Lecture Notes in Mathematics

A collection of informal reports and seminars
Edited by A. Dold, Heidelberg and B. Eckmann, Zürich

168

The Steenrod Algebra and Its Applications:

A Conference to Celebrate N.E. Steenrod's Sixtieth Birthday

Proceedings of the Conference held at the Battelle Memorial Institute, Columbus, Ohio
March 30th-April 4th, 1970

Edited by F.P. Peterson, MIT, Cambridge, MA/USA

Springer-Verlag
Berlin · Heidelberg · New York 1970

ISBN 3-540-05300-X Springer-Verlag Berlin · Heidelberg · New York
ISBN 0-387-05300-X Springer-Verlag New York · Heidelberg · Berlin

Offsetdruck: Julius Beltz, Weinheim/Bergstr.

Preface

This volume consists of lectures given by some of N. E. Steenrod's students and close associates at a conference held this past spring in honor of his sixtieth birthday. His influence on the development of algebraic topology and on our own work has been enormous, and we are all very happy that we have been able to work with him. All of these papers are gratefully dedicated to N. E. Steenrod.

We would like to thank the Battelle Memorial Institute very much for sponsoring this conference and for their help in making it a great success.

F. P. Peterson

Cambridge, Mass., June 15, 1970

Table of Contents

Contributors' Addresses

Jose Adem
Centro de Investigacion del I.P.N.
Apartado Postal 14-740
Mexico 14, D.F., Mexico

Peter J. Freyd
Department of Mathematics
University of Pennsylvania
Philadelphia, Pennsylvania 19104, USA

S. Gitler
Centro de Investigacion del I.P.N.
Apartado Postal 14-740
Mexico 14, D.F., Mexico

Peter Hilton
Department of Mathematics
Cornell University
Ithaca, New York 14850, USA

P. Holm
Department of Mathematics
University of Oslo
Blindern, Oslo, Norway

Wu-chung Hsiang
Department of Mathematics
Yale University
New Haven, Connecticut 06520, USA

Wu-yi Hsiang
Department of Mathematics
University of California
Berkeley, California 94720, USA

Sufian Y. Husseini
Department of Mathematics
University of Wisconsin
Madison, Wisconsin 53706, USA

Ioan M. James
Mathematical Institute
24-29 St. Giles
Oxford, England

Kee Yuen Lam
Department of Mathematics
University of British Columbia
Vancouver 8, British Columbia, Canada

Saunders MacLane
Department of Mathematics
University of Chicago
Chicago, Illinois 60637, USA

J. Peter May
Department of Mathematics
University of Chicago
Chicago, Illinois 60637, USA

John C. Moore
Fine Hall
Princeton University
Princeton, New Jersey 08540, USA

Franklin P. Peterson
Department of Mathematics
Massachusetts Institute of Technology
Cambridge, Massachusetts 02139, USA

Ted Petrie
Institute for Defense Analysis
100 Prospect Street
Princeton, New Jersey 08540

Joseph Roitberg
Department of Mathematics
State University of New York at Stony Brook
Stony Brook, New York 11790, USA

Paul A. Schweitzer, S.J.
Institute for Advanced Study
Princeton, New Jersey 08540, USA

Edwin Spanier
Department of Mathematics
University of California
Berkeley, California 94720, USA

Emery Thomas
Department of Mathematics
University of California
Berkeley, California 94720, USA

George W. Whitehead
Department of Mathematics
Massachusetts Institute of Technology
Cambridge, Massachusetts 02139, USA

Alexander Zabrodsky
Department of Mathematics
University of Illinois at Chicago Circle
Chicago, Illinois 60680, USA

THE WORK OF NORMAN E. STEENROD IN ALGEBRAIC TOPOLOGY: AN APPRECIATION

by George W. Whitehead

When one thinks of Steenrod's work in algebraic topology, it is the algebra of operators which bears his name and which is the subject of this conference that first comes to mind. There are, however, at least two other aspects of his work which have had a profound and lasting influence on the development of our subject.

The first of these is concerned with the foundations. The fifty years following the appearance of Poincaré's fundamental memoir saw great progress in the development of algebraic topology. The fundamental theorems of the subject - the invariance theorem, the duality theorems of Poincaré and Alexander, the universal coefficient and Künneth theorems, the Lefschetz fixed-point theorem - had all been proved, at least for finite complexes. The intersection theory for algebraic varieties had been extended, first to manifolds, then, with the invention of cohomology groups, to arbitrary complexes. Nevertheless, by the early forties the subject was in a chaotic state. Partly in a quest for greater insight, partly in order to extend the range of validity of the basic theorems, there had arisen a plethora of homology theories - the singular theories due to Alexander-Veblen and to Lefschetz, the Vietoris and Čech theories, as well as many minor variants. Thus, while there were many <u>homology theories</u> there was not as yet a <u>theory of homology</u>. Indeed, many of the concepts which are routine today, while appearing implicitly in much of the literature, had never been explicitly formulated. The time was ripe to find a framework in which the above-mentioned results could be placed in order to determine their interconnections, as well as to ascertain their relative importance.

This task was accomplished by Steenrod and Eilenberg, who announced, in 1945, their system of axioms for homology theory. What was especially impressive was that a subject as complicated as homology theory could be characterized by properties of such beauty and simplicity. But the first impact of their work was not so much in the explicit results as in their whole philosophy. The conscious recognition of the functorial properties of the concepts involved, as well as the explicit use of diagrams as a tool in constructing proofs, had thoroughly permeated the subject by the time of appearance of their book "Foundations of Algebraic Topology" in 1952.

Inspection of their axioms reveals that the seventh "Dimension axiom" has an entirely different character from the first six, the latter being of a very general character, while the former is very specific, asserting as it does, that the homology groups of a point vanish in all non-zero dimensions. That it is accorded the same status as the others is no doubt due to the fact that no interesting examples of non-standard theories were known at that time. In any case, a great deal of the work does not depend on the dimension axiom.

With the great advances in homotopy theory in the '50's and '60's, there arose numerous examples of "extraordinary theories" - theories satisfying the first six axioms; among these were the stable homotopy and cohomotopy groups of Spanier and Whitehead; the K-theories of Atiyah and Hirzebruch; the bordism theories of Atiyah and Conner-Floyd and their more recent generalizations. It is a tribute to the insight of Eilenberg and Steenrod that these theories, whose existence was undreamt-of in 1945, fit so beautifully into their framework.

Another beautiful application was the theorem of Dold and Thom. The symmetric group $S(n)$ acts on the n-fold Cartesian power X^n of a space X by permuting the factors; the orbit space is the n-fold symmetric power $SP^n(X)$. There are imbeddings $SP^n(X) \subset SP^{n+1}(X)$;

thus one can form the infinite symmetric product $SP^\infty(X)$. Dold and Thom showed that the groups $\pi_q(SP^\infty X)$ satisfy the axioms and hence

$$\pi_q(SP^\infty X) \approx H_q(X;Z) .$$

For example, $SP^\infty(S^n)$ is an Eilenberg-MacLane space $K(Z,n)$.

The other aspect of Steenrod's work mentioned above is the theory of fibre bundles. Steenrod has always had a strong interest in differential geometry; indeed, one of his earliest papers, written jointly with Myers, established one of the fundamental results of global differential geometry - the group of isometries of a Riemannian manifold is a Lie group. Thus it was to be expected that Steenrod would have something to say about the young and rapidly growing subject of fibre bundles.

One of the first problems in the subject is that of the existence of a cross-section. This problem is attacked by a stepwise extension process parallel to that used in ordinary obstruction theory. However, the coefficient groups for the obstruction vary from point to point; the various groups are isomorphic, but the isomorphism between the groups at two different points depends on a homotopy class of paths joining them. In other words, the coefficients form a <u>bundle of groups</u>. During the 1940's, Steenrod introduced the notion of homology and cohomology with coefficients in a bundle of groups ("local coefficients"). This theory, which extended and clarified the work of Reidemeister on "homotopy chains", provided the proper setting for obstructions, not only in bundle theory, but also for mappings into non-simple spaces. Moreover, it gave the first satisfactory formulation of Poincaré duality for non-orientable manifolds.

One of the most vital notions in bundle theory is that of universal bundle (equivalently, of classifying space). The importance of the Grassmann manifold $G(k,\ell)$ of k-planes in $R^{k+\ell}$ was first recognized by Whitney, who proved that every (k-1)-sphere bundle over a complex of dimension $\leqq \ell$ is induced by a map of its base space into $G(k,\ell)$. In 1944 Steenrod proved the decisive result in this direction: the space $G(k,\ell)$ is a classifying space for (k-1)-sphere bundles over a base complex of dimension $\leqq \ell$, i.e., for such a space X the homotopy classes of maps of X into $G(k,\ell)$ are in one-to-one correspondence with the isomorphism classes of S^{k-1}-bundles over X. There is a standard bundle $B(k,\ell)$ over $G(k,\ell)$, whose total space is the set of all pairs (π,x), where π is a k-plane in $R^{k+\ell}$ and x a unit vector in π. The above correspondence associates with each map $f: X \to G(k,\ell)$ the induced bundle $f^*B(k,\ell)$. The total space of the principal associated bundle is the Stiefel manifold $V_{k+\ell,k}$ of oriented k-frames in $R^{k+\ell}$; and the crucial fact used in the proof is that $V_{k+\ell,k}$ is $(\ell-1)$-connected. Once this was realized, the generalization to bundles whose group is an arbitrary compact Lie broup was not difficult, and was found by Steenrod and several other authors independently. Later developments included the proof by Milnor of a classifying space for an arbitrary topological group; while in recent years the classifying spaces BO,BU,... for stable vector bundles have been of enormous importance.

These noteworthy contributions to bundle theory occurred in an era during which the subject was growing with great rapidity, and was in a confused state. A systematic account of the subject was badly needed, and this need was amply met with the appearance in 1951 of Steenrod's book "The Topology of Fibre Bundles". But the importance of the book was not limited to its treatment of bundle theory. It must be remembered that, while homotopy groups had been in existence for 16 years and obstruction theory for 12, neither topic had received a treatment in book form. Steenrod's book gave a very clear, if succinct, treatment of both topics and thus served as an introduction to homotopy theory to a whole generation of young topologists.

And now the time has come to turn our attention to the Steenrod algebra. The problem of classifying the maps of an m-complex K into the n-sphere S^n had long occupied topologists. In 1933 Hopf gave the solution for $m = n$; these results were reformulated in terms of cohomology and thus greatly simplified by Whitney in 1937. In 1931 Hopf showed that $\pi_3(S^2)$ was non-zero; that it was infinite cyclic was established by Hurewicz in 1935. In 1937 Freudenthal proved his fundamental <u>suspension</u> <u>theorem</u> and showed that, for $n \geq 3$, $\pi_{n+1}(S^n)$ is a cyclic group of order two generated by the suspension of the Hopf map. In 1941 Pontryagin classified the maps of K^3 into S^2 ; his classification involved the relatively new cup-products of Alexander-Čech-Whitney.

The next outstanding problem was the classification of maps of K^{n+1} into S^n for $n \geq 3$. That this problem was quite subtle is evinced by the fact that at least two very distinguished mathematicians had announced solutions which turned out to be incorrect.

In 1947 Steenrod solved this problem. His solution was interesting, not only <u>per</u> <u>se</u>, but by virtue of the new operations in terms of which the solution was expressed. These were, of course, the celebrated Steenrod squares. They were presented as generalizations of the cup-square; and Steenrod's first construction was by cochain formulas generalizing the Alexander-Čech-Whitney formula for the cup-product. While this gave a simple and effective procedure for calculating the squares, it was not at all clear how to generalize the construction to obtain reduced $n^{\underline{th}}$ powers. It was not long before Steenrod realized that it was the Lefschetz approach to cup-products via chain approximations to the diagonal, rather than that of Alexander-Čech-Whitney, which yielded a fruitful generalization, and he soon succeeded in constructing the higher reduced powers.

The power of the new operations soon became apparent. The Cartan formulas for $Sq^i(u \cup v)$ allowed one to calculate the squares in truncated projective spaces; and this had deep consequences for the old problem: how many vector fields can be found on S^n which are linearly independent at each point? Using the above results, Steenrod and J. H. C. Whitehead were able to show that, if k is the exponent of the largest power of two dividing $n+1$, then S^n does not admit a tangent 2^k-frame. This was an enormous step forward; previously this was known to be true only for $k=0,1$.

The reduced powers are examples of cohomology operations, i.e., natural transformations of one cohomology functor into another. Moreover, they are stable operations, in the sense that they are defined in every dimension and commute with suspension. The set of all stable operations in mod p cohomology forms an algebra $\mathcal{A}_p$, which was soon to be known as the Steenrod algebra. In 1952 Serre showed that $\mathcal{A}_2$ was generated by the squares, and exhibited an additive basis composed of certain iterated squares. In 1954, Cartan proved an analogous result for odd primes; besides the reduced p^{th} powers $\mathcal{P}^i$, one additional operation, the Bockstein β_p, was needed.

In the meantime Adem used Steenrod's approach to find relations among the squares, In particular, he proved that Sq^i is decomposable if i is not a power of 2. This had a most important application: there is no map of S^{2n-1} into S^n of Hopf invariant one unless n is a power of 2. This again was a great step forward; previously it was known only that n had to be divisible by 4 $(n>2)$.

Adem also showed that his relations gave rise to secondary cohomology operations; these differed from the old ones in that they were not everywhere defined (the domain was the kernel of a certain primary operation) and not single-valued (the range was the cokernel of another primary operation). Nevertheless, they were sufficiently powerful to prove that the iterated Hopf maps η^2, ν^2, σ^2 were stably non-trivial.

The analogous relations among the reduced $p^{\underline{th}}$ powers were found independently by Adem and Cartan a year or so later.

Meanwhile, Steenrod had not been idle. His new approach to the subject revealed deep connections with the Eilenberg-MacLane homology of groups. Specifically, he showed that each element of $H_q(\pi,G)$, where π is a subgroup of the symmetric group $S(n)$ of degree n , gives rise to an operation. These operations included the old ones (which arose from a cyclic transitive subgroup of the symmetric group) and more - the generalized Pontryagin powers $\mathfrak{P}_p$ of Thomas were also included. Moreover, Steenrod and Thomas showed that all operations derived from permutation groups by Steenrod's procedure were generated by the $\mathcal{P}^i$ and the $\mathfrak{P}_p$ with the aid of the primitive operations of addition, cup-product, coefficient group homomorphisms, and Bocksteins. Later Moore, Dold, and Nakamure showed that <u>all</u> operations are obtained in this way (at least if the coefficient groups are finitely generated).

Further applications now followed thick and fast. It is hardly necessary to detail to this audience how thoroughly the Steenrod algebra has permeated our work in recent years. To mention only a few examples: the structure of the cohomology of the Thom spectrum $M(G)$ as an $\mathcal{A}$-module was crucial in the determination of the various bordism rings by Thom, Wall, Milnor, Anderson-Brown-Peterson; Milnor's observation that the Cartan formulas make $\mathcal{A}$ into a Hopf algebra and his investigation of its structure have greatly deepened our insight; the introduction of homological-algebraic methods by Adams has revealed the crucial importance of the cohomology of $\mathcal{A}$ in stable homotopy; and Steenrod's own work on unstable $\mathcal{A}$-modules led Massey and Peterson to their unstable version of the Adams spectral sequence, one of the most promising of our tools in studying unstable homotopy theory.

Some other aspects of Steenrod's work that deserve mention are:

1) the notion of direct limit (of a system of abelian groups) made its first appearance in his Ph.D. thesis;

2) his 1940 paper on regular cycles in compact metric spaces was a forerunner of Borel-Moore homology theory;

3) his calculation, in the same paper, of the cohomology groups of the complement of a solenoid in R^3 led Eilenberg and MacLane to study the relation between group extensions and homology, thereby inaugurating a long and fruitful collaboration;

4) in 1941, Hurewicz and Steenrod introduced the concept of fibre space as opposed to fibre bundle, formulated the notion of covering homotopy, and proved the basic homotopy properties of fibre maps.

There is still another facet of Steenrod's work, and perhaps not the least important, which is evinced by the very existence of this conference. The weekly conferences in his office, and the long hours of patient guidance he spent with us, not only on our maiden efforts at research, but also on the exposition of our results will be long and gratefully remembered by all of us.

BIBLIOGRAPHY OF N. E. STEENROD

I. Papers

1. Finite arc-sums, Fund. Math. 23 (1934), 38-53.

2. Characterizations of certain finite curve-sums, Am. Jour. Math. 56 (1934), 558-568.

3. On universal homology groups, Proc. Nat. Acad. Sci. U. S. A. 21 (1935), 482-484.

4. Universal homology groups, Am. Jour. Math. 58 (1936), 661-701.

5. Remark on weakly convergent cycles, Fund. Math. 31 (1932), 135-136.

6. (with J. H. Roberts) Monotone transformations of two-dimensional manifolds, Ann. of Math. 39 (1938), 851-862.

7. (with S. B. Myers) The group of isometries of a Riemannian manifold, Ann. of Math. 40 (1939), 400-416.

8. Regular cycles of compact metric spaces, Ann. of Math. 41 (1940), 833-851.

9. Regular cycles of compact metric spaces. Lectures in Topology, pp. 43-55. University of Michigan Press, Ann Arbor, Mich., 1941.

10. (with W. Hurewicz) Homotopy relations in fibre spaces, Proc. Nat. Acad. Sci. U. S. A. 27 (1941), 60-64.

11. Topological methods for the construction of tensor functions, Ann. of Math. 43 (1942), 116-131.

12. Homology with local coefficients, Ann. of Math. 44 (1943), 610-627.

13. The classification of sphere bundles, Ann. of Math. 45 (1944), 294-311.

14. (with S. Eilenberg) Axiomatic approach to homology theory, Proc. Nat. Acad. Sci. U. S. A. 31 (1945), 117-120.

15. Products of cocycles and extensions of mappings, Ann. of Math. 48 (1947), 290-320.

16. Cohomology invariants of mappings, Proc. Nat. Acad. Sci. U. S. A. 33 (1947), 124-128.

17. Cohomology invariants of mappings, Ann. of Math. 50 (1949), 954-988.

18. (with J. H. C. Whitehead) Vector fields on the n-sphere, Proc. Nat. Acad. Sci. U. S. A. 37 (1951), 58-63.

19. Reduced powers of cohomology classes, Ann. of Math. 56 (1952), 47-67.

20. Homology groups of symmetric groups and reduced power operations, Proc. Nat. Acad. Sci. U. S. A. 39 (1953), 213-217.

21. Cyclic reduced powers of cohomology classes, Proc. Nat. Acad. Sci. U. S. A. 39 (1953), 217-233.

22. The work and influence of Professor S. Lefschetz in algebraic topology. Algebraic Geometry and Topology, A symposium in honor of S. Lefschetz, pp. 24-43. Princeton University Press, Princeton, N. J., 1957.

23. Cohomology operations derived from the symmetric group, Comm. Math. Helv. 31 (1957), 195-218.

24. (with P. E. Thomas) Cohomology operations derived from cyclic groups, Comm. Math. Helv. 32 (1957), 129-152.

25. Cohomology operations. Symposium Internacional de Topologia Algebraica, pp. 165-185. Universidad Nacional Autonoma de Mexico and UNESCO, Mexico City, 1958.

26. Cohomology operations, and obstructions to extending continuous functions (mimeographed), A.M.S. Colloquium Lectures, August, 1957.

27. The cohomology algebra of a space, Ens. Math. (2) 7 (1961), 153-178.

28. (with M. Rothenburg) The cohomology of classifying spaces of H-spaces, Bull. Am. Math. Soc. 71 (1965), 872-875.

29. A convenient category of topological spaces, Mich. Math. Jour. 14 (1967), 133-152.

30. Milgram's classifying space of a topological group, Topology 7 (1968), 349-368.

II. Books

1. The Topology of Fibre Bundles, Princeton University Press, Princeton, N. J., 1951.

2. (with S. Eilenberg) Foundations of Algebraic Topology. Princeton University Press, Princeton, N. J., 1952.

3. Cohomology Operations. Lectures by N. E. Steenrod. Written and revised by D. B. A. Epstein. Annals of Mathematics Studies, No. 50. Princeton University Press, Princeton, N. J., 1962.

4. (with W. G. Chinn) First Concepts of Topology. The geometry of mappings of segments, curves, circles, and disks. New Philosophical Library, vol. 18, Random House, New York; The L. W. Singer Co., Syracuse, N. Y., 1966.

5. Homology of Cell Complexes, by George E. Cooke and Ross L. Finney (based on lectures by Norman E. Steenrod). Princeton University Press, Princeton, N. J. and University of Tokyo Press, Tokyo, 1967.

III.

1. Reviews of Papers in Algebraic and Differential Topology, Topological Groups, and Homological Algebra, Classified by N. E. Steenrod, American Mathematical Society, Providence, R. I., 1968.

ON NONSINGULAR BILINEAR MAPS

BY JOSE ADEM

1. INTRODUCTION

Let $n\xi_k$ denote the n-fold Whitney sum of the Hopf bundle ξ_k over the real projective space P^k. As it is known ([2],[4]), the bundle $n\xi_k$ has s independent sections if and only if there exists a nonsingular skew-linear map $f: R^{k+1} \times R^s \to R^n$. In terms of $\operatorname{gd} n\xi_k$, the geometrical dimension of $n\xi_k$, this is also equivalent to assure that $\operatorname{gd} n\xi_k \leq n-s$. An interesting problem is to find out when f can be replaced by a bilinear map.

Let K be the space of Cayley numbers and H the space of quaternions. For all integer $n \geq 1$, we construct maps $K^2 \times K^{2n} \to K^{2n+1}$ and $H^3 \times H^{4n} \to K^{2n+1}$, that can be regarded as a generalization of the constructions given by Lam in [3]. These maps give raise to several nonsingular bilinear maps $R^r \times R^s \to R^t$, with $r \leq 16$.

Many of our maps can be used to solve the problem mentioned above. For example, for all $k \geq 0$, we obtain

$$R^{13} \times R^{13+16k} \longrightarrow R^{19+16k},$$

$$R^{12} \times R^{15+16k} \longrightarrow R^{21+16k},$$

$$R^{12} \times R^{11+16k} \longrightarrow R^{17+16k}.$$

Observe that the restriction of the first map to $R^{12} \times R^{13+16k}$, can be combined with the other two, in the single formula

(1.1) $$R^{12} \times R^{11+2t+16k} \longrightarrow R^{17+2t+16k},$$

that holds for $t = 0,1,2$ and all $k \geq 0$. Now, an unpublished work of Gitler shows that $gd(17+2t+16k)\xi_{11} = 6$, for the following cases: $t = 0$ and $k = 1,2$; $t = 1$ and $k = 0,2$; $t = 2$ and $k = 0,2$. So, it follows from (1.1) that, in each of these cases, the maximal number of sections is associated with a bilinear map.

Finally, I believe that the maps of this paper, together with those constructed in [3], [5] and [6], give essentially all the possible nonsingular bilinear maps of the form $R^{12} \times R^{s} \to R^{n}$, when $n \leq 64$.

2. THE MAP $K^2 \times K^{2n} \to K^{2n+1}$

Let K be the Cayley algebra over R, where R is the field of real numbers. For $n \geq 1$, we will construct a nonsingular bilinear map $K^2 \times K^{2n} \to K^{2n+1}$, with a commutator in one of its components. This map can be regarded as a generalization of [3; Th. 1] and [1; Th. 3.6].

Let $u \in K^2$ and $v \in K^{2n}$, where $u = (x_1,x_2)$, $v = (y_1,\dots,y_{2n})$, with x_i, y_j Cayley numbers. Set

(2.1) $$\Phi_1(u,v) = x_1y_1 + \bar{y}_{2n}x_2,$$

(2.2) $$\Phi_{2t}(u,v) = y_{2t}x_1 - x_2\bar{y}_{2t-1}, \text{ for } 1 \leq t \leq n,$$

(2.3) $$\Phi_{2t+1}(u,v) = x_1y_{2t+1} - \bar{y}_{2t}x_2, \text{ for } 1 \leq t \leq n-1,$$

(2.4) $$\Phi_{2n+1}(u,v) = x_1y_1 - y_1x_1.$$

Here, the product is the Cayley multiplication and the bar means the conjugate. Now, define

$$f(u,v) = [\Phi_1(u,v),\ldots,\Phi_{2n+1}(u,v)].$$

Theorem 2.5. *The bilinear map* $f: K^2 \times K^{2n} \to K^{2n+1}$, *where* $n \geq 1$, *is nonsingular. Moreover, by suitable restrictions,* f *induces the following nonsingular bilinear maps, where* $k \geq 0$:

(2.6) $R^{16} \times R^{16+16k} \to R^{23+16k}$, (2.7) $R^{13} \times R^{13+16k} \to R^{19+16k}$,

(2.8) $R^{11} \times R^{11+16k} \to R^{17+16k}$, (2.9) $R^{10} \times R^{10+16k} \to R^{16+16k}$,

(2.10) $R^{10} \times R^{16+16k} \to R^{22+16k}$, (2.11) $R^{16} \times R^{10+16k} \to R^{22+16k}$,

(2.12) $R^{11} \times R^{15+16k} \to R^{21+16k}$, (2.13) $R^{15} \times R^{11+16k} \to R^{21+16k}$,

(2.14) $R^{10} \times R^{14+16k} \to R^{20+16k}$, (2.15) $R^{14} \times R^{10+16k} \to R^{20+16k}$,

(2.16) $R^{9} \times R^{16+16k} \to R^{16+16k}$, (2.17) $R^{16} \times R^{9+16k} \to R^{16+16k}$.

Proof. Suppose that $f(u,v) = 0$, then we obtain the following $2n+1$ equations that x_i, y_j must satisfy: $\Phi_t(u,v) = 0$, with $1 \leq t \leq 2n+1$. To prove that f is nonsigular we consider three different cases.

First case: if $x_1 = 0$ and y_1 any number. In this case, we have $x_2\bar{y}_{2t-1} = 0$ and $\bar{y}_{2t}x_2 = 0$ for $1 \leq t \leq n$. Therefore, wither $u = 0$ or $v = 0$.

<u>Second case</u>: if $x_1 \neq 0$ and $y_1 = 0$. Here, we have that $y_t = 0$ implies $y_{t+1} = 0$. In fact, by substitution of $y_t = 0$ in $\Phi_{t+1}(u,v) = 0$, we get $y_{t+1}x_1 = 0$ if t is odd or $x_1 y_{t+1} = 0$ if t is even. Since $x_1 \neq 0$, it follows that $y_{t+1} = 0$. Therefore, starting with $y_1 = 0$, by induction we obtain in this case, that $v = 0$.

<u>Third case</u>: if $x_1 \neq 0$ and $y_1 \neq 0$. From (2.1), we have

$$x_1 y_1 = -\bar{y}_{2n} x_2, \tag{2.18}$$

then, it follows that $x_2 \neq 0$, $y_{2n} \neq 0$. From (2.2) and (2.3), we have

$$y_{2t} x_1 = x_2 \bar{y}_{2t-1}, \tag{2.19}$$

$$x_1 y_{2t+1} = \bar{y}_{2t} x_2, \tag{2.20}$$

and from these we obtain, that $y_k \neq 0$ implies that $y_{k+1} \neq 0$. Consequently, in this case all the x_i, y_j must be different from zero.

Before arriving to a contradiction, we will prove by induction the following. For all $1 \leq k \leq 2n-2$, we have that

$$y_k n(y_2) = y_{k+2} n(y_1), \tag{2.21}$$

where $n(y_i) = y_i \bar{y}_i = \bar{y}_i y_i$ is the norm of y_i.

From (2.19) and (2.20) with $t = 1$, we get

$$y_2 x_1 = x_2 \bar{y}_1, \tag{2.22}$$

(2.23) $$x_1y_3 = \bar{y}_2x_2.$$

Observe that (2.4) implies that x_1 and y_1 commute. Then, right multiplying (2.22) by y_1 and using the associative property [1; (2.3)], give

$$(y_2x_1)y_1 = y_2(x_1y_1) = x_2n(y_1),$$

and now, left multiplying by $\bar{y}_2$, yields

(2.24) $$x_1y_1n(y_2) = \bar{y}_2x_2n(y_1).$$

Substitution of (2.23) in this gives

$$x_1y_1n(y_2) = x_1y_3n(y_1),$$

and cancelling the factor x_1, proves (2.21) for $k = 1$.

Now, suppose that (2.21) holds for all $1 \leq i < k$. If $k = 2t$ is even, using the induction hypothesis, we have that

$$x_2\bar{y}_{2t-1}n(y_2) = x_2\bar{y}_{2t+1}n(y_1),$$

and from this and (2.19) we obtain that

$$y_{2t}x_1n(y_2) = y_{2t+2}x_1n(y_1),$$

so, by cancelling the factor x_1, we get (2.21) for this case. If $k = 2t+1$ is odd, a similar argument works using (2.20), and this completes the induction step.

Using (2.21), it follows trivially that

$$y_2n(y_2)^{n-1} = y_{2n}n(y_1)^{n-1},$$

and by substitution of this in (2.24), we get

$$x_1 y_1 n(y_2)^n = \bar{y}_{2n} x_2 n(y_1)^n .$$

Finally, combining this last result with (2.18), we obtain

$$x_1 y_1 [n(y_1)^n + n(y_2)^n] = 0,$$

and this is a contradiction to the assumption that $x_1 \neq 0$ and $y_1 \neq 0$. This ends the proof that f is nonsingular.

Since the component $\Phi_{2n+1}(u,v)$ of $f(u,v)$ is the commutator

$$x_1 y_1 - y_1 x_1 ,$$

the maps (2.6-17) are obtained by restricting x_1 and y_1 as is done in [1; p. 99]. The four extra cases (2.11), (2.13), (2.15) and (2.17) that we are getting here, are obtained by respectively interchanging the restrictions of x_1 and y_1 in (2.10), (2.12), (2.14) and (2.16).

3. THE MAP $H^3 \times H^{4n} \to K^{2n+1}$

Let H denote the space of quaternions. We regard the elements of K as ordered pairs of quaternions and we recall that given Cayley numbers $x = (a_1, a_2)$, $y = (b_1, b_2)$ where a_1, a_2, b_1, b_2 are quaternions, the product xy is given by

(3.1) $$xy = (a_1 b_1 - \bar{b}_2 a_2, b_2 a_1 + a_2 \bar{b}_1),$$

and the conjugate $\bar{x}$ of x by $\bar{x} = (\bar{a}_1, -a_2)$.

Let $u \in H^3$ and $v \in H^{4n}$, where $u = (a_1,a_2,a_3)$, $v = (b_1,b_2,\ldots,b_{4n})$, with a_i, b_j quaternions, and assume that the real part of b_2 is <u>always</u> zero. This is equivalent to suppose that

(3.2) $$b_2 = -\bar{b}_2.$$

Define the Cayley numbers $x = (a_2,a_3)$ and $y_t = (b_{2t-1},b_{2t})$ for $1 \leq t \leq 2n$, and set

(3.3) $$\Psi_1(u,v) = (0,b_2a_1) + \bar{y}_{2n}x,$$

(3.4) $$\Psi_{2t}(u,v) = (a_1b_{4t-1},a_1b_{4t}) - x\bar{y}_{2t-1}, \text{ for } 1 \leq t \leq n,$$

(3.5) $$\Psi_{2t+1}(u,v) = (a_1b_{4t+1},b_{4t+2}a_1) - \bar{y}_{2t}x, \text{ for } 1 \leq t \leq n-1,$$

(3.6) $$\Psi_{2n+1}(u,v) = (a_1b_1,\bar{b}_2a_1 + \bar{a}_1b_2).$$

Now, define

$$g(u,v) = [\Psi_1(u,v),\ldots,\Psi_{2n+1}(u,v)].$$

<u>Theorem</u> 3.7. <u>The bilinear map</u> $g: H^3 \times H^{4n} \longrightarrow K^{2n+1}$, <u>where</u> $n \geq 1$, <u>is nonsingular</u>. <u>Furthermore, for all integer</u> $k \geq 0$, g <u>gives rise to the following nonsingular bilinear maps</u>:

(3.8) $$R^{12} \times R^{15+16k} \longrightarrow R^{21+16k},$$

(3.9) $$R^{12} \times R^{11+16k} \longrightarrow R^{17+16k}.$$

These maps improve (2.8) and (2.12).

<u>Proof</u>. If $g(u,v) = 0$, then we get the equations $\Psi_t(u,v) = 0$ that a_i, b_j must satisfy. For $t = 2n+1$, the resulting equation

is equivalent with

(3.10) $$a_1 b_1 = 0,$$

(3.11) $$\bar{b}_2 a_1 + \bar{a}_1 b_2 = 0.$$

From (3.10) it follows that either $a_1 = 0$ or $b_1 = 0$. As before, to prove that g is nonsingular we consider different cases.

<u>First case</u>: if $a_1 = 0$ and b_1 any number. From (3.3), (3.4) and (3.5) we obtain $x\bar{y}_{2t-1} = 0$ and $\bar{y}_{2t}x = 0$ for $t = 1, \ldots, n$. Then, either $x = 0$ or $y_k = 0$ for $k = 1, \ldots, 2n$. Consequently, $u = 0$ or $v = 0$.

<u>Second case</u>: if $a_1 \neq 0$, $b_1 = 0$ and $b_2 = 0$. Since $y_1 = (b_1, b_2)$, in this case $y_1 = 0$. We will prove that $y_k = 0$ implies $y_{k+1} = 0$. If $k = 2t-1$ is odd, from (3.4) we get $a_1 b_{4t-1} = 0$ and $a_1 b_{4t} = 0$. Since $a_1 \neq 0$, it follows that $b_{4t-1} = b_{4t} = 0$ therefore $y_{2t} = 0$. Using (3.5), the same argument works if k is even. Then $v = 0$ in this case.

<u>Third case</u>: if $a_1 \neq 0$, $b_1 = 0$ and $b_2 \neq 0$. Before we arrive to a contradiction with the assumption that $g(u,v) = 0$, we will prove the following. For all $0 \leq k \leq n-1$ we have

(3.12) $$b_{4k+1} = 0,$$

(3.13) $$b_{4k+2} \neq 0,$$

(3.14) $$b_{4k+2} = -\bar{b}_{4k+2},$$

(3.15) $$\bar{b}_{4k+2}a_1 + \bar{a}_1 b_{4k+2} = 0,$$

and for $1 \leq k \leq n-1$, we have

(3.16) $$y_{2k} \neq 0,$$

(3.17) $$b_{4k+2} = b_2\lambda_k,$$

where $\lambda_k > 0$ is the real number given by

(3.18) $$\lambda_k = \prod_{i=1}^{k} n(y_{2i})n(b_{4i-2})^{-1}.$$

The proof is by induction. If $k = 0$, from (3.2) and (3.11), it follows that (3.12-15) hold. Since $b_2 a_1 \neq 0$, from (3.3) we have that $\bar{y}_{2n}x \neq 0$, and this implies that $y_{2n} \neq 0$ and $x \neq 0$. Now, assume that (3.12-18) hold for $i < k$. From (3.4) and (3.5), we get

$$(a_1 b_{4k-1}, a_1 b_{4k}) - (a_2, a_3)(\bar{b}_{4k-3}, -b_{4k-2}) = 0,$$

$$(a_1 b_{4k+1}, b_{4k+2}a_1) - (\bar{b}_{4k-1}, -b_{4k})(a_2, a_3) = 0.$$

Multiplying as in (3.1) and using the hypothesis $b_{4k-3} = 0$, we obtain

(3.19) $$a_1 b_{4k-1} - \bar{b}_{4k-2}a_3 = 0,$$

(3.20) $$a_1 b_{4k} + b_{4k-2}a_2 = 0,$$

(3.21) $$a_1 b_{4k+1} - \bar{b}_{4k-1}a_2 - \bar{a}_3 b_{4k} = 0,$$

(3.22) $$b_{4k+2}a_1 - a_3\bar{b}_{4k-1} + b_{4k}\bar{a}_2 = 0.$$

From $x = (a_2,a_3) \neq (0,0)$, the hypothesis $b_{4k-2} \neq 0$ and from (3.19), (3.20), it follows that $y_{2k} = (b_{4k-1},b_{4k}) \neq (0,0)$, according with (3.16).

We will prove that

$$\bar{b}_{4t-1}a_2 + \bar{a}_3 b_{4t} = 0. \tag{3.23}$$

Clearly, (3.19) and (3.20) imply that $a_3 = 0$ iff $b_{4t-1} = 0$ and $a_2 = 0$ iff $b_{4t} = 0$. Then, to establish (3.23) we may suppose that these four elements are different from zero. Left and right multiplying (3.19), respectively, by b_{4k-2} and $\bar{b}_{4k-1}$, gives

$$b_{4k-2}a_1 n(b_{4k-1}) = a_3\bar{b}_{4k-1}n(b_{4k-2}),$$

and from $b_{4k-2} = -\bar{b}_{4k-2}$, we get

$$\bar{b}_{4k-2}a_1 n(b_{4k-1}) = -a_3\bar{b}_{4k-1}n(b_{4k-2}).$$

The fact that $\bar{b}_{4k-2}a_1 = -\bar{a}_1 b_{4k-2}$ implies that $a_3\bar{b}_{4k-1} = -b_{4k-1}\bar{a}_3$. Therefore, we obtain

$$\bar{b}_{4k-2}a_1 n(b_{4k-1}) = b_{4k-1}\bar{a}_3 n(b_{4k-2}).$$

Now, left and right multiplying (3.20), respectively, by $\bar{b}_{4t-2}$ and $\bar{b}_{4k}$, gives

$$\bar{b}_{4k-2}a_1 n(b_{4k}) = -a_2\bar{b}_{4k}n(b_{4k-2}),$$

and from this and the above equation we get

$$a_2\bar{b}_{4k}n(b_{4k-1}) + b_{4k-1}\bar{a}_3 n(b_{4k}) = 0.$$

Left and right multiplying this result, respectively, by $\bar{b}_{4k-1}$ and b_{4k}, establishes (3.23).

If we substitute (3.23) in (3.21), we get that $a_1 b_{4k+1} = 0$, and since $a_1 \neq 0$, it follows that $b_{4k+1} = 0$, according with (3.12). Then, substitute $b_{4k+1} = 0$ in (3.5), to obtain

$$(0, b_{4k+2} a_1) = \bar{y}_{2k} x,$$

and since we have established already that $y_{2k} \neq 0$ and $x \neq 0$, it follows that $b_{4k+2} \neq 0$, according with (3.13).

As we have remarked before $a_3 \neq 0$ iff $b_{4k-1} \neq 0$. To complete the induction step we consider two cases.

If $a_3 \neq 0$, then right multiplying (3.22) by b_{4k-1}, and using (3.19) and the conjugate of (3.23), gives

$$[b_{4k+2}\bar{b}_{4k-2} - n(b_{4k-1}) - n(b_{4t})]a_3 = 0,$$

therefore,

$$b_{4k+2}\bar{b}_{4k-2} = n(b_{4k-1}) + n(b_{4k}) = n(y_{2k})$$

and from this, we have

$$b_{4k+2} = b_{4k-2} n(y_{2k}) n(b_{4k-2})^{-1}. \tag{3.24}$$

If $a_3 = 0$, then $b_{4t-1} = 0$ and, since $x = (a_2, a_3) \neq (0,0)$, we have that $a_2 \neq 0$ and $b_{4t} \neq 0$. The equation (3.22) becomes

$$b_{4k+2} a_1 + b_{4k} \bar{a}_2 = 0,$$

and right multiplying it by $\bar{b}_{4k-2}$, and then using $a_1 \bar{b}_{4k-2} =$

$-b_{4k-2}\bar{a}_1$ gives

$$-b_{4k+2}b_{4k-2}\bar{a}_1 + b_{4k}\bar{a}_2\bar{b}_{4k-2} = 0.$$

From the conjugate of (3.20), we have that $\bar{a}_2\bar{b}_{4k-2} = -\bar{b}_{4k}\bar{a}_1$ and since $b_{4k-2} = -\bar{b}_{4k-2}$, we obtain

$$[b_{4k+2}\bar{b}_{4k-2} - n(b_{4k})]\bar{a}_1 = 0,$$

therefore, (3.24) also holds in this case.

Now, if we substitute the induction hypothesis $b_{4k-2} = b_2\lambda_{k-1}$ in (3.24), it follows (3.17). From (3.11), this also implies that (3.15) holds.

Finally, for $k = 1$, observe that all the arguments work without the induction hypothesis, and that (3.24) becomes $b_6 = b_2\lambda_1$. This ends the proof and establishes (3.12-18).

We are now ready to settle the third case. From (3.3), we have

$$(0, b_2a_1) + (\bar{b}_{4n-1}, -b_{4n})(a_2, a_3) = 0,$$

and this gives,

$$\bar{b}_{4n-1}a_2 + \bar{a}_3b_{4n} = 0, \tag{3.25}$$

$$b_2a_1 + a_3\bar{b}_{4n-1} - b_{4n}\bar{a}_2 = 0. \tag{3.26}$$

Right multiplying (3.26) by b_{4n-1} and then using $\bar{a}_2b_{4n-1} = -\bar{b}_{4n}a_3$ and $a_1b_{4n-1} = \bar{b}_{4n-2}a_3$, that come, respectively, from the conjugate of (3.25) and from (3.19) with $k = n$, we get

$$[b_2\bar{b}_{4n-2}+n(b_{4n-1})+n(b_{4n})]a_3 = 0.$$

Then, using (3.17), we have $(\lambda_o = 1)$,

$$[\lambda_{n-1}n(b_2)+n(y_{2n})]a_3 = 0,$$

therefore, $a_3 = 0$. Consequently, $b_{4n-1} = 0$.

Now, by substitution of these values in (3.26), we obtain $b_2a_1-b_{4n}\bar{a}_2 = 0$, and, since from (3.2), (3.11), it follows that $b_2a_1 = -\bar{b}_2a_1 = \bar{a}_1b_2$, we have $\bar{a}_1b_2-b_{4n}\bar{a}_2 = 0$. Its conjugate gives

$$\bar{b}_2a_1-a_2\bar{b}_{4n} = 0.$$

Right multiplying this by b_{4n}, and using $a_1b_{4n} = -b_{4n-2}a_2$ that follows from (3.20) with $k = n$, we get

$$[\bar{b}_2b_{4n-2}+n(b_{4n})]a_2 = [\lambda_{n-1}n(b_2)+n(b_{4n})]a_2 = 0.$$

Hence, $a_2 = 0$, and this is a contradiction to the fact that $x = (a_2,a_3) \neq (0,0)$. Therefore, $g(u,v) = 0$ is not possible in the third case, and this ends the proof that g is nonsingular.

Since $\bar{b}_2a_1+\bar{a}_1b_2$ is real, the component $\Psi_{2n+1}(u,v)$ of $g(u,v)$ lies in a 5-dimensional subspace of K. Then, recalling that the real part of b_2 is zero, with $n = k+1$, we can regard g as the map (3.8). The map (3.9) follows from this by taking the restriction $b_1 = 0$.

Centro de Investigación del IPN.

REFERENCES

[1] J. Adem, Some immersions associated with bilinear maps, Bol. Soc. Mat. Mex. 13, 95-104 (1968).

[2] S. Gitler, The projective Stiefel manifolds - II. Applications, Topology, 7, 47-53 (1968).

[3] K.Y. Lam, Construction of nonsingular bilinear maps, Topology, 6, 423-26 (1967).

[4] ________, On bilinear and skew-linear mpas that are nonsingular, Quart. J. Math. Oxford Ser. 2, 281-88 (1968).

[5] ________, Construction of some nonsingular bilinear maps, Bol. Soc. Mat. Mex. 13, 88-94 (1968).

[6] R.J. Milgram, Immersing projective spaces, Ann. of Math. 85, 473-82 (1967).

HOMOTOPY IS NOT CONCRETE

by

Peter Freyd

Theorem

Let $\mathcal{T}$ *be a category of base-pointed topological spaces including all finite-dimensional CW-complexes. Let* $T: \mathcal{T} \to \mathcal{S}$ *be any set-valued functor which is homotopy-invariant. There exists* $f: X \to Y$ *such that* f *is not null-homotopic, but* $T(f) = T(*)$, *where* $*$ *is null-homotopic.*

Corollary

Let K *be any cardinal number. There exist finite-dimensional CW-complexes* X, Y *and a map* $f: X \to Y$ *not null-homotopic but such that for any* $X' \subset X$, X' *having less than* K *cells,* $f|X'$ *is null-homotopic.*

The corollary follows from the theorem as follows: let Z be the wedge of all CW-complexes having less than K cells. The theorem says that there must exist $f: X \to Y$ not null-homotopic such that $[Z,X] \xrightarrow{[Z,f]} [Z,Y]$ is constant. For any $X' \subset X$, X' having less than K cells, there exists $X' \to Z \to X' = 1_{X'}$ from which we may conclude that $[X',X] \xrightarrow{[X',f]} [X',Y]$ is constant and in particular $f|X'$ is null-homotopic.

Let $\mathcal{H}$ be the "homotopy category" obtained from $\mathcal{T}$. Its objects are the objects of $\mathcal{T}$, its maps are homotopy-classes of maps. The theorem says that $\mathcal{H}$ may not be faithfully embedded in the category of sets, or in the language of Kurosh, $\mathcal{H}$ is not "concrete." There is no interpretation of the objects of $\mathcal{H}$ so that the maps may be interpreted as functions (in a functorial way, at least). $\mathcal{H}$ has always been the best example of an abstract category, historically and philosophically. Now we know that it was of necessity abstract, mathematically.

The theorem says a bit more: $\mathcal{H}$ has a zero-object, that is an object O such that for any X there is a unique $O \to X$ and a unique $X \to O$, and consequently for any X, Y a unique $X \to O \to Y$, the "zero-map" from X to Y.

We shall shortly restrict our attention to zero-preserving functors between categories with zero. Instead of functors into the category of sets $\mathcal{S}$, we'll consider functors into the category of base-pointed sets $\mathcal{S}_*$ and only those functors which preserve zero. But first:

<u>Proposition</u>

<u>If</u> $\mathcal{A}$ <u>is a category with zero, and</u> $T: \mathcal{A} \to \mathcal{S}$ <u>any functor then there exists</u> $T_*: \mathcal{A} \to \mathcal{S}_*$ <u>preserving zero such that for all</u> $f, g: A \to B$ <u>in</u> $\mathcal{A}$, $T(f) = T(g) \Leftrightarrow T_*(f) = T_*(g)$.

<u>Proof</u>

Let $F: \mathcal{S} \to \mathcal{G}$, the category of abelian groups, be the functor which assigns free groups. F is faithful, hence $T(f) = T(g) \Leftrightarrow FT(f) = FT(g)$. Let Z be the constant functor valued $FT(0)$. There exists transformations $Z \to FT \to Z = 1_Z$, and FT splits as $H \oplus Z$ (remember that the category of functors from $\mathcal{A}$ to $\mathcal{G}$ is an abelian category). H preserves zero and $FT(f) = FT(g) \Leftrightarrow H(f) = H(g)$. Let U: (Abelian Groups) $\to \mathcal{S}_*$ be the forgetful functor and define $T_* = UH$. ∎

In light of this proposition the main theorem is equivalent with

<u>For any zero-preserving functor</u> $T: \mathcal{H} \to \mathcal{S}_*$, <u>there exists</u> f <u>in</u> $\mathcal{H}$, $f \neq 0$, $T(f) = 0$.

$\mathcal{H}$ is thus worse than non-concrete: not only must any $T: \mathcal{H} \to \mathcal{S}_*$ confuse two distinct maps, it must confuse two maps one of which is a zero-map. Such failure to be concrete is easier to work with than the more general. We will say that a functor $T: \mathcal{C} \to \mathcal{S}_*$ is FAITHFUL-At-ZERO if $T(f) = 0 \Rightarrow f = 0$, and $\mathcal{C}$ is CONCRETE-At-ZERO if there exists $T: \mathcal{C} \to \mathcal{S}_*$ faithful at zero. The last proposition says that concrete implies concrete-at-zero. We wish to show that $\mathcal{H}$ is not concrete-at-zero; we shall isolate a property that any concrete-at-zero category must possess and then demonstrate its failure in $\mathcal{H}$.

We shall work for awhile in an arbitrary category $\mathcal{C}$ with zero. Given $A \in \mathcal{C}$ we may define an equivalence relation on the monomorphisms into A: $(B_1 \to A) \equiv (B_2 \to A)$ if there exists an isomorphism $B_1 \xrightarrow{\approx} B_2$ such that

$$\begin{array}{ccc} B_1 & \searrow & \\ \downarrow & & A \\ B_2 & \nearrow & \end{array}.$$

A <u>subobject</u> of A is defined to be an equivalence class of monomorphisms. A kernel of a map $A \to B$ is usually defined as a monomorphism into A satisfying the well-known universal property. We note here that "the" kernel of $A \to B$ may be defined as a subobject, removing completely the ambiguity. (Every monomorphism in the equivalence class must of necessity be a kernel). A NORMAL SUBOBJECT is one that appears as a kernel. The following will be a corollary of a later theorem:

If $\mathcal{C}$ is concrete-at-zero then every object in A *has only a set of normal subobjects. Moreover, if every map in $\mathcal{C}$ has a kernel then the converse holds.*

This theorem, as it stands, is not useful for $\mathcal{H}$. There are very few kernels, indeed there are very few monomorphisms, in $\mathcal{H}$. We therefore introduce another equivalence relation, this time on *all* the maps into a fixed object A.

$X \to A \equiv X' \to A$ if for all $A \to Y$, $X \to A \to Y = 0 \Leftrightarrow X' \to A \to Y = 0$.

We shall call the equivalence classes, ABSTRACT NORMAL SUBOBJECTS of A. The connection with normal subobjects is this:

Proposition

If $\mathcal{C}$ has kernels and cokernels for every map then each abstract normal subobject contains a unique normal subobject.

Proof

One may first check that two normal monomorphisms are equivalent in the previous (the "subobject" sense) iff they are equivalent in the new sense (the "abstract subobject" sense). If $\mathcal{C}$ has kernels and cokernels then given $f: X \to A$ we note that $\mathrm{Ker}(\mathrm{Cok}(f))$ is equivalent to f. ∎

Theorem

$\mathcal{C}$ is concrete-at-zero iff every object has only a set of abstract normal subobjects.

Proof

If $T: \mathcal{C} \to \mathcal{S}_*$ is faithful-at-zero and $X \to A \not\equiv X' \to A$ then $T(X) \to T(A) \not\equiv T(X') \to T(A)$ by a direct verification. Thus by choosing a representative from each abstract normal subobject of A we obtain a one-to-one function into the set of abstract normal subobjects of $T(A)$. The latter are in natural one-to-one correspondence with the subobjects of $T(A)$ which in turn are in natural one-to-one correspondence with the pointed subsets of $T(A)$.

For the converse, define $T\colon \mathcal{G} \to \mathcal{S}_*$ by letting $T(A)$ be the set of abstract normal subobjects of A. Given $f\colon A \to B$ note that if $X \to A \equiv X' \to A$ then $X \to A \to B \equiv X' \to A \to B$; thus $A \to B$ induces a function $T(A) \to T(B)$ clearly seen to be functorial. If $T(A) \to T(B)$ were constant then $A \xrightarrow{1} A \to B \equiv 0 \to A \to B$ and from $A \xrightarrow{1} A \to B \xrightarrow{1} B = 0 \Leftrightarrow 0 \to A \to B \xrightarrow{1} B = 0$ we conclude that $A \to B = 0$. ▮

The previous assertion that if $\mathcal{G}$ has kernels then the fact that every object has only a set of normal subobjects is equivalent with concreteness-at-zero, may be seen by looking at $\mathcal{G}^{op}$ and noticing that in general $X \to A \equiv X' \to A \Leftrightarrow \mathrm{Cok}(X \to A) = \mathrm{Cok}(X' \to A)$. This would yield $T\colon \mathcal{G}^{op} \to \mathcal{S}_*$. However, the contravariant functor represented by the two-point set is faithful and we would obtain $\mathcal{G} \to \mathcal{S}_*$.

A WEAK-KERNEL of $A \to Y$ is a map $X \to A$ such that

1) $X \to A \to Y = 0$

2) If $Z \to A \to Y = 0$ then there exists $\begin{array}{c} Z \\ \downarrow \\ X \end{array} \begin{array}{c} \searrow \\ \nearrow \end{array} A$

(no uniqueness condition).

If both $X \to A$ and $X' \to A$ are weak-kernels (of possibly different things) then $X \to A \equiv X' \to A$ iff there exist

$$\begin{array}{c} X \\ \downarrow \\ X' \end{array} \begin{array}{c} \searrow \\ \nearrow \end{array} A \quad \text{and} \quad \begin{array}{c} X' \\ \downarrow \\ X \end{array} \begin{array}{c} \searrow \\ \nearrow \end{array} A$$

by direct application of the definition of $\equiv$ and weak-kernels.

In $\mathcal{H}$ we have many weak-cokernels. Indeed, the suspension of any map is always such. We are directed, therefore, to look at the dual side, keeping in mind that a contravariant $T: \mathcal{H} \to \mathcal{S}_*$ faithful-at-zero yields a covariant functor if followed by the faithful $(-,2)$. We wish to find a space A, a class of maps $\{A \to X_i\}_{i \in I}$ all of which are weak-cokernels such that for $i \neq j$ not both $A \begin{smallmatrix} \nearrow & X_i \\ & \downarrow \\ \searrow & X_j \end{smallmatrix}$ and $A \begin{smallmatrix} \nearrow & X_j \\ & \downarrow \\ \searrow & X_i \end{smallmatrix}$ exist.

From the theory of abelian groups:

<u>Lemma</u>

<u>For any positive integer</u> p, <u>there exists a family of</u> p-<u>primary torsion abelian groups</u> $\{G_\alpha\}$, α <u>running through the ordinal numbers</u>, <u>and for each</u> G_α <u>a special element</u> $x_\alpha \in G_\alpha$ <u>such that</u>

$$x_\alpha \neq 0$$
$$px_\alpha = 0$$

<u>for any homomorphism</u> $f: G_\beta \to G_\alpha$, $\beta > \alpha$, $f(x_\beta) = 0$.

<u>Proof</u>

We recall the theory of "height" in torsion groups. Let $\mathcal{G}_p$ be the category of p-primary torsion abelian groups, let I be its identity functor. For each ordinal α we define a subfunctor inductively by

$$I_0 = I$$
$$I_{\alpha+1} = \mathrm{Image}(I_\alpha \xrightarrow{p} I_\alpha)$$
$$I_\beta = \cap_{\alpha<\beta} I_\beta \qquad \beta \text{ a limit ordinal.}$$

We must show that this descending sequence continues to descend forever. Given α we shall find G_α such that $I_\alpha(G_\alpha) \neq 0$, $I_{\alpha+1}(G_\alpha) = 0$. By letting x_α be a non-zero element in $I_\alpha(G_\alpha)$ we will achieve the announced end, because if $\beta > \alpha$, then $I_\beta(G_\beta) \to I_\beta(G_\alpha)$ but $I_\beta(G_\alpha) \subset I_{\alpha+1}(G_\alpha) = 0$.

Given α let W_α be the set of finite words of ordinals $\langle\gamma_1\gamma_2\cdots\gamma_n\rangle$ where $\gamma_1 < \gamma_2 < \ldots < \gamma_n \leq \alpha$, including the empty word $\langle\ \rangle$. Let G_α be the group generated by W_α subject to the relations $p\langle\gamma_1\cdots\gamma_n\rangle = \langle\gamma_2\gamma_3\cdots\gamma_n\rangle$, $\langle\ \rangle = 0$. Then G_α is p-primary torsion. Note that every non-zero element in G_α is expressible uniquely as something of the form $a_1w_1 + \ldots + a_nw_n$, $a_i = 1,2,\ldots,p-1$ $w_i \in W - \{\langle\ \rangle\}$. We may then show, inductively that $I_\gamma(G_\alpha)$ is generated by elements of the form $\langle\gamma_1\gamma_2\cdots\gamma_n\rangle$ where $\gamma \leq \gamma_1$. Hence $I_\alpha(G_\alpha) \simeq Z_p$, $I_{\alpha+1}(G_\alpha) = 0$. ∎

Let $M(G)$ be the Moore space, $H_1(MG) \simeq G$. Choose a prime p, a generator x for $H_1(M(Z_p))$ and for each α a map $f_\alpha\colon M(Z_p) \to M(G_\alpha)$ such that $(H(f))(x) = x_\alpha$.

For $\beta > \alpha$ there is no

$$\begin{array}{ccc} & \xrightarrow{\Sigma(f_\beta)} & \Sigma M(G_\beta) \\ \Sigma M(Z_p) & & \downarrow \\ & \xrightarrow{\Sigma(f_\alpha)} & \Sigma M(G_\alpha) \end{array}$$

because application of H_2 would contradict the choice of x_α, G_α. Each $\Sigma(f_\alpha)$ is a weak-cokernel. Hence each $\Sigma(f_\alpha)$ represents a different abstract normal quotient object. Hence $\Sigma(M(Z_p))$ has more than a set of abstract normal quotient objects. Hence $\mathcal{H}$ is not concrete-at-zero.

We may be more specific: for any $n \geq 0$ consider the

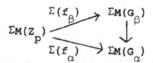

mapping-cone sequence

$$\Sigma^n M(Z_p) \to \Sigma^n M(G_\alpha) \to \Sigma^n M(G_\alpha/Z_p) \xrightarrow{g_\alpha} \Sigma^{n+1} M(Z_p) \to \Sigma^{n+1} M(G_\alpha)$$

(The mapping cone of $\Sigma^n M(Z_p) \to \Sigma^n M(G_\alpha)$ is a Moore space because $Z_p \to G_\alpha$ is monomorphic.) For $\beta > \alpha$

$$\Sigma^n M(G_\beta/Z_p) \xrightarrow{g_\beta} \Sigma^{n+1} M(Z_p) \to \Sigma^{n+1} M(G_\alpha) \neq 0.$$

Let $T: \mathcal{H} \to \mathcal{S}_*$ be any functor. Let $\beta > \alpha$ be such that $T(g_\beta)$ and $T(g_\alpha)$ have the same image in $T\,\Sigma^{n+1} M(Z_p)$. Then because $T(\Sigma^{n+1} f_\alpha \cdot g_\alpha) = 0$ it must be the case that $T(\Sigma\, f_\alpha \cdot g_\beta) = 0$.

Note that for each n we have shown that the homotopy category of (n+3)-dimensional, n-connected CW-complexes is not concrete-at-zero. With $n \geq 1$ we know that it is not the basepoints that prevent concreteness. With $n = 3$ we know that the stable category is not concrete. We have shaved it as fine as possible: the homotopy category of (n+2)-dimensional, n-connected CW-compleses is concrete, at least for $n \geq 2$.

On Concreteness in General

When we move away from zero, the notion of normal subobjects is not enough. A regular subobject is one that appears as an equalizer. We accordingly define yet another equivalence relation on maps into A.

$$X \to A \equiv X' \to A \quad \text{iff for all} \quad f, g: A \to Y$$

$$X \to A \xrightarrow{f} Y = X \to A \xrightarrow{g} Y \Leftrightarrow X' \to A \xrightarrow{f} Y = X' \to A \xrightarrow{g} Y.$$

The equivalence classes will be called ABSTRACT REGULAR

SUBOBJECTS. A necessary condition for concreteness is that every object have only a set of abstract regular subobjects, and I have just recently proved that for categories with finite products this is a sufficient condition. For categories without products a different condition is available, discovered by John Isbell [2]: fixing A, B define

$$\begin{matrix} & & A \\ & \nearrow & \\ X & & \\ & \searrow & \\ & & B \end{matrix} \;\equiv\; \begin{matrix} & & A \\ & \nearrow & \\ X' & & \\ & \searrow & \\ & & B \end{matrix} \quad \text{iff for all} \quad \begin{matrix} A & & \\ & \searrow & \\ & & Y \\ & \nearrow & \\ B & & \end{matrix}$$

$$\begin{matrix} & & A & & \\ & \nearrow & & \searrow & \\ X & & & & Y \\ & \searrow & & \nearrow & \\ & & B & & \end{matrix} \quad \text{commutes} \Leftrightarrow \begin{matrix} & & A & & \\ & \nearrow & & \searrow & \\ X' & & & & Y \\ & \searrow & & \nearrow & \\ & & B & & \end{matrix} \quad \text{commutes.}$$

The condition, then, is that for any A, B only a set of equivalence classes arise. This condition allows us to formally adjoin finite products in a way to get the abstract-regular-subobject condition. That condition allows us to formally adjoin equalizers (while preserving the products) to get the condition that every object has only a set of regular subobjects. Now the hard part. A long rather arduous construction takes place in the category of set-valued functors.

In a previous paper [1], I showed that the category-of-small-categories-and-natural-equivalence-classes-of-functors is not concrete. I also gave an unenlightening proof that the category of groups-and-conjugacy-classes-of-homomorphisms <u>is</u> concrete, a fact rather easily seen from the sufficiency of the abstract-regular-subobject condition. Also, the characterization therein of those $\mathfrak{a}$ for which the category of "petty" functors from $\mathfrak{a}$ is concrete becomes much easier. The Eckmam-Hilton analogue of homotopy in abelian categories usually yields non-concrete categories, as do the notions of homotopy on chain complexes.

This rather esoteric side of category theory has its highest moments in the work of the Prague School. Hedrlin, Kucera, and all, have shown that every concrete category is _fully_ enbeddable in the category of semigroups (or graphs, or topological spaces and local homeomorphisms, or Hopf-algebras, or the category of discrete representation of a certain 5-element nomoid). Non-concreteness easily says that there is no fruitful algebraic representation. They have shown that the alternative is the necessary existence of a _full_ and faithful representation.

University of Pennsylvania

Bibliography

[1] P.J. Freyd, _On the concreteness of certain categories_, Volume for Rome conference, March 1969.

[2] J. Isbell, _Two set-theoretical theorems in categories_, Fundamenta Math., 53 (1963), pp. 43-49.

THE K THEORY OF STIEFEL MANIFOLDS

BY

S. GITLER AND KEE YUEN LAM

1. INTRODUCTION

Hodgkin in [6] has shown that for a compact connected Lie group G with $\pi_1(G)$ torsion free, $K^*(G)$ is an exterior algebra on the basic representations of G, and is primitively generated. In particular, $K^*(Spin(m))$ has such a description. In this paper, we go on to determine $K^*(SO(m))$ and the K theory of Stiefel manifolds $V_{m,r}$. The answers turn out to be quite simple. For example, as a ring, $K^*(SO(2n)) = E \otimes A$, where E is an exterior algebra, A is a Z_2-graded ring with $A^o = K^o(RP^{2n-1})$ and $A^1 = Z$. We refer to theorems (4.8), (4.9) for complete statements.

Our methods are, in fact, quite simple-minded. We first settle $K^*(SO(m))$ by explicitly constructing all its basic generators. (§4,§5). It turns out that, unlike Spin(m), some generators of $K^*(SO(m))$ are necessarily non-primitive. We then determine $K^*(V_{m,r})$ by induction on r, using the Gysin sequence for $S^{m-r} \to V_{m,r} \to V_{m,r-1}$ as the main tool. Usually the Gysin; sequence does not determine all the ring structure in $K^*(V_{m,r})$, but a look at its image in $K^*(SO(m))$ will.

To carry out such a program it is necessary to study the representation rings of spinor groups and induced homomorphisms between these rings. We collect the pertinent information in §2. Elements in these representation rings are used to describe the basic elements in $K^*(V_{m,r})$ via the so called α and β constructions. These constructions are explained in §3.

2. REPRESENTATION RINGS OF SPINOR GROUPS

In this section we recall the structure of the complex representation ring $R(m) = R(Spin(m))$. Let T_m be the maximal torus in $Spin(m)$. Then $R(m)$ is the subring in $R(T_m)$ invariant under the Weyl group W_m. We separate into cases $m = 2n$ and $m = 2n+1$.

For $m = 2n$,

$$R(T_{2n}) = Z[u_1^2,\ldots,u_n^2,u_1^{-2},\ldots,u_n^{-2},(u_1\ldots u_n)], \tag{2.1}$$

and W_{2n} is generated by permutations of $\{u_1,\ldots,u_n\}$ and even number of inversions $u_i \to u_i^{-1}$. Let

$$\Delta_{2n}^{\pm} = \Sigma\, u_1^{\varepsilon_1}\ldots u_n^{\varepsilon_n},$$

where the sum ranges over all sequences $(\varepsilon_1,\ldots,\varepsilon_n)$ with $\varepsilon_i = \pm 1$, $\Pi\varepsilon_i = +1$ (for Δ_{2n}^{+}) or $\Pi\varepsilon_i = -1$ (for Δ_{2n}^{-}).

For $m = 2n+1$, $R(T_{2n})$ is again given by (2.1), but W_{2n+1} is now generated by permutations of $\{u_1,\ldots,u_n\}$ and transformations $u_i \to u_i^{\varepsilon_i}$, $\varepsilon_i = \pm 1$. Define Δ_{2n+1} by

$$\Delta_{2n+1} = \Pi_{i=1}^{n}(u_i+u_i^{-1}).$$

<u>Theorem</u> (2.2) (See [7]) <u>We have</u>

$$R(2n) = Z[\lambda_1,\dots,\lambda_{n-2},\Delta_{2n}^{+},\Delta_{2n}^{-}],$$

$$R(2n+1) = Z[\lambda_1,\dots,\lambda_{n-1},\Delta_{2n+1}],$$

where λ_i *is the representation* $Spin(m) \xrightarrow{p} SO(m) \xrightarrow{\lambda_i} U(\binom{m}{i})$. *Moreover, in* $R(2n)$,

$$(\Delta_{2n}^{\pm})^2 = \lambda_n^{\pm} + \lambda_{n-2} + \lambda_{n-4} + \cdots$$

$$\Delta_{2n}^{+}\Delta_{2n}^{-} = \lambda_{n-1} + \lambda_{n-3} + \lambda_{n-5} + \cdots$$

while in $R(2n+1)$,

$$(\Delta_{2n+1})^2 = \lambda_n + \lambda_{n-1} + \cdots + \lambda_1 + 1.$$

Here $\lambda_n^{\pm}$ *are the two representations into which* λ_n *splits*.

Theorem (2.2) is not convenient when we consider restriction homomorphisms such as $R(m) \to R(m-r)$. For this purpose we introduce the "K-Pontryagin classes" π_i in $R(m)$, first considered in [2]. Define $\pi_i \in R(m)$ to be the ith elementary symmetric function on $u_1^2 + u_1^{-2} - 2,\dots,u_n^2 + u_n^{-2} - 2$. Then the π_i's and λ_i's are mutually expressible as integral linear combinations of each other. In fact, their precise relationship is given by

$$1+\lambda_1 t+\dots+\lambda_m t^m = 1+\pi_1 s+\dots+\pi_{[m/2]} s^{[m/2]},$$

where $s = t(1+t)^{-2}$. From this, and some amount of computation, one can get

Theorem (2.3). *Let* $\delta_{2n}^{\pm} = \Delta_{2n}^{\pm} - 2^{n-1}$, $\delta_{2n+1} = \Delta_{2n+1} - 2^n$.

Then

$$R(2n) = Z[\pi_1,\ldots,\pi_{n-2},\delta_{2n}^+,\delta_{2n}^-],$$

$$R(2n+1) = Z[\pi_1,\ldots,\pi_{n-1},\delta_{2n+1}].$$

Furthermore, in $R(2n)$

$$(\Delta_{2n}^+ - \Delta_{2n}^-)^2 = \pi_n, \quad (\Delta_{2n}^+ + \Delta_{2n}^-)^2 = \pi_n',$$

where π_n' is defined to be $\pi_n+4\pi_{n-1}+16\pi_{n-2}+\ldots+2^{2n}$; while in $R(2n+1)$,

$$\Delta_{2n+1}^2 = \pi_n'.$$

Theorem (2.4). Under the homomorphism induced by inclusion, $\delta_{2n}^{\pm} \to \delta_{2n-1}$, $\delta_{2n+1} \to \delta_{2n} = \delta_{2n}^+ + \delta_{2n}^-$. Likewise, if $j^*:R(m) \to R(m-r)$ is so induced, then

$$j^*(\pi_i) = \pi_i, \quad 1 \leq i \leq [\tfrac{1}{2}(m-r)]$$

$$j^*(\pi_i) = 0, \quad i > [\tfrac{1}{2}(m-r)].$$

From these, it is not hard to read out the kernel and cokernel of j^*.

We can now state the structure of $K^*(\mathrm{Spin}(m))$. Each $\rho \in R(G)$ determines an element $\beta\rho \in K^1(G)$ (see [6] or §3). Then

Theorem (2.5) (Hodgkin).

$$K^*(\mathrm{Spin}(2n) = E(\beta\pi_1,\ldots,\beta\pi_{n-2},\beta\delta_{2n}^+,\beta\chi_{2n}),$$

$$K^*(\mathrm{Spin}(2n+1)) = E(\beta\pi_1,\ldots,\beta\pi_{n-1},\beta\delta_{2n+1}),$$

where $\chi_{2n} = \delta^+_{2n} - \delta^-_{2n}$.

This theorem will be the basis for the rest of our work.

3. THE α AND β CONSTRUCTIONS ([3], [6])

Let $j:H \to K$ be an inclusion of (compact) Lie groups and $j^*:R(K) \to R(H)$ the induced homomorphism between their representation rings. Suppose $\rho \in R(K)$ belongs to ker j^*. Then we can write $\rho = \rho_1 - \rho_2$ where $\rho_1, \rho_2 : K \to U(n)$ are such that they have the same restrictions to H. The map $K/H \to U(n)$ given by $kH \to \rho_1(k)\rho_2(k^{-1})$ is well-defined, and represents an element $\beta\rho$ in $K^1(K/H)$. This element is said to be obtained from ρ "by the β construction". It depends, in general, on the choice of ρ_1 and ρ_2.

Let $\mu \in R(H)$ be a (virtual) representation of H. Then we can associate a (virtual) bundle $\alpha_o(\mu)$ to the principal bundle $H \to EH \to BH$. We regard $\alpha_o(\mu)$ as an element in $K^o(BH)$. In the same way, using $H \to K \to K/H$, we get $\alpha(\mu) \in K^o(K/H)$. These elements are said to be obtained from μ "by the α construction". Clearly, $\alpha:R(H) \to K^o(K/H)$ is a ring homomorphism, and $\alpha(\mu) = \dim \mu$ if $\mu \in \mathrm{Im}\, j^*$.

Both the α and β constructions are natural, each in an appropriate sense which will not be detailed here.

We go on to describe a relative version of the α construction.

To simplify matters somewhat, in this section we suppose that H is a compact connected Lie group then $K^1(BH) = 0$ [4 ; (4.8)]. Regard $B_j : BH \to BK$ as inclusion. Then there is a commutative diagram of exact sequences

$$\begin{array}{ccccccc}
0 & \longrightarrow & \ker j^* & \longrightarrow & R(K) & \xrightarrow{j^*} & R(H) \\
 & & \Big\downarrow \alpha_0 & & \Big\downarrow \alpha_0 & & \Big\downarrow \alpha_0 \\
0 = K'(BH) & \xrightarrow{\delta} & K^0(BK,BH) & \longrightarrow & K^0(BK) & \xrightarrow{B_j^*} & K^0(BH),
\end{array}$$

where the first vertical α_0 is defined in the obvious way. For ω in $\ker j^*$, $\alpha_0(\omega) \in K^0(BK,BH)$ is said to be obtained from ω "by the relative α construction". If $\omega' \in R(K)$ is any element, then $\omega\omega' \in \ker j^*$, and clearly $\alpha_0(\omega\omega') = \alpha_0(\omega)\alpha_0(\omega')$.

More generally, whenever $K/H \to E_p \xrightarrow{p} X_p$ is a fibration associated with some principal K-bundle over X_p, one can obviously use universality to define an element $\alpha_p(\omega)$ in $K^0(X_p, E_p)$ (regard p as an inclusion !), "obtained by relative α construction". Of particular interest is when $H \xrightarrow{j} K \xrightarrow{i} G$ is a sequence of Lie group inclusions, and $K/H \to G/H \xrightarrow{p} G/K$ the naturally induced bundle. For this p, we give a more explicit description of $\alpha_p(\omega)$: write $\omega = \omega_1 - \omega_2$ where $\omega_1, \omega_2 : K \to U(n)$ restrict to the same representation $\omega_0 : H \to U(n)$. Then the bundles $\alpha(\omega_1)$, $\alpha(\omega_2)$ over G/K are isomorphic when restricted to G/H. Indeed, both restrict to $\alpha(\omega_0)$, and the identity map of $\alpha(\omega_0)$ provides an explicit isomorphism σ. Then $\alpha_p(\omega)$ is the "difference element" represented by the exact sequence over G/H:

$$0 \longrightarrow \alpha(\omega_1) \xrightarrow{\sigma} \alpha(\omega_2) \longrightarrow 0,$$

in the sense of [3, § 9].

Suppose now $\omega = i^*(\eta)$ for a certain $\eta \in R(G)$. Since $j^*i^*(\eta) = 0$, η gives arise to $\beta\eta \in K^1(G/H)$. The relation between $\beta\eta$ and $\alpha_p(\omega)$ is given by

Proposition (3.1). *For any* $\eta \in \text{kern}\,(R(G) \to R(H))$, *we have* $\delta(\beta\eta) = -\alpha_p(i^*\eta)$, *where* δ *is the coboundary in the exact sequence of the pair* $(G/K, G/H)$.

Proof. Write $\eta = \eta_1 - \eta_2$ where η_1, η_2 $G \to U(n)$ restricts to the same $\omega_o : H \to U(n)$. Let $i^*(\eta) = \omega$, $i^*(\eta_i) = \omega_i$. The coboundary $\delta(\beta\eta)$ can be represented as a "difference element " by the exact sequence

$$0 \longrightarrow \varepsilon^n \xrightarrow{\tau} \varepsilon^n \longrightarrow 0,$$

where ε^n is the n dimensional trivial bundle over G/K, and τ is the following isomorphism defined over G/H:

$$\tau : (gH, c) \longrightarrow (gH, \eta_1(g)\eta_2(g^{-1})c)\,; \quad g \in G,\ c \in C^n.$$

On the other hand, each $\alpha(\omega_i)$ is trivial over G/K. An explicit trivialisation is

$$\nu_i : \frac{G \times C^n}{(g,c) \sim (gk^{-1}, \omega_i(k)c)} \xrightarrow{\approx} (G/K) \times C^n, \quad (k \in K)$$

defined by $\nu_i[g,c] = (gK, \eta_i(g)c)$. It is now easy to check that the following diagram is commutative over G/H:

$$\begin{array}{ccccccc} 0 & \longrightarrow & \alpha(\omega_1) & \xrightarrow{\sigma} & \alpha(\omega_2) & \longrightarrow & 0 \\ & & \downarrow{\scriptstyle \nu_1} & & \downarrow{\scriptstyle \nu_2} & & \\ 0 & \longrightarrow & \epsilon^n & \xrightarrow{\tau^{-1}} & \epsilon^n & \longrightarrow & 0. \end{array}$$

From this it follows that $\delta(\beta\eta) = -\alpha_p(i^*(\eta))$, as claimed.

In this paper, we shall exploit proposition (3.1) in the following way. Take $H = Spin(2q-1)$, $K = Spin(2q)$ so that p has S^{2q-1} as fiber. One can then consider the Gysin sequence of p:

$$\begin{array}{ccc} K^*(G/K) & \xrightarrow{p^*} & K^*(G/H) \\ {\scriptstyle \phi}\nwarrow & & \swarrow{\scriptstyle \psi} \\ & K^*(G/K). & \end{array}$$

Here ϕ is "cupping with the Euler class" and ψ is the coboundary δ followed by the inverse Thom isomorphism $\tau^{-1}: K^*(G/K, G/H) \to K^*(G/K)$. Suppose, as before, that $j^*i^*(\eta) = 0$. Proposition (3.1) can now be used to compute $\psi(\beta\eta)$:

<u>Proposition</u> (3.2). $\psi(\beta\eta) = -\alpha(\zeta) \in K^0(G/K)$, where $\zeta = (\delta^+_{2q} - \delta^-_{2q})^{-1} i^*(\eta)$.

<u>Proof</u>. Since $\ker j^*$ is the ideal generated by $\delta^+_{2q} - \delta^-_{2q}$, ζ is well-defined. The Thom class $U_p \in K^0(G/K, G/H)$ is given by

$\alpha_p(\delta_{2q}^+ - \delta_{2q}^-)$, as can be seen from the universal fibration $BH \to BK$ and naturality of the relative α construction. Hence by (3.1)

$$\psi(\beta\eta) = -T^{-1}(\alpha_p(i^*(\eta))) = -T^{-1}(\alpha_p(\delta_{2q}^+ - \delta_{2q}^-)\alpha(\zeta))$$

$$= -T^{-1}(U_p \cdot \alpha(\zeta)) = -\alpha(\zeta).$$

Finally we shall explain another relation between the α and β constructions, generalising proposition (4.1) of [6] up to a sign difference. Again consider the inclusion $H \xrightarrow{j} K$. Suppose $\rho = \rho_1 - \rho_2$ belongs to kern j^*, as in the beginning paragraph of this section. Corresponding to the fibration $K/H \to BH \xrightarrow{B_j} BK$, form the diagram

(3.3)
$$\begin{array}{ccccccc} 0 = K'(BH) & \longrightarrow & K'(K/H) & \xrightarrow{\delta} & K^0(BH, K/H) & \longrightarrow & K^0(BH) \\ & & & & \uparrow{\scriptstyle \bar{B}_j^*} & & \uparrow{\scriptstyle B_j^*} \\ & & & & K^0(BK, *) & \xrightarrow{\approx} & K^0(BK). \end{array}$$

It is harmless to identify the two groups at the bottom. As usual, define the <u>suspension homomorphism</u>

$$\sigma\colon \ker\ B_j^* \longrightarrow K'(K/H)$$

by $\chi \to \delta^{-1}\bar{B}_j^*(\chi)$. Since $K'(BH)$ is assumed zero, σ is well-defined. We assert

<u>Proposition</u> (3,4). $\beta\rho = \sigma(\alpha_o(\rho))$.

<u>Proof</u>. We must show that $\delta(\beta\rho) = \bar{B}_j^*(\alpha_o(\rho))$. This is done

by explicit checking. Let EK be a contractible space on which K acts freely. We can take BK = EK/K, BH = EK/H. Choose a base point $e \in EK$ above *, and identify K with $eK \subset EK$. The bundles $\alpha_o(\rho_i)$ over BK and $B_j^*\alpha_o(\rho_i)$ over BH, i = 1,2, have total spaces given by

$$\frac{EK \times C^n}{(x,c) \sim (xk^{-1}, \rho_i(k)c)} \quad \text{and}$$

$$\frac{EK \times C^n}{(x,c) \sim (xh^{-1}, \rho_i(h)c)} \qquad ((x,c) \in EK \times C^n; k \in K; h \in H)$$

respectively. Consider now the commutative diagram of exact bundle maps:

$$\begin{array}{ccccccccc} 0 & \longrightarrow & B_j^*(\alpha_o(\rho_1)) & \xrightarrow{f_1} & B_j^*(\alpha_o(\rho_2)) & \longrightarrow & 0 \\ & & \downarrow & & \downarrow & & \\ 0 & \longrightarrow & \alpha_o(\rho_1) & \xrightarrow{f_o} & \alpha_o(\rho_2) & \longrightarrow & 0. \end{array}$$

Here, the vertical maps are induced by the identity of $EK \times C^n$; f_o is defined over *; f_1 is defined over K/H (identified as the fiber above *). Both f_o and f_1 are given, in terms of representatives, by $(ek,c) \mapsto (ek, \rho_2(k^{-1})\rho_1(k)c)$. The diagram shows how, when represented as difference elements, $\alpha_o(\rho)$ is lifted to $\bar{B}_j^*(\alpha_o(\rho))$.

The two bundles on the top row are in fact identical, because $\rho_1/H = \rho_2/H$. We trivialise the restriction of either of them to K/H by

$$\nu : \frac{eK \times C^n}{(ek,c) \sim (ekh^{-1}, \rho_i(h)c)} \longrightarrow (K/H) \times C^n, \quad (k \in K, h \in H)$$

with $\nu[ek,c] = (kH, \rho_1(k)c)$. If is now straight forward to check that the following is a commutative diagram over K/H:

$$\begin{array}{ccccccc} 0 & \longrightarrow & B_j^*(\alpha_o(\rho_1)) & \xrightarrow{f_1} & B_j^*(\alpha_o(\rho_2)) & \longrightarrow & 0 \\ & & \downarrow \nu & & \downarrow \nu & & \\ 0 & \longrightarrow & \epsilon^n & \xrightarrow{f_2} & \epsilon^n & \longrightarrow & 0 \end{array}$$

where f_2 is given by $f_2(kH,c) = (kH, \rho_1(k)\rho_2(k^{-1})c)$. By the properties of difference elements, this shows that $\delta(\beta\rho) = \bar{B}_j^*(\alpha_o(\rho))$.

Two useful consequences of this proposition are:

1) When $K^1(BH) = 0$, $\beta\rho$ is independent of the representation $\rho = \rho_1 - \rho_2$. This is clear from (3.3).

2) If $\rho, \bar{\rho} \in \ker j^*$, then $\beta(\rho\bar{\rho}) = 0$. This follows from general properties of suspension homomorphism.

4. BASIC ELEMENTS IN $K^*(V_{m,r})$

In this section we describe the basic elements which generate $K^*(V_{m,r})$ as a ring. Think of $V_{m,r}$ as

Spin(m)/Spin(m-r), and consider the restriction homomorphism

$$j^*:R(m) \to R(m-r)$$

where j denotes the inclusion map. The basic elements are defined via α and β constructions.

1) Define $\tau_{m,r} \in \widetilde{K}^0(V_{m,r})$ to be $\alpha(\delta_{m-r})$. Recall that $2^t\delta_{m-r} \in \operatorname{Im} j^*$, where

$$(4.1) \qquad t = \begin{cases} r/2-1 & \text{if } m,r \text{ are both even,} \\ [r/2] & \text{otherwise.} \end{cases}$$

Hence $\tau_{m,r}$ is a torsion element with order dividing 2^t. Consideration of its image under the homomorphism induced by the embedding $RP^{m-1}/RP^{m-r-1} \to V_{m,r}$ whows that the order is exactly 2^t. (See [1].) Moreover, since $\delta^2_{m-r} \equiv -2^{[\frac{1}{2}(m-r)]+1}\delta_{m-r} \bmod \operatorname{Im} j^*$, we have

$$(4.2) \qquad \tau^2_{m,r} = -2^{[\frac{1}{2}(m-r)]+1}\tau_{m,r}.$$

2) When $m-r$ is even, there is an additional element $\gamma_{m-r} \in \widetilde{K}^0(V_{m,r})$ given by $\gamma_{m-r} = \alpha(\delta^+_{m-r})$. This is not a torsion element, because it restricts to the generator of $\widetilde{K}^0(S^{m-r})$ in the fibration $S^{m-r} \to V_{m,r} \to V_{m,r-1}$. Ir shows, incidentally, that this fibration is totally noncohomologous to zero in K theory. Moreover, since

$$(\delta^+_{m-r})^2 = \delta^+_{m-r}(\delta_{m-r}-\delta^-_{m-r}) \equiv (\delta^+_{m-r}+2^{2(m-r)-1})\delta_{m-r} (\mathrm{mod}\ \mathrm{Im}\, j^*) ,$$

we have

$$(4.3) \qquad \gamma^2_{m-r} = (\gamma_{m-r}+2^{\frac{1}{2}(m-r)-1})\tau_{m,r}.$$

3) According to §2, the representations

$$\left\{\pi_\ell \,\middle|\, \left[\frac{m-r}{2}\right] + 1 \leq \ell \leq \left[\frac{m-3}{2}\right]\right\}$$

lie in ker j^*, hence the β construction gives elements $\beta\pi$ in $K^1(V_{m,r})$, for each ℓ in the indicated range.

4) If m is even, $\chi_m = \delta^+_m - \delta^-_m$ also lies in ker j^*, giving $\beta\chi_m$ in $K^1(V_{m,r})$. If m is odd, there is no such element.

5) Finally, a last generator in ker j^* is

$$u_{m,r} = \begin{cases} (\Delta^+_m)^2 - 2^{2t}\pi'_{[\frac{1}{2}(m-r)]} & \text{for } m \text{ even} \\ (\Delta_m)^2 - 2^{2t}\pi'_{[\frac{1}{2}(m-r)]} & \text{for } m \text{ odd.} \end{cases}$$

This again produces $\beta u_{m,r} \in K'(V_{m,r})$. However, concerning this element we have

Proposition (4.4). βu_{mr}, is divisible by 2^t in $K^1(V_{m,r})$, t being given by (4.1).

Proof. Let us take m even, the argument being the same for odd m. As in proposition (3.4), $\beta u_{m,r} = \sigma(\alpha_o(u_{m,r}))$, where σ is the suspension homomorphism. Let $\sigma_{(t)}$ denote the suspension homomorphism in K theory with coefficients Z_{2^t}, and ρ_t the mod 2^t reduction. Then

$$\rho_t \beta u_{m,r} = \sigma_{(t)} \rho_t \alpha_o(u_{m,r}) = \sigma_{(t)} \rho_t \alpha_o(\delta_m^+)^2.$$

But, in K theory with Z_{2^t} coefficients, $\rho_t \alpha_o(\delta_m^+)$ is already a suspendable element. Since the product of two suspendable elements suspends to zero, it follows that $\rho_t \beta u_{m,r} = 0$, or $\beta u_{m,r}$ is divisible by 2^t, as claimed.

Having proved proposition (4.4), we can now define a basic element $\theta_{m,r}$ in $K^1(V_{m,r})$ by $2^t \theta_{m,r} = \beta u_{m,r}$. Unlike the previous elements, $\theta_{m,r}$ is obviously not unique.

From the naturality of the α and β constructions, it goes without saying that under the homomorphisms induced by the fibration $V_{m,r+1} \xrightarrow{p} V_{m,r}$ (or $\mathrm{Spin}(m) \xrightarrow{\tilde{p}} V_{m,r}$), the elements $\beta\pi$, $\beta\chi_m$ correspond. Furthermore, $p^*(\tau_{m,r}) = b\tau_{m,r+1}$, where b is 1 or 2 according as $m-r$ is odd or even. If $m-r$ is even, it is also clear from the representation rings that $p^*(\gamma_{m-r}) = \tau_{m,r+1}$. However, due to the non-uniqueness of $\theta_{m,r}$, there is no definite relation between $p^*(\theta_{m,r})$ and $\theta_{m,r+1}$, although it is true that

$$(4.5) \qquad 2^t p^*(\theta_{m,r}) = \begin{cases} 2^t \theta_{m,r+1} & \text{if } m-r \text{ is odd,} \\ 2^{t+1}\theta_{m,r+1} - 2^{2t}\beta\pi_{\frac{1}{2}(m-r)} & \text{if } m-r \text{ is even;} \end{cases}$$

(4.6) $$\tilde{p}^*(\theta_{m,r}) = 2^{-t}\beta u_{m,r} \in K^1(\mathrm{Spin}(m)).$$

The following general result will be useful in determining the ring structure of $K^*(V_{m,r})$:

<u>Lemma</u> (4.7). <u>Let</u> X <u>be a finite</u> CW <u>complex and</u> $x \in K^1(X)$, <u>then</u> $x^2 = 0$.

<u>Proof</u>. x can be represented by a map $f_x: X \to U$. Which factors as $X \xrightarrow{g_x} U(N) \xrightarrow{j} U$. Then $g_x^*[j] = x$, where $[j] \in K^1(U(N))$ satisfies $[j]^2 = 0$ because $K^*(U(N))$ is torsion free. The lemma follows.

We are now ready to state our result on $K^*(V_{m,r})$. For simplicity, first take $m = 2n$. Denote by $A_{2n,2k-1}$ the Z_2-graded ring with $Z \oplus Z_{2^{k-1}}$ in degree zero and Z in degree one, generated by 1, $\tau_{2n,2k-1}$ and $\theta_{2n,2k-1}$ respectively, with ring structure prescribed by

$$\tau^2_{2n,2k-1} = -2^{n-k+1}\tau_{2n,2k-1}; \; \theta^2_{2n,2k-1} = 0 \text{ and}$$

$$\theta_{2n,2k-1}\tau_{2n,2k-1} = 0,$$

(compare (4.2), (4.7)). Also, write $E(Y)$ for the exterior algebra generated by a set Y.

<u>Theorem</u> (4.8). i) $K^*(V_{2n,1}) = E(\beta\chi_{2n})$.

ii) <u>For</u> $k \geq 2$, <u>the element</u> $\theta_{2n,2k-1}$ <u>in</u> $K^1(V_{2n,2k-1})$ <u>can</u>

be appropriately choosen to satisfy $\theta_{2n,2k-1}\tau_{2n,2k-1} = 0$, so that as a ring,

$$K^*(V_{2n,2k-1}) = E(\beta\pi_{n-k+1},\ldots,\beta\pi_{n-2},\beta\chi_{2n}) \otimes A_{2n,2k-1}.$$

iii) For $k \geq 1$, $K^*(V_{2n,2k}) = K^*(V_{2n,2k-1}) \otimes K^*(S^{2n-2k})$ as a $K^*(V_{2n,2k-1})$-module. Moreover, γ_{2n-2k} corresponds to the generator of $\tilde{K}^o(S^{2n-2k})$. The ring structure is completed by

$$\gamma^2_{2n-2k} = (\gamma_{2n-2k}+2^{n-k-1})\tau_{2n,2k}.$$

(compare (4.3)).

The next theorem gives the analogous result on $K^*(V_{m,r})$ when $m = 2n+1$. Its proof is similar and omitted.

Theorem (4.9). i) $K^*(V_{2n+1,1}) = K^*(S^{2n})$.

ii) For $k \geq 1$, $K^*(V_{2n+1,2k}) = E(\beta\pi_{n-k+1},\ldots,\beta\pi_{n-1})\otimes A_{2n+1,2k}$, where $A_{2n+1,2k}$ is the subring generated by 1, $\dot{\tau}_{2n+1,2k}$, and an appropriately choosen $\theta_{2n+1,2k}$ satisfying $\theta_{2n+1,2k}\tau_{2n+1,2k} = 0$.

iii) $K^*(V_{2n+1,2k+1}) = K^*(V_{2n+1,2k}) \otimes K^*(S^{2n-2k})$ as a $K^*(V_{2n+1,2k})$-module, with ring structure completed by

$$\gamma^2_{2n-2k} = (\gamma_{2n-2k}+2^{n-k-1})\tau_{2n+1,2k+1}.$$

5. K THEORY OF ORTHOGONAL GROUPS

The purpose of this section is to prove theorem (4.8) in the special case of $V_{2n,2n-1} = SO(2n)$, and to make a few observations on the Hopf algebra structure.

We begin by observing that $\tau_{2n,2n-1} \in \widetilde{K}^0(SO(2n))$ is represented by $t_{2n}-1$, where t_{2n} is the complexified line bundle associated with the double covering $p:Spin(2n) \to SO(2n)$. Its order is 2^{n-1} and $\tau^2_{2n,2n-1} = -2\tau_{2n,2n-1}$. Next, we present a different (but later shown to be equivalent) description of $\theta_{2n,2n-1}$. Consider the diagram

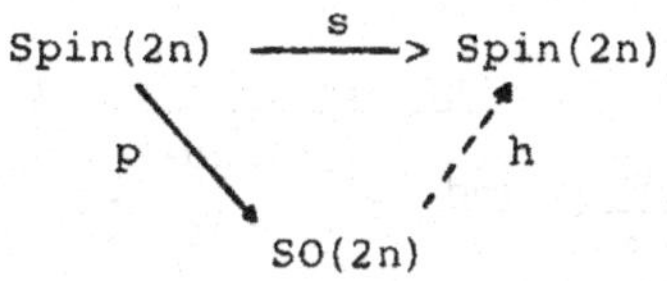

in which s is the "squaring" map, and h the unique well-defined map making the diagram commutative. <u>Define</u> $\theta_{2n,2n-1}$ <u>in</u> $K^1(SO(2n))$ <u>to be</u> $h^*(\beta\delta^+_{2n})$. More explicitly, $\theta_{2n,2n-1}$ is represented by the map $x \to \Delta^+_{2n}(\bar{x})^2$ of $SO(2n)$ into $U(2^{n-1})$, where $\bar{x} \in Spin(2n)$ is such that $p(\bar{x}) = x$. Clearly, $p^*(\theta_{2n,2n-1}) = 2\beta(\delta^+_{2n})$.

Let $b:SO(2n) \to RP^\infty$ be the classifying map for P. Up to homotopy type, one can think of $Spin(2n) \xrightarrow{p} SO(2n) \xrightarrow{b} RP^\infty$ as a fibration with $Spin(2n)$ as fiber. The local coefficient system $\{K^*(Spin(2n))\}$ is trivial, being induced from a map of RP^∞ into $BSpin(2n)$. The Atiyah-Hirzebruch spectral sequence

[4] for this fibration has(*)

$$E_2 = H^*(RP^{\infty};Z) \otimes K^*(\mathrm{Spin}(2n)).$$

Let $z \in H^2(RP^{\infty};Z) = Z_2$ be the generator. In the spectral sequence, $\tau_{2n,2n-1}$ is represented by z. Since $\tau^q_{2n,2n-1} = 0$ iff $q \geq n$, we deduce that

$$d_2 = \dots = d_{2n-1} = 0, \quad d_{2n}(\beta\delta^+_{2n}) = z^n,$$

in this spectral sequence. Thus $\beta\pi_1,\dots,\beta\pi_{n-2}, \beta\chi_{2n}$ and $2\beta\delta^+_{2n}$ in $K^*(\mathrm{Spin}(2n))$ survive to E_{∞}, while $\beta\delta^+_{2n}$ dos not. Moreover, counting torsion elements in E_{2n+1} gives

Proposition (5.1). The order of the torsion subgroup of $K^*(SO(2n))$ does not exceed $2^{(n-1)2^{n-1}}$.

We now prove that, additively at least,

$$K^*(SO(2n)) = E(\beta\pi_1,\dots,\beta\pi_{n-2},\beta\chi_{2n}) \otimes A_{2n,2n-1}.$$

It is already clear that the monomials

$$\left\{(\beta\pi_1)^{\varepsilon_1}\dots(\beta\pi_{n-2})^{\varepsilon_{n-2}}(\beta\chi_{2n})^{\varepsilon_{n-1}}(\theta_{2n,2n-1})^{\varepsilon_n}\right\}_{\varepsilon_i=0,1}$$

give a basis of $K^*(SO(2n))/\mathrm{Torsion}$. In particular, $K^*(SO(2n))$ contains $E = E(\beta\pi_1,\dots,\beta\pi_{n-2},\beta\chi_{2n})$ as a subring. If we can show

(*) We are indebted to R. Held and U. Suter for pointing out the usefulness of this spectral sequence, at this point, resulting in simplifications of our original argument.

that the torsion subgroup $E\tau_{2n,2n-1}$ contains exactly $2^{(n-1)2^{n-1}}$ elements, we would have established the desired additive structure and at the same time proved that $E_{2n+1} = E_{\infty}$ in the spectral sequence. This is done in

Lemma (5.2). *For any* $e \in E$, $e\tau_{2n,2n-1} \neq 0$ *unless* e *is divisible by* 2^{n-1}.

Proof. It is quite easy to see that, under the action map

$$\tilde{\mu}: Spin(2n) \times SO(2n) \longrightarrow SO(2n),$$

$\tilde{\mu}^*(y) = p^*(y) \otimes 1 + 1 \otimes y$ for y = any generator of E, while $\tilde{\mu}^*(\tau_{2n,2n-1}) = 1 \otimes \tau_{2n,2n-1}$. Consequently

(5.3) $\quad \tilde{\mu}(e\tau_{2n,2n-1}) = p^*(e) \otimes \tau_{2n,2n-1}$ + terms not cancelling the first $\neq 0$,

provided that e is not divisible by 2^{n-1}. This proves the lemma.

To complete the ring structure there is just one more thing to show, namely that $\theta_{2n,2n-1}\tau_{2n,2n-1} = 0$. For this we need

Proposition (5.4). *Let* $\phi: SO(2n) \longrightarrow U(2^{n-1})$ *preserve base points and represent an element* $[\phi] \in K^1(SO(2n))$. *Then* $[\phi]t_{2n} \in K^1(SO(2n))$ *is represented by the (well-defined) map*

$$\phi': SO(2n) \longrightarrow U(2^{n-1})$$

given by $\phi'(x) = \Delta^+_{2n}(\bar{x})\phi(x)\Delta^+_{2n}(\bar{x})^{-1}$, where $\bar{x} \in \mathrm{Spin}(2n)$, $P(\bar{x}) = x$.

Proof. Recall that t_{2n} is the complexified line bundle associated with p. We know that $2^{n-1}t_{2n}$ is trivial. Its total space is $\mathrm{Spin}(2n) \times_T C^{2^{n-1}}$, where T acts by $T(\bar{x},c) = (-\bar{x},-c)$. An explicit trivialisation is

$$\lambda : \mathrm{Spin}(2n) \times_T C^{2^{n-1}} \xrightarrow{\approx} SO(2n) \times C^{2^{n-1}},$$

with $\lambda[\bar{x},c] = (p(\bar{x}), \Delta^+_{2n}(\bar{x})c)$. On the other hand, $[\phi]$ can be regarded as an element in

$$K^O(\Sigma SO(2n)) = K^O(SO(2n) \times I, SO(2n) \times \partial I \cup e \times I).$$

We represent $[\phi]$ as a "difference element" by the sequence of bundle maps:

$$0 \longrightarrow \varepsilon^{2^{n-1}} \xrightarrow{\sigma} \varepsilon^{2^{n-1}} \longrightarrow 0,$$

where $\varepsilon^{2^{n-1}}$ has product total space $(SO(2n) \times I)\ C^{2^{n-1}}$, and σ is a bundle isomorphism, defined over $SO(2n) \times \partial I \cup e \times I$, as follows:

$$\begin{cases} \sigma \text{ is the identity over } SO(2n) \times I \cup e \times I; \\ \sigma(x \times 0, c) = (x \times 0, \phi(x)c) \quad \text{over } SO(2n) \times 0. \end{cases}$$

Extend t_{2n} throughout $SO(2n) \times I$ in the obvious way. In the diagram

$$
\begin{array}{ccccccccc}
0 & \longrightarrow & \varepsilon^{2^{n-1}} \otimes t_{2n} & \xrightarrow{\sigma\otimes 1} & \varepsilon^{2^{n-1}} \otimes t_{2n} & \longrightarrow & 0 \\
 & & \downarrow \lambda & & \downarrow \lambda & & \\
0 & \longrightarrow & \varepsilon^{2^{n-1}} & \xrightarrow{\sigma'} & \varepsilon^{2^{n-1}} & \longrightarrow & 0,
\end{array}
$$

the top row represents $[\phi]t_{2n}$ as a difference element, λ is the trivialisation of $\varepsilon^{2^{n-1}} \otimes t_{2n} (=2^{n-1}t_{2n})$, and σ' is a bundle isomorphism defined over $SO(2n) \times \partial I \cup e \times I$, ensuring commutativity. A little diagram chasing confirms that

σ' is the identity over $SO(2n) \times 1 \cup e \times I$,

$\sigma'(x \times 0, c) = (x \times 0, \Delta^+_{2n}(\bar{x})\phi(x)\Delta^+_{2n}(\bar{x})^{-1}c)$ over $SO(2n) \times 0$.

This shows that $[\phi]t_{2n}$ is indeed represented by the map ϕ'.

<u>Corollary</u> (5.5). $\theta_{2n,2n-1}\tau_{2n,2n-1} = 0$ <u>in</u> $K^*(SO(2n))$.

<u>Proof</u>. Take ϕ to be the map $x \to \Delta^+_{2n}(\bar{x})^2$ in this proposition.

The ring structure of $K^*(SO(2n))$ is now completed. In fact, one can obtain some additional information, such as

<u>Corollary</u> (5.6). <u>Under the action map</u> $\tilde{\mu}$,

$$\tilde{\mu}^*(\theta_{2n,2n-1}) = P^*(\theta_{2n,2n-1}) \otimes 1 + 1 \otimes \theta_{2n,2n-1} + \beta\delta^+_{2n} \otimes \tau_{2n,2n-1}.$$

<u>Proof</u>. $\tilde{\mu}^*(\theta_{2n,2n-1})$ is represented by a map ψ of $Spin(2n) \times SO(2n)$ into $U(2^{n-1})$. If $\bar{x}, \bar{y} \in Spin(2n)$, $p(\bar{y}) = y$,

then

$$\psi(\bar{x},y) = (\Delta_{2n}^{+}(\bar{x}\bar{y}))^2$$

$$= \Delta_{2n}^{+}(\bar{x})(\Delta_{2n}^{+}(\bar{y})\Delta_{2n}^{+}(\bar{x})\Delta_{2n}^{+}(\bar{y})^{-1})\Delta_{2n}^{+}(\bar{y})^2 .$$

The maps $(\bar{x},y) \to \Delta_{2n}^{+}(\bar{x})$, $(\bar{x},y) \to \Delta_{2n}^{+}(\bar{y})^2$ represent, respectively, the elements $\beta\delta_{2n}^{+} \otimes 1$ and $1 \otimes \theta_{2n,2n-1}$ in $K^1(\mathrm{Spin}(2n) \times SO(2n))$. By an argument quite similar to that of proposition (5.4), the map

$$(\bar{x},y) \to \Delta_{2n}^{+}(\bar{y})\Delta_{2n}^{+}(\bar{x})\Delta_{2n}^{+}(\bar{y})^{-1}$$

can be shown to represent $\beta\delta_{2n}^{+} \otimes t_{2n}$. Since $P^*(\theta_{2n,2n-1})$ is equal to $2\beta\delta_{2n}^{+}$, the conclusion follows.

This corollary says, in particular, that $\theta_{2n,2n-1}$ cannot be primitive under the multiplication μ of $SO(2n)$. As for the other generators, we have

Proposition (5.7). The generators $\beta\pi_\ell$ $(1 \leq \ell \leq n-2)$ and $\tau_{2n,2n-1}$ are primitive under μ. However, $\beta(\chi_{2n})$ is not. In fact

$$\text{(5.)} \qquad \mu^*(\beta\chi_{2n}) = \beta\chi_{2n} \otimes 1 + 1 \otimes \beta\chi_{2n} + \tau_{2n,2n-1} \otimes \beta\chi_{2n} .$$

Proof. The first assertion is easy because each π_ℓ is a virtual representation of $SO(2n)$. The element $\mu^*(\beta\chi_{2n}) - \beta\chi_{2n} \otimes 1$ is represented by a map ψ of $SO(2n) \times SO(2n)$ into $U(2^{n-1})$. If $\bar{x},\bar{y} \in \mathrm{Spin}(2n)$, $p(\bar{x}) = x$, $p(\bar{y}) = y$, then

$$\psi(x,y) = (\Delta_{2n}^{+}(\bar{x}\bar{y})\Delta_{2n}^{-}(\bar{x}\bar{y})^{-1})(\Delta_{2n}^{+}(\bar{x})\Delta_{2n}^{-}(\bar{x})^{-1})^{-1}$$

$$= \Delta_{2n}^{+}(\bar{x})(\Delta_{2n}^{+}(\bar{y})\Delta_{2n}^{-}(\bar{y})^{-1})\Delta_{2n}^{+}(\bar{x})^{-1}.$$

The map $(x,y) \to \Delta_{2n}^{+}(\bar{y})\Delta_{2n}^{-}(\bar{y})^{-1}$ represents the element $1 \otimes \beta\chi_{2n}$. Again an argument similar to that of proposition (5.4) shows that ψ represents $t_{2n} \otimes \beta\chi_{2n}$. Hence (5.8).

Summarising, we now know that $K^*(SO(2n))$ is not primitively generated, and that μ is not homotopy commutative.

The following proposition brings together the two descriptions of $\theta_{2n,2n-1}$, thus establishing theorem (4.8) completely in the case of $SO(2n)$.

<u>Proposition</u> (5.9). <u>The</u> $\theta_{2n,2n-1}$ <u>defined in this section satisfies</u>

$$2^{n-1}\theta_{2n,2n-1} = \beta u_{2n,2n-1}.$$

<u>Proof</u>. Recall that $u_{2n,2n-1} = (\Delta_{2n}^{+})^2 - 2^{2n-2} = (\delta_{2n}^{+})^2 + 2^n\delta_{2n}^{+}$. In $K^1(Spin(2n))$ we have

$$P^*(\beta u_{2n,2n-1}) = 2^n\beta\delta_{2n}^{+} = 2^{n-1}P^*(\theta_{2n,2n-1});$$

therefore $\beta u_{2n,2n-1} - 2^{n-1}\theta_{2n,2n-1}$ is a torsion element in $K^1(SO(2n))$. Under the action $\tilde{\mu}$, $\beta u_{2n,2n-1}$ is primitive, $2^{n-1}\theta_{2n,2n-1}$ is also primitive by (5.6), while (5.3) shows that no torsion element in $K^1(SO(2n))$ could be primitive. Hence $\beta u_{2n,2n-1} = 2^{n-1}\theta_{2n,2n-1}$.

We conclude this section by remarking that, analogously,

$$K^*(SO(2n+1)) = E(\beta\pi_1, \ldots, \beta\pi_{n-1}) \otimes A_{2n+1,2n},$$

$A_{2n+1,2n}$ being the ring generated by 1, $\tau_{2n+1,2n}$ and $\theta_{2n+1,2n}$. Here, $\theta_{2n,2n+1}$ is represented by $x \mapsto \Delta_{2n+1}(\bar{x})^2$, and satisfies $2^n\theta_{2n+1,2n} = \beta(\Delta^2_{2n+1} - 2^{2n})$. The proof is entirely similar. There are also similar remarks concerning the action and the multiplication map. We omit the details.

6. PROOF OF THEOREM (4.8)

In this section we give a proof of theorem (4.8) based on the preparations laid down in §3-§5. Part i) of the theorem is well known. Assume inductively that $K^*(V_{2n,2k-1})$ is given by ii), (or i) in case $k = 1$). As explained in (2) of 4, the fibration $S^{2n-2k} \longrightarrow V_{2n,2k} \longrightarrow V_{2n,2k-1}$ is totally non-cohomologous to zero. Hence (see [5]) the structure of $K^*(V_{2n,2k})$ is given by iii). To proceed to $K^*(V_{2n,2k+1})$, consider the Gysin sequence for the fibration $S^{2n-2k-1} \longrightarrow V_{2n,2k+1} \xrightarrow{p} V_{2n,2k}$:

$$\begin{array}{ccc} K^*(V_{2n,2k}) & \longrightarrow & K^*(V_{2n,2k+1}) \\ {\scriptstyle\phi}\nwarrow & & \swarrow{\scriptstyle\psi} \\ & K^*(V_{2n,2k}) & \end{array}$$

Here, ϕ is "cupping with the Euler class $\chi \in K^0(V_{2n,2k})$". Since the Euler class for the universal fibration $S^{2n-2k-1} \longrightarrow BSpin(2n-2k-1) \longrightarrow BSpin(2n-2k)$ is given by $\alpha_0(\delta^+_{2n-2k} - \delta^-_{2n-2k})$ in $K^0(BSpin(2n-2k))$,

(6.2) $$\chi = \alpha(\delta^{+}_{2n-2k}-\delta^{-}_{2n-2k}) = \alpha(2\delta^{+}_{2n-2k}-\delta_{2n-2k}) = 2\gamma_{2n-2k}-\tau_{2n,2k}.$$

In the somewhat exceptional case $k = 1$, $K^*(V_{2n,2})$ is free, $\tau_{2n,2}$ should be interpreted as zero and $\psi(\theta_{2n,3}) = -\gamma_{2n-2}$ (see (6.8) below). This information suffices to resolve the extension problem in the Gysin sequence, giving that, additively,

$$K^*(V_{2n,3}) = E(\beta\chi_{2n}) \otimes A_{2n,3}.$$

Furthermore, $\bar{p}^*: K^*(V_{2n,3}) \to K^*(SO(2n))$ is a monomorphism, whence the relation $\theta_{2n,3}\tau_{2n,3} = 0$ can be verified. (Compare the more detailed argument in (6.16) below). This completes the ring structure of $K^*(V_{2n,3})$.

Having settled $V_{2n,3}$, we now take $k \geq 2$ in our inductive argument. Abbreviate by E^* the exterior subalgebra in $K^*(V_{2n,2k})$ generated by $\beta\pi_{n-k+1},\ldots,\beta\pi_{n-2}$ and $\beta\chi_{2n}$. Then (6.1) is a sequence of E^*-modules and E^*-homomorphisms. Since we know ϕ completely, we can break (6.1) up as

(6.3) $$0 \longrightarrow C^* \xrightarrow{p^*} K^*(V_{2n,2k+1}) \xrightarrow{\psi} N^* \longrightarrow 0,$$

where $N^* = \ker\phi$, $C^* = \operatorname{coker}\phi$. After certain amount of computation, the following descriptions for N^* and C^* as modules over E^* emerge:

<u>Lemma</u> (6.4). $N^* \approx E^* \otimes \bar{N}^*$, <u>where</u> $\bar{N}^*$ <u>is the subgroup of</u> $K^*(V_{2n,2k})$ <u>with</u> $\bar{N}^0 = Z \oplus Z_2 \oplus Z_{2^{k-1}}$, <u>generated by</u> χ, $2^{k-2}\tau_{2n,2k}$ <u>and</u> $(\gamma_{2n-2k}+2^{n-k})\tau_{2n,2k}$, <u>and</u> $\bar{N}^1 = Z$ <u>generated by</u> $\theta_{2n,2k}\gamma_{2n-2k}$;

$C^* \approx E^* \otimes \bar{C}^*$; $\bar{C}^*$ being the group with $\bar{C}^0 = Z \oplus Z_{2^k}$ generated by $[1]$ and $[\gamma_{2n-2k}]$, and $\bar{C}^1 = Z \oplus Z_2$ generated by $[\theta_{2n,2k}]$ and $[\theta_{2n,2k}\gamma_{2n-2k}]$.

Consider now the elements $\beta\pi_{n-k}$, $\beta\pi_{n-k}\tau_{2n,2k+1}$ and $\beta u_{2n,2k+1}$ in $K^1(V_{2n,2k+1})$. We can use proposition (3.2) to compute their images under ψ, using $P^*\gamma_{2n-2k} = \tau_{2n,2k+1}$:

$$(6.5)\qquad \begin{cases} \psi(\beta\pi_{n-k}) = -\chi = -2\gamma_{2n-2k} + \tau_{2n,2k}, \\ \\ \psi(\beta\pi_{n-k}\tau_{2n,2k+1}) = -\chi\gamma_{2n-2k} = -(\gamma_{2n-2k}+2^{n-k})\tau_{2n,2k}; \end{cases}$$

$$(6.6)\qquad \psi(\beta u_{2n,2k+1}) = -2^{2k-2}\chi = -2^{2k-1}\gamma_{2n-2k}.$$

Next recall the element $\theta_{2n,2k+1}$ in $K^1(V_{2n,2k+1})$ introduced by

$$(6.7)\qquad 2^k\theta_{2n,2k+1} = \beta u_{2n,2k+1}.$$

On account of (6.6),

$$(6.8)\qquad \psi(\theta_{2n,2k+1}) \equiv -2^{k-1}\gamma_{2n-2k} + \text{a torsion element in } N^0.$$

By lemma (6.4), this torsion element must have the form

$$2^{k-2}e_o\tau_{2n,2k} + e_o'(\gamma_{2n-2k}+2^{n-k})\tau_{2n,2k},$$

where $e_o, e_o' \in E^0$. Taking $\theta_{2n,2k+1}+e_o'(\beta\pi_{n-k})\tau_{2n,2k+1}$ as a new choice of $\theta_{2n,2k+1}$, we do not change (6.7), but simplify (6.8) to

(6.9) $$\psi(\theta_{2n,2k+1}) = -2^{k-1}\gamma_{2n-2k}+2^{k-2}e_o\tau_{2n,2k}.$$

If we introduce

(6.10) $$\bar{\theta}_{2n,2k+1} = \theta_{2n,2k+1}-2^{k-2}\beta\pi_{n-k},$$

then we get the equivalent equation

(6.11) $$\psi(\bar{\theta}_{2n,2k+1}) = -2^{k-2}(1-e_o)\tau_{2n,2k}.$$

Summarizing, we have broken up the Gysin sequence in (6.3) and computed ψ in (6.5) and (6.11). We now need some lemmas to solve the extension problem posed by (6.3).

Lemma A. The element $(1-e_o)$ in (6.11) is invertible mod 2.

Proof. Invertibility mod 2 amounts to the fact that, when $1-e_o$ is written as a sum of monomials in E^*, the constant term is odd. Assume otherwise. Let ω be the product of the generators of E^*. Then $\omega(1-e_o)$ is either zero, or an even multiple of ω. In any case $\psi(\omega\bar{\theta}_{2n,2k+1}) = 0$. By exactness $\omega\bar{\theta}_{2n,2k+1} \in \text{Im}\, p^*$. On the other hand, simple calculations show that the lift of $\omega\bar{\theta}_{2n,2k+1}$ to $K'(\text{Spin}(2n))$ can never come from $K^1(V_{2n,2k})$ (see (4.6)). This contrdiction establishes lemma A.

Remark. Since we are interested in $2^{k-2}(1-e_o)\tau_{2n,2k}$ rather than in $(1-e_o)$, we might as well suppose that $(1-e_o)$ is invertible in E^*.

Lemma B. Concerning $p^*(\theta_{2n,2k})$ we have

1) $2p^*(\theta_{2n,2k})\tau_{2n,2k+1} = 0.$

2) $\theta_{2n,2k+1}$ <u>can be suitably choosen so that</u>

(6.12) $$p^*(\theta_{2n,2k}) = 2\bar{\theta}_{2n,2k+1}(1+\varepsilon_o\tau_{2n,2k+1}),,$$

where $\bar{\theta}_{2n,2k+1}$ is given by (6.10) and ε_o is some element in E^o.

<u>Proof</u>. 1) is immediate because $\tau_{2n\ 2k+1} = p^*(\gamma_{2n-2k})$, and $[\theta_{2n,2k}\gamma_{2n-2k}]$ generates a Z_2 in C^*. To prove (2), take the second half of (4.5), divide by 2^{k-1} to get

(6.13) $$p^*(\theta_{2n,2k}) = 2\bar{\theta}_{2n,2k+1}+T$$

in $K^1(V_{2n,2k+1})$, where T is annihilated by 2^{k-1}. By (6.11), $\psi(T) = 0$. By exactness, one can write

$$T = p^*(\varepsilon_1[\gamma_{2n-2k}]+\varepsilon_o[\theta_{2n,2k}\gamma_{2n-2k}])$$

$$= \varepsilon_1\tau_{2n,2k+1}+\varepsilon_o p^*(\theta_{2n,2k})\tau_{2n,2k+1},$$

for some $\varepsilon_1 \in E^1$, $\varepsilon_o \in E^o$. For T to be annihilated by 2^{k-1}, $\varepsilon_1 \in E^1$ must be even. Take $\theta_{2n,2k+1}+(\varepsilon_1/2)\tau_{2n,2k+1}$ as a revised choice for $\theta_{2n,2k+1}$, and correspondingly redefine $\bar{\theta}_{2n,2k+1}$ as in (6.10). This does not cha ge (6.7) or (6.11), but simplifies (6.13) in the following way:

(6.14) $$p^*(\theta_{2n,2k}) = 2\bar{\theta}_{2n,2k+1}+\varepsilon_o p^*(\theta_{2n,2k})\tau_{2n,2k+1}$$

$$= 2\bar{\theta}_{2n,2k+1}+\varepsilon_o(2\bar{\theta}_{2n,2k+1}+\varepsilon_o p^*(\theta_{2n,2k})\tau_{2n,2k+1})\tau_{2n,2k+1}$$

$$= 2\bar{\theta}_{2n,2k+1}(1+\varepsilon_o\tau_{2n,2k+1}) \quad \text{by 1)}.$$

This completes lemma B.

The element $2\bar{\theta}_{2n,2k+1}\tau_{2n,2k+1}$ occurs in lemma B. We need another expression for it, given in

<u>Lemma</u> C. <u>There is an</u> $e_1 \in E^1$ <u>such that</u>

$$2\bar{\theta}_{2n,2k+1}\tau_{2n,2k+1} = (2^{k-1}(1-e_o)\beta\pi_{n-k}+2e_1)\tau_{2n,2k+1}.$$

<u>Proof</u>. Both $\bar{\theta}_{2n,2k+1}\tau_{2n,2k+1}$ and $2^{k-2}(1-e_o)\beta\pi_{n-k}\tau_{2n,2k+1}$ are checked to have the same image under ψ, using (6.11) and (6.5). By exactness, ther exist $\bar{e}_o \in E^o$, $e_1 \in E^1$, such that

$$\bar{\theta}_{2n,2k+1}\tau_{2n,2k+1} = 2^{k-2}(1-e_o)\beta\pi_{n-k}\tau_{2n,2k+1}$$

$$+ p^*(e_1[\gamma_{2n-2k}]+\bar{e}_o[\theta_{2n,2k}\gamma_{2n-2k}]).$$

The lemma is proved by multiplying this equation by 2, noting $2[\theta_{2n,2k}\gamma_{2n-2k}] = 0$.

<u>Proposition</u> (6.15). <u>The following elements</u>

$$\{1,\tau_{2n,2k+1},\beta\pi_{n-k},\beta\pi_{n-k}\tau_{2n,2k+1},\theta_{2n,2k+1},\beta\pi_{n-k}\theta_{2n,2k+1}\}$$

<u>generate</u> $K^*(V_{2n,2k+1})$ <u>as an</u> E^*-<u>module</u>.

<u>Proof</u>. Denote by S the E^*-module generated by these elements. Refer to (6.3). Lemmas B and C can be used to check

that S contains $p^*(C^*)$. It suffices to show $\psi(S) = N^*$. Now N^* has the following four generators over E^*:

$$\{\chi, 2^{k-2}\tau_{2n,2k}, (\gamma_{2n-2k}+2^{n-k})\tau_{2n,2k}, \theta_{2n,2k}\gamma_{2n-2k}\}.$$

By (6.5), the first and third are already in $\psi(S)$. By lemma A, we can multiply (6.11) by $(1-e_o)^{-1}$ to show that the second is also in $\psi(S)$. It remains to settle $\theta_{2n,2k}\gamma_{2n-2k}$.

Write (6.14) in the form

$$p^*((1-\epsilon_o\gamma_{2n-2k})\theta_{2n,2k}) = 2\bar{\theta}_{2n,2k+1};$$

then use the fact that ψ is a $K^*(V_{2n,2k})$-homomorphism to obtain

$$\psi(2\beta\pi_{n-k}\bar{\theta}_{2n,2k+1}) = (-2\gamma_{2n-2k}+\tau_{2n,2k})(1-\epsilon_o\gamma_{2n-2k})\theta_{2n,2k}$$

$$= -2\theta_{2n,2k}\gamma_{2n-2k}.$$

Hence $\psi(\beta\pi_{n-k}\bar{\theta}_{2n,2k+1})$ is equal to $-\theta_{2n,2k}\gamma_{2n-2k}$ modulo elements annihilated by 2. Since $\beta\pi_{n-k}\bar{\theta}_{2n,2k+1}$ is in S, this suffices to show $\theta_{2n,2k}\gamma_{2n-2k} \in \psi(S)$. Therefore $S = K^*(V_{2n,2k+1})$ as claimed.

We are now ready to complete the inductive step by giving the structure of $K^*(V_{2n,2k+1})$. This is done by noting that:

<u>Proposition</u> (6.16). $\bar{p}:SO(2n) \to V_{2n,2k+1}$ <u>induces a monomorphism of</u> $K^*(V_{2n,2k+1})$ <u>into</u> $K^*(SO(2n))$.

In fact, $\bar{p}^*$ is completely determined by $\bar{p}^*(\tau_{2n,2k+1}) = 2^{n-k-1}\tau_{2n,2n-1}$, $\bar{p}^*(\beta\pi_{n-k}) = \beta\pi_{n-k}$, and

$$\bar{p}^*(\theta_{2n,2k+1}) = \bar{p}^*(\beta u_{2n,2k+1}) \text{ divided by } 2^k$$
$$= 2^{n-k-1}(\theta_{2n,2n-1}+T_{2n,2k+1})-2^k\beta\pi'_{n-k+1},$$

where a torsion element $T_{2n,2k+1}$ is introduced because division by 2^k in $K^*(SO(2n))$ is not unique. It is now straight forward to check, using proposition (6.15), that $\bar{p}^*$ is a monomorphism.

Having checked this, it is routine to verify that

$$K^*(V_{2n,2k+1}) = E(\beta\pi_{n-k},\ldots,\beta\pi_{n-2},\beta\chi_{2n}) \quad A_{2n,2k+1}$$

as abelian groups. Actually $K^*(V_{2n,2k+1})$ is so given as a ring, except that the relation $\theta_{2n,2k+1}\tau_{2n,2k+1} = 0$ must be established. As long as $k \leq [n/2]$, this is easy because

$$\bar{p}^*(\theta_{2n,2k+1}\tau_{2n,2k+1}) = 0 \quad \text{for} \quad k \leq [n/2].$$

For $k \geq [n/2]+1$, however, we only have

$$2^{2k-n}\bar{p}^*(\theta_{2n,2k+1}\tau_{2n,2k+1}) = 0, \tag{6.17}$$

and a more careful choice of $\theta_{2n,2k+1}$ is called for. To this end, write

$$\theta_{2n,2k+1}\tau_{2n,2k+1} = Q\tau_{2n,2k+1},$$

where Q is a polynomial in the exterior generators. (6.17) implies that $Q = 2^{n-k}Q_1$, and we can write $Q\tau_{2n,2k+1} = -Q_1\tau^2_{2n,2k+1}$. Now take $\theta_{2n,2k+1}+Q_1\tau_{2n,2k+1}$ as a new choice of $\theta_{2n,2k+1}$. This does not affect (6.7), but ensures that

$$\theta_{2n,2k+1}\tau_{2n,2k+1} = 0.$$

The inductive proof of theorem (4.8) is now completed.

Centro de Investigación del IPN, México

University of British Columbia, Canada

REFERENCES

[1] F. Adams, Vector fields on spheres, Ann. of Math. 75(1962), 603-32.

[2] D. Anderson, E. Brown and F. Peterson, SO-cobordism, KO-characteristic numbers and the Kervaire invariant, Ann. of Math. 83(1966),54-67.

[3] M. Atiyah, R, Bott and A. Shapiro. Clifford modules, Topology 3(1964), 3-38.

[4] M. Atiyah and F. Hirzebruch. Vector bundles on homogeneous spaces, Proceedings of Symposia Pure Mathematics of the A.M.S. 3(1961), 7-38.

[5] A. Dold, Relations between ordinary and extraordinary homology, Colloquium on Algebraic topology, Aarhus (1962), 2-9.

[6] L. Hodgkin, On the K-theory of Lie groups, Topology 6(1967), 1-36.

[7] D. Husemoller, Fibre bundles, McGraw Hill Book Co. New York, N.Y, 1966.

ON THE CLASSIFICATION OF TORSION-FREE H-SPACES OF RANK 2

by

Peter Hilton and Joseph Roitberg

1. Introduction

If X is a finite CW-complex which is an H-space, then, by a classical theorem of Hopf,

$$(1.1) \qquad H^*(X; Q) \cong H^*(S^{n_1} \times \ldots \times S^{n_k}; Q), \; n_i \text{ odd.}$$

We then say

$$\text{type } X = (n_1, n_2, \ldots, n_k)$$
$$\text{rank } X = k.$$

We assume knowledge of rank 1 H-spaces and study H-spaces of rank 2. By studying the cohomology of the projective plane of X, it may be proved [1,4,7] that if X is an H-space of rank 2 without 2-torsion then

$$(1.2) \quad \text{type } X = (1,1), (1,3), (1,7), (3,3), (3,5), (3,7), \text{ or } (7,7).$$

Indeed all these types occur, viz

(1,1)	$S^1 \times S^1$
(1,3)	$S^1 \times S^3$
(1,7)	$S^1 \times S^7$
(3,3)	$S^3 \times S^3$
(3,5)	$SU(3)$
(3,7)	$S^3 \times S^7$, $Sp(2)$
(7,7)	$S^7 \times S^7$.

Notice that the two examples given of type (3,7) are principal S^3-bundles over S^7. Such bundles are classified by $\pi_6(S^3) = Z_{12}$, generated by ω, the element measuring the noncommutativity of the quaternionic multiplication on S^3. If we write such a bundle as

$$S^3 \rightarrow E_{kw} \rightarrow S^7$$

then plainly $E_{kw} \cong E_{-kw}$, so we must consider $0 \leq k \leq 6$. The cases $k = 0,1$ are $S^3 \times S^7$, $Sp(2)$ respectively. Methods due to Zabrodsky (see also [6,10]) show that E_{3w}, E_{4w}, E_{5w} are H-spaces and that E_{2w} is an H-space iff E_{6w} is an H-space. Zabrodsky has announced (unpublished) that, in fact, neither is an H-space; nevertheless, we will add them to our previous list to provide

	(1,1)	$S^1 \times S^1$
	(1,3)	$S^1 \times S^3$
List A	(1,7)	$S^1 \times S^7$
	(3,3)	$S^3 \times S^3$
	(3,5)	$SU(3)$
	(3,7)	$S^3 \times S^7$, $Sp(2)$, E_{2w}, E_{3w}, E_{4w}, E_{5w}, E_{6w}
	(7,7)	$S^7 \times S^7$.

Theorem 1 (i) Every homotopy type of torsion-free rank 2 H-spaces is represented in List A.

(ii) Each homotopy type, with possibly one exception, contains exactly one PL-manifold.

(The exception is, of course, $S^1 \times S^3$).

Remark It seems probable that we can replace torsion-free by 2-torsion-free. However, we must exclude 2-torsion as the example of the exceptional Lie group G_2 shows. By using Zabrodsky's mixing techniques we may construct three 'fake' G_2's, which will be distinguished from G_2 by 3-component and 5- sompenent.

Theorem 1(i) has been proved independently by M. L. Curtis,

G. Mislin, and E. Thomas and by Zaboodsky (unpublished). Details of this work will appear elsewhere [6].

2. Sketch of proof of Theorem 1

Let us write type X = (q,n), $q \leq n$. We may immediately dispose of the cases q = 1 by invoking

Proposition 2.1. *Let X be an H-space with $\pi_1 X$ free abelian of rank k. Then $X \simeq \tilde{X} \times T^k$, where T^k is the k-dimensional torus.*

For we have, for any X, a fibration

$$\tilde{X} \to X \to K(\pi_1 X, 1);$$

and since X is an H-space and $K(\pi, X, 1) = T^k$ it follows that there is a cross-section and hence that the proposition holds.

We may now write (excluding the cases disposed of)

$$X \simeq S^q \cup_\alpha e^n \cup_\beta e^{n+q} \qquad (2.2)$$

since X is 1-connected. We will invoke three facts about H-spaces:

(2.3) Whitehead products vanish in an H-space;

(2.4) H-spaces satisfy Poincare duality [2];

(2.5) H-spaces are stably reducible (i.e., $\Sigma^N \beta = 0$, for some N) [3].

Proposition 2.2 *If $\alpha = 0$, then $X \simeq S^q \times S^n$.*

For $\pi_{n+q-1}(S^q \vee S^n)$ contains a cyclic infinite summand generated by the Whitehead product $[\iota', \iota'']$. Thus $\beta = \pm [\iota', \iota'']$ by (2.3).

This disposes of types (3,3) and (7,7). We look at type (3,5) so that

$$X \simeq S^3 \cup_\alpha e^5 \cup_\beta e^8$$

Since $S^3 \times S^5$ is not an H-space, it follows from Proposition 2.2 that $\alpha \neq 0 \in \pi_4(S^3)$. Thus we must compare X with

$$SU(3) \simeq S^3 \cup_\alpha e^5 \cup_{\beta'} e^8.$$

From the sequence

$$(2.6) \qquad \pi_7(S^3) \xrightarrow{i} \pi_7(S^3 \cup_\alpha e^5) \xrightarrow{j} \pi_7(S^3 \cup_\alpha e^5, S^3)$$

we infer from (2.4), (2.5) and a theorem of James [8] that

$$j(\beta) = \pm [\sigma, \iota], \; j(\beta') = \pm [\sigma, \iota],$$

where $[\sigma, \iota]$ is the relative Whitehead product of σ, generating $\pi_5(S^3 \cup_\alpha e^5, S^3)$, and ι, generating $\pi_3(S^3)$. Since we may, of course, replace β by $-\beta$ without changing the homotopy type, we may write $j(\beta) = j(\beta')$, so that $\beta = \beta'$, since, in (2.6), i is the zero homomorphism.

A very similar argument applies to the case of type (3,7). Let us write

$$\begin{aligned} C_k &= S^3 \cup_{kw} e^7 \\ X_k &= C_k \cup_\beta e^{10} \\ E_{kw} &= C_k \cup_{\beta'} e^{10} \end{aligned}$$

We will show that $X_k \simeq E_{kw}$ if X_k is an H-space by showing that $\beta = \pm\beta'$. Notice that E_{kw} satisfies Poincare duality and is stably reducible whether or not it is an H-space, so that, adjusting signs if necessary we may again conclude from

$$(2.7) \qquad \pi_9(S^3) \xrightarrow{i} \pi_9(C_k) \xrightarrow{j} \pi_9(C_k, S^3)$$

that $j(\beta) = j(\beta') = [\sigma, \iota]$. Now $\pi_9(S^3) = Z_3$, generated by

$S^9 \xrightarrow{\Sigma^3 w} S^6 \xrightarrow{w} S^3$. It thus follows that i is the zero homomorphism if $3 \nmid k$, and $\beta = \beta'$. The case $k = 0$ is covered by Proposition 2.2 so that it remains to consider $k = 3,6$.

For $k = 3$ we know that E_{3w} is an H-space. Also, in the sequence

$$\pi_7(C_3) \xrightarrow{j} \pi_7(C_3, S^3) \xrightarrow{\partial} \pi_6(S^3),$$

$\partial\sigma = 3w$, so that there exists τ with $j(\tau) = 4\sigma$. Then $j[\tau, \iota] = 4[\sigma, \iota]$. It follows from (2.3) that $4\beta = 4\beta'(= [\tau, \iota])$, so that $\beta = \beta'$ since $\beta - \beta' \in Z_3$.

Finally let $k = 6$; we retain the notation above for $k = 3$ and use bars to indicate corresponding elements in C_6, X_6, E_{6w}. We have the extra difficulty here that we are not entitled to suppose E_{6w} an H-space. Thus

(2.8) $$X_6 = C_6 \cup_{\bar\beta} e^{10}, \quad E_{6w} = C_6 \cup_{\bar\beta'} e^{10}$$

and we prove as above that $2\bar\beta = [\bar\tau, \iota]$. It is thus clearly sufficient to show that $2\bar\beta' = [\bar\tau, \iota]$. We have a map

(2.9) $$\begin{array}{ccccc} S^3 & \to & E_{6w} & \to & S^7 \\ & & f & & 2 \\ S^3 & \to & E_{3w} & \to & S^7 \end{array}$$

From (2.9) we readily infer that we may choose $\bar\tau, \tau$ so that $f_*(\bar\tau) = \tau$, whence $f_*[\bar\tau, \iota] = [\tau, \iota]$. Now $2\bar\beta' = [\bar\tau, \iota] + \bar{i}\lambda$, $\lambda \in \pi_9(S^3)$. From (2.8) and (2.9) we deduce $f_*(\bar\beta') = m\beta'$ for some integer m, so that $2m\beta' = [\tau, \iota] + i\lambda$; comparing this with $4\beta' = [\tau, \iota]$ we have $m = 2$ (which was clear on other grounds) and $i\lambda = 0$. Since $i : \pi_9(S^3) \to \pi_9(C_3)$ is monic, we conclude that $\lambda = 0$

and the proof of Theorem 1(i) is complete. As to Theorem 1(ii), the case $(q,n) = (1,1)$ is trivial and the case $(q,n) = (1,7)$ follows from a theorem of Browder-Levine. For the remaining cases we invoke Sullivan's formulation of simply-connected surgery [11], according to which a homotopy equivalence $h : N \to M$ may be deformed to a PL-equivalence if a certain obstruction $C_h : M_o \to F/PL$ is nullhomotopic. Here M_o is M with a disk removed. Now in our cases M_o may be deformation-retracted onto $S^q \cup_\alpha e^n$, q, n odd, so that any map $M_o \to F/PL$ is nullhomotopic.

3. Quasifibrations and orthogonal bundles

Similar techniques to those which establish Theorem 1(i) may be used to study quasifibrations. Let X be a closed, smooth manifold with

$$(3.1) \qquad X \simeq S^3 \cup_\alpha e^n \cup_\beta e^{n+3}, \qquad n = 5 \text{ or } 7.$$

By studying the stable tangent bundle $\tau \in [X, BO]$, one readily shows that X is a π-manifold, so that $j(\beta) = [\sigma, \iota]$. It follows that the pinching map $S^3 \cup_\alpha e^n \to S^n$ extends to e^{n+3},

$$p : X, S^3 \to S^n, pt,$$

and that p is a quasi-fibration, by a theorem of Sasao [9]. A count of homotopy types (3.1) yields 3 if $n = 5$, 10 if $n = 7$. However, the classical results of James and Whitehead (applied by Curtis and Mislin (unpublished) if $n = 7$) show that this is precisely the number of homotopy types of total spaces of orthogonal S^3-bundles. Thus, in this case, every quasifibration is equivalent (in this sense) to an orthogonal bundle.

That this is now so, in general, may be seen from the example of S^2-quasifibrations over S^n, $n \geq 3$, with cross-section. Such total spaces X have the form

$$(3.2) \qquad X_\theta \simeq (S^2 \vee S^n) \cup_{[\iota', \iota'']+\theta} e^{n+2}, \quad \theta \in \pi_{n+1}(S^2).$$

If $\theta \neq 0$ then X_θ cannot have the homotopy type of an orthogonal bundle. For if

$$S^2 \to X \to S^n$$

is an orthogonal buncld then $X \simeq S^2 \cup_\alpha e^n \cup_\beta e^{n+2}$ and, if $X \simeq X_\theta$ then $\alpha = 0$. However, this implies the existence of a cross-section which, in turn, implies $X \simeq S^2 \times S^n$. On the other hand, one may show that $X_\theta \simeq X_{\theta'}$ iff $\theta = \pm \theta'$.

Let us take $n = 5$. Then $\pi_6(S^2) = Z_{12}$ and there are 7 possible values for $|\theta|$, of which one ($\theta=0$) yields an orthogonal bundle. Since $\Sigma\, \pi_6(S^2) = 0$, the remaining 6 homotopy types yield π-manifolds which quasifibre (with cross-section) over S^5 with quasifibre S^2, but are not total spaces of orthogonal bundles. Now **Helmut Kneser** has shown that the group of homeomorphisms of S^2 deformation-retracts onto the orthogonal group. Thus the group of these 6 quasifibrations may not even be reduced to the group of homeomorphisms of S^2.

Bibliography

1. J.F. Adams, H-spaces with few cells, Topology 1 (1962), 67-72.

2. W. Browder, Torsion in H-spaces, Ann. of Math. 74 (1961), 24-51.

3. W. Browder and E. Spanier, H-spaces and duality, Pac. Journ. Math. 12 (1962), 411-414.

4. R.R. Douglas and F. Sigrist, Sphere-bundles over spheres and H-spaces, Topology 8 (1969), 115-118.

5. P.J. Hilton and J. Roitberg, On principal S^3-bundles over spheres, Ann. of Math. 90 (1969), 91-107.

6. P.J. Hilton and J. Roitberg, On the classification problem for H-spaces of rank 2, Comm. Math. Helv. (1970).

7. J.R. Hubbuck, Generalized cohomology operations and H-spaces of low rank, Trans. Amer. Math. Soc. 141 (1969), 335-360.

8. I.M. James, Note on cup-products, Proc. Amer. Math. Soc. 8 (1957), 374-383.

9. S. Sasao, On homotopy groups of certain complexes, Journ. Fac. Sci., Univ. Tokyo, Sec. 1, Vol. 8 (1960), 605-630.

10. J.D. Stasheff, Manifolds of the homotopy type of (non-Lie) groups, Bull. Amer. Math. Soc. 75 (1969), 998-1000.

11. D. Sullivan, Thesis, Princeton University (1965).

Cornell University, Ithaca, N. Y.

SUNY at Stonybrook

COMPACT PERTURBATIONS AND DEGREE THEORY

by

P. Holm* and E. Spanier*

University of Oslo and University of California, Berkeley

This note is concerned with an extension of the Geba-Granas theory of compact fields in a Banach space [3]. It is related to the degree theory on Banach manifolds formulated by Smale [6] and Elworthy-Tromba [1] in much the same way as the Geba-Granas theory [3] is related to the Leray-Schauder degree [4], [5]. Given an arbitrary Banach space E and a paracompact Hausdorff space X we consider equivalence classes of proper "singularity free" maps $X \to E$, equivalent maps being compact perturbations of each other. If X is a manifold some of these classes admit a degree function.

In section 1 we discuss the extension of the Geba-Granas theory and show that the compact homotopy classes of maps of a pair (X, A) into a pair $(E-W', E-W)$ are "stable" homotopy classes of maps of pairs (X^m, A^m) into (E^m-W', E^m-W) where E^m is finite dimensional. In section 2 we show how a degree function can be defined on compact homotopy classes of maps which are compact perturbations of Fredholm maps of index 0 of (X, A) into $(E, E-0)$. This degree function uses a limit argument applied to maps into finite dimensional subspaces of E where the degree is defined by homological methods. There are no assumptions on the Banach space E.

1. <u>Extension of the Geba-Granas theory</u>. Let X be a paracompact Hausdorff space and E a Banach space. A map

---

*Research partially supported by the National Science Foundation.

$K : X \to E$ is <u>compact</u> (or <u>finite dimensional</u>) if $\operatorname{im} K$ has compact (or finite dimensional) closure in E. A map $f : X \to E$ is a <u>compact perturbation</u> (or <u>finite dimensional perturbation</u>) of a map $\varphi : X \to E$ if $f = \varphi + K$ where $K : X \to E$ is a compact (or finite dimensional) map. A map $f : X \to E$ is <u>proper</u> if $f^{-1}C$ is a compact subset X for every compact subset $C \subset E$. A map $f : X \to E$ is <u>σ-proper</u> if there is a countable collection of closed subsets $\{F_i\}$ of X such that $X = \cup F_i$ and, for each $i, f|F_i : F_i \to E$ is a proper map.

<u>Remark</u>. A compact perturbation of a proper (or σ-proper) map is proper (or σ-proper).

Let $W \subset E$ be a closed subset contained in a finite dimensional subspace of E and let $W' \subset W$ be closed. Let A be a closed subset of X and $\varphi : X \to E$ a continuous map. A <u>φ-map</u> $f : X \to E$ is a compact perturbation f of φ such that $f(X) \subset E-W'$ and $f(A) \subset E-W$, and a <u>φ-homotopy</u> $h : X{\times}I \to E$ is a compact perturbation h of $\varphi \circ pr : X{\times}I \to X \to E$ such that $h(X{\times}I) \subset E-W'$ and $h(A{\times}I) \subset E-W$. For such a map h we write $h = \varphi + H$ instead of $h = \varphi \circ pr + H$. At this stage we do not assume φ itself to be a φ-map (i.e. φ does not necessarily map (X,A) into $(E-W', E-W)$).

Given a map $f : X \to E$ we form the closed subsets $W_t = W_t(f) \subset W$ with

$$W_t = \{w' \in W' | \operatorname{dist}(w', f(X)) \leq t\} \cup \{w \in W | \operatorname{dist}(w, f(A)) \leq t\}$$

where t varies over the non-negative real numbers. We say that f is <u>(W,W')-bounded</u> if W_t is a bounded subset of E for all $t \geq 0$. If either $f(X)$ is bounded or W is bounded, then f is (W,W')-bounded, so the definition covers both of these cases.

<u>Lemma 1.1</u>. If φ is (W,W')-bounded and f is a compact perturbation of φ, then f is (W,W')-bounded.

<u>Proof</u>. Let $f = \varphi + K$ with $K : X \to E$ a compact map. If f

were not (W,W')-bounded, then either (1) there is a sequence $\{w_i'\}$ in W' with $\|w_i'\| > i$ and a sequence $\{x_i\}$ in X with $\operatorname{dist}(w_i', \varphi(x_i) + K(x_i)) \leq t_o$ for some $t_o \geq 0$, or (2) there is a sequence $\{w_i\}$ in W with $\|w_i\| > i$ and a sequence $\{a_i\}$ in A with $\operatorname{dist}(w_i, \varphi(a_i) + K(a_i)) \leq t_o$ for some $t_o \geq 0$. However, if c is a bound for $\{\|K(x)\|\}_{x\in X}$, we would have $\operatorname{dist}(w_i', \varphi(x_i)) \leq t_o + c$ in case (1) (or $\operatorname{dist}(w_i, \varphi(a_i)) \leq t_o + c$ in case (2)). Therefore, $w_i' \in W_{t_o+c}(\varphi)$ (or $w_i \in W_{t_o+c}(\varphi)$) contradicting the (W,W')-boundedness of φ in either case.

Henceforth, we assume φ is a proper (W,W')-bounded map.

Lemma 1.2. Let $h : X\times I \to E$ be a φ-homotopy. Then $h(X\times I)$ is bounded away from W' and $h(A\times I)$ is bounded away from W.

Proof. Assume $h(X\times I)$ is not bounded away from W'. Then there exist sequences $\{x_i\}$ in X, $\{t_i\} \in I$, $\{w_i'\} \in W'$ such that $\operatorname{dist}(w_i', h(x_i,t_i)) \leq \frac{1}{i}$. Since φ is (W,W')-bounded, so is $\varphi \circ pr : X\times I \to E$, and therefore, by lemma 1.1, so is h. It follows that $\{w_i'\}$ is in a bounded, hence compact, subset of W'. We may assume $\{w_i'\}$ converges to a point $w' \in W'$. Then $h(x_i,t_i)$ converges to w'. Since h is a compact perturbation of the proper map $\varphi \circ pr$, it is proper and so $\{(x_i,t_i)\}$ may be assumed to converge to a point $(x,t) \in X\times I$. By continuity, $h(x,t) = w'$ contradicting $h(X\times I) \subset E-W'$.

A similar argument shows that $h(A\times I)$ is bounded away from W.

Corollary 1.3. If $f : X \to E$ is a φ-map then $f(X)$ is bounded away from W' and $f(A)$ is bounded away from W.

Proof. The composite $f \circ pr : X\times I \to E$ is a φ-homotopy. By lemma 1.2, $f(X) = (f \circ pr)(X\times I)$ is bounded away from W' and $f(A) = (f \circ pr)(A\times I)$ is bounded away from W.

Corollary 1.4. Let $f : X \to E$ be a φ-map. Then f is φ-homotopic to a φ-map f' which is a finite dimensional perturbation of φ. Two compact finite dimensional perturbations of φ which are φ-homotopic are homotopic by a compact finite dimensional perturbation of $\varphi \cdot pr$.

Proof. Let $f = \varphi + K$ with K a compact map. By Corollary 1.3 there is $\epsilon > 0$ such that $dist(W', f(X)) > \epsilon$ and $dist(W, f(A)) > \epsilon$. Let $K' : X \to E$ be a compact finite dimensional map which ϵ-approximates K (such maps exist as is shown in [5]). Then $f' = \varphi + K'$ is a compact finite dimensional perturbation of φ and f' maps (X,A) into $(E-W', E-W)$. Therefore, f' is a φ-map. The map $h : X\times I \to E$ defined by $h(x,t) = (1-t)f(x) + tf'(x)$ maps $(X\times I, A\times I)$ into $(E-W', E-W)$. Furthermore, $h(x,t) - \varphi(x) = (1-t)(f(x) - \varphi(x)) + t(f'(x) - \varphi(x) = (1-t)K(x) + tK'(x)$. Therefore, $h - \varphi \cdot pr$ is a compact map, so h is a φ-homotopy from f to f'.

Suppose f_o, f_1 are φ-maps which are compact finite dimensional perturbations of φ and there is a φ-homotopy $h = \varphi + H$ from f_o to f_1. By lemma 1.2 there is $\epsilon > 0$ such that $dist(W', h(X\times I)) > \epsilon$ and $dist(W, h(A\times I)) > \epsilon$. As before let $H' : X\times I \to E$ be a compact finite dimensional map which ϵ-approximates H. Then $h' = \varphi + H'$ is an ϵ-approximation to h so maps $(X\times I, A\times I)$ into $(E-W', E-W)$, and is a φ-homotopy from f_o' to f_1' (where $f_o'(x) = \varphi(x) + H'(x,o)$, $f_1'(x) = \varphi(x) + H'(x,1)$). Since f_o, f_o' are compact finite dimensional perturbations of φ, the homotopy $h_o(x,t) = (1-t)f_o(x) + tf_o'(x)$ is a compact finite dimensional perturbation of $\varphi \cdot pr$ and maps $(X\times I, A\times I)$ into $(E-W', E-W)$. Similarly the homotopy $h_1(x,t) = (1-t)f_1'(x) + tf_1(x)$ is a compact finite dimensional perturbation of $\varphi \cdot pr$ and is a φ-homotopy from f_1' to f_1. The composite of the φ-homotopies, h_o, h', and h_1 is a compact finite dimensional perturbation of

$\varphi \cdot pr$ which is a φ-homotopy from f_o to f_1.

Let $\{E_j\}$ be a directed filtration of E by finite dimensional subspaces such that every finite dimensional subspace of E is contained in some E_j. Without loss of generality we may assume $W \subset E_j$ for all j. For each j let (X_j, A_j) $= (X \cap \varphi^{-1}E_j, A \cap \varphi^{-1}E_j)$ and let $\varphi_j : X_j \to E_j$ be the map defined by φ. Then φ_j is a proper (W,W')-bounded map. We have the following lemma (compare with [3, Theorems (2.1), (2.2)]).

Lemma 1.5. Let $f_j : X_j \to E_j$ be a φ-map. Then there is an extension of f_j to a φ-map $f = \varphi + K$ with $K(X) \subset E_j$.

Let $f_j, f_j' : X_j \to E_j$ be φ_j-homotopic φ_j-maps with extensions to φ-maps $f, f' : X \to E$ which differ from φ by compact maps into E_j. If $h_j : X_j \times I \to E_j$ is a φ_j-homotopy from f_j to f_j', then there is an extension of h_j to a φ-homotopy $h = \varphi + H$ from f to f' with $H(X \times I) \subset E_j$.

Proof. Let $f_j = \varphi_j + K_j$ where $K_j : X_j \to E_j$ is a compact (hence, bounded) map. Let $K : X \to E_j \subset E$ be a compact extension of K_j (which exists by Tietze's theorem) and define $f = \varphi + K$. We show f maps (X,A) into $(E-W, E-W)$. In fact, $f(x) \in W' \subset E_j$ implies $\varphi(x) \in E_j$ so that $x \in X_j$ and $f(x) = f_j(x) \in E_j - W'$, a contradiction. Similarly $f(a) \in W$ for $a \in A$ implies $a \in A_j$ so $f(a) = f_j(a) \in E_j - W$, again a contradiction. Therefore, f is a φ-map which is an extension of f_j with $\operatorname{im}(f - \varphi) = \operatorname{im} K \subset E_j$.

For the second part let $h_j = \varphi_j + H_j$ with $H_j : X_j \times I \to E_j$ a compact map and let $f = \varphi + K$, $f' = \varphi + K'$ with $\operatorname{im} K \subset E_j$, $\operatorname{im} K' \subset E_j$. Define

$$H' : (X \times 0) \cup (X_j \times I) \cup (X \times I) \to E_j$$

so that $H'(x,o) = K(x)$, $H'(x,1) = K'(x)$, and $H'(x,t) = H_j(x,t)$ for $x \in X_j$, $t \in I$. Then $\operatorname{im} H'$ is contained in a compact subset of E_j, so, by Tietze's theorem, there is a compact map $H : X \times I \to E_j$ which is an extension of H'. Define $h = \varphi + H$. Then, as above, h maps $(X\times I, A\times I)$ into $(E-W', E-W)$ so is a φ-homotopy from f to f' and $\mathrm{im}(h - \varphi) = \mathrm{im}\, H \subset E_j$.

We introduce a "suspension operation" in the sets of finite dimensional φ-homotopy classes. Let $[X,A; E-W', E-W]_\varphi$ be the set of φ-homotopy classes of φ-maps and let $[X_j, A_j; E_j-W', E_j-W]_{\varphi_j}$ be the set of φ_j-homotopy classes of φ_j-maps. From lemma 1.5 we obtain a canonical map

$$S_j : [X_j, A_j; E_j-W', E_j-W]_{\varphi_j} \to [X,A; E-W', E-W]_\varphi$$

such that $S_j[f_j] = [f]$ where f is a φ-map which is an extension of f_j such that $\mathrm{im}(f - \varphi) \subset E_j$. Similarly if $i \leq j$ by choosing $X = X_j$, $A = A_j$, $E = E_j$, and $\varphi = \varphi_j$ we obtain a canonical map

$$S_i^j : [X_i, A_i; E_i-W', E_i-W]_{\varphi_i} \to [X_j, A_j; E_j-W' E_j-W]_{\varphi_j} .$$

From lemma 1.5 it follows that $S_i = S_j \circ S_i^j$, and if $i \leq j \leq k$ then $S_i^k = S_j^k \circ S_i^j$. Therefore, the family $\{[X_j, A_j; E_j-W', E_j-W]_{\varphi_j}, S_i^j\}$ is a direct system, and the maps $\{S_j\}$ define uniquely a map

$$S : \underrightarrow{\lim}\{[X_j, A_j; E_j-W', E_j-W]_{\varphi_j}\} \to [X,A; E-W', E-w]_\varphi$$

such that, for every $[f_j] \in [X_j, A_j; E_j-W', E_j-W]_{\varphi_j}$, $S\{[f_j]\} = S_j[f_j]$.

The following generalizes the main result in [3; Theorems 2,3] which treats the case where X is a closed bounded subset of E, A is empty, and φ is the inclusion map.

Theorem 1.6. The map

$$S : \varinjlim\{[X_j, A_j;\ E_j - W',\ E_j - W]_{\varphi_j}\} \to [X, A;\ E - W',\ E - W]_{\varphi}$$

is a bijection.

Proof. The first part of Corollary 1.4 implies that S is surjective, and the second part of Corollary 1.4 implies that S is injective.

In case φ is bounded, the properness of φ implies that the spaces X_j are compact for each j. In this case every continuous map $(X_j, A_j) \to (E_j - W',\ E_j - W)$ is a φ_j-map, and every continuous homotopy $(X_j \times I,\ A_j \times I) \to (E_j - W',\ E_j - W)$ is a φ_j-homotopy. Thus, we obtain the following:

Corollary 1.7. If φ is bounded, there is a canonical bijection

$$S : \varinjlim\{[X_j, A_j;\ E_j - W',\ E_j - W]\} \approx [X, A;\ E - W',\ E - W]_{\varphi} .$$

Specializing further to the case where A is empty and $W' = \{0\}$ we have:

Corollary 1.8. If φ is bounded, there is a canonical bijection $\varinjlim\{\pi^{d_j}(X_j)\} \approx [X; E - \{0\}]_{\varphi}$, where $d_j = \dim E_j - 1$.

Proof. Since S^{d_j} is a deformation retract of $E_j - \{0\}$, it follows that $[X_j;\ E_j - \{0\}] \approx [X_j, S^{d_j}] = \pi^{d_j}(X_j)$, and the result is a consequence of Corollary 1.7.

2. Applications to degree theory. We recall some of the basic definitions and properties connected with the degree of a Fredholm map (see [1] for a more complete discussion).

Let $L(E)$ be the Banach algebra of bounded linear operators on E and $GL(E)$ the multiplicative subgroup of invertible elements. Let $c(E)$ be the completely continuous operators and

$L_c(E)$ and $GL_c(E)$ the subsets of $L(E)$ and $GL(E)$, respectively, of operators of the form $I + T$, $T \in c(E)$. Then $GL_c(E)$ is a subgroup of $GL(E)$. It is known [1] that $GL_c(E)$ has two components. The component containing the identity will be denoted by $SL_c(E)$ and the other component will be denoted by $SL_c^-(E)$.

Given a Banach manifold M a c-structure on M is an admissible atlas $\{(U_i, \varphi_i)\}$ maximal with respect to the property that for every i, j the differential $d(\varphi_j \circ \varphi_i^{-1})$ at every point where it is defined lies in $GL_c(E)$. The c-structure is orientable if it admits a subatlas for which the differentials lie in $SL_c(E)$. An orientation is a subatlas maximal with respect to this property. Observe that every finite dimensional manifold has a unique c-structure and that orientability in this case has its usual meaning.

A smooth map $f : M \to N$ between c-manifolds (i.e. manifolds with given c-structures) modelled on E is a c-map if for every local representative $\psi_j \circ f \circ \varphi_i^{-1}$ of f the differential $d(\psi_j \circ f \circ \varphi_i^{-1})$ at every point where it is defined is in $L_c(E)$. This implies that f is a Fredholm map of index 0 (i.e. the differential df at every point is a linear transformation with finite dimensional kernel and finite dimensional cokernel of the same dimension as the kernel).

If f is a proper c-map between oriented manifolds M, N with N connected, the oriented degree of f is defined as follows. By the Sard-Smale theorem [6; Theorem 1.3] there exist regular values $y \in N$ of f. Then $f^{-1}(y) \subset M$ is a finite set of points. The sum of these with suitable signs is defined to be the degree of f

$$\deg f = \sum_{x \in f^{-1}(y)} \operatorname{sgn} df_x$$

where $\operatorname{sgn} df_x = +1$ or -1 according to whether the differential of a local representative of f , $d(\psi_j \circ f \circ \varphi_i^{-1})$ at $\varphi_i(x)$ (which is in $GL_c(E)$ because x is a regular point of f) , lies in $SL_c(E)$ or $SL_c^-(E)$. This definition of degree extends the finite dimensional one (cf. [5]).

Suppose now that $N = U$ an open subset of E with its canonical c-structure and that $f : M \to U$ is Fredholm of index 0 . By a result of Elworthy and Tromba [1] there is a unique c-structure $c_f = \{(U_i,\varphi_i)\}$ on M with respect to which f is a c-map. The map f is said to be orientable if c_f is orientable. If f is proper and oriented (i.e. c_f is oriented) the degree of f is defined and up to sign is invariant under proper Fredholm homotopies.

Referring to the notation of section 1 we now assume that (X,A) is a smooth relative manifold modelled on E (i.e. A is a closed subset of X and $X-A$ is a smooth manifold modelled on E) and that $(W,W') = (\{0\},\emptyset)$. We also assume that $\varphi : X \to E$ is proper and maps A into $E-0$ so that φ is itself a φ-map. Then the set of φ-maps is an equivalence class with respect to the relation "compact perturbation of" in the set of all proper maps $(X,A) \to (E,E-0)$. The relation of φ-homotopy (i.e. "compactly homotopic to") is a still finer equivalence relation in this set of proper maps. We will define a degree on some of the perturbation classes, namely those which admit a smooth Fredholm representative of degree 0 , which is invariant under compact homotopies.

From now on we assume that $\varphi : (X,A) \to (E,E-0)$ is proper, Fredholm of index 0 and oriented (i.e., $\varphi : X \to E$ is proper and $\varphi|X-A$ is a smooth Fredholm map of index 0 and oriented). The degree of φ with respect to the origin, denoted by $\deg(\varphi,A,0)$ is defined as follows. Let U be the (open) connected component of $E - \varphi(A)$ containing the origin 0 and let $V = \varphi^{-1}(U)$. Then V is an open subset of X contained in $X-A$ and $\varphi : V \to U$ is a proper oriented Fredholm map of degree 0. Hence, it has a degree and by definition this degree is defined to be the degree of φ with respect to 0 and denoted by $\deg(\varphi,A,0)$. It is invariant under smooth compact homotopies $(X,A)\times I \to (E,E-0)$. Note that if $B \subset E$ is a small open ball around the origin (an open neighborhood in $E - \varphi(A)$ would suffice) and $A' = X - \varphi^{-1}(B)$, then $\deg(\varphi,A,0) = \deg(\varphi,A',0)$. Thus, for degree purposes we may assume that $X-A = \varphi^{-1}(B)$ which makes $\varphi|X-A : X-A \to B$ a proper map and $A = \varphi^{-1}(\varphi(A))$. This will be used repeatedly in the sequel.

Lemma 2.1. Let (X,A) be a relative manifold modelled on a Banach space E and let $\varphi : (X,A) \to (E,E-0)$ be a proper continuous map such that $\varphi|X-A$ is Fredholm. Let $\{E_i\}$ be a finite collection of finite dimensional subspaces of E. If W is a sufficiently small neighborhood of the origin in E there are points $q \in E$ arbitrarily near the origin such that $\varphi - q$ is transverse regular to all E_i on $\varphi^{-1}(W)$.

Proof. For each i let E_i' be a complement of E_i in E such that there is a split short exact sequence of Banach spaces

$$0 \to E_i \to E \xrightarrow{p_i} E_i' \to 0$$

and a canonical isomorphism $E_i \times E_i' \approx E$. Since φ is proper, $\varphi(A)$ is closed so bounded away from 0. Let W_i, W_i' be open neighborhoods of the origin in E_i, E_i', respectively, such that $W_i \times W_i' \subset E - \varphi(A)$. Then for each i the composite

$$\varphi^{-1}(W_i \times W_i') \xrightarrow{\varphi} W_i \times W_i' \xrightarrow{pr} W_i'$$

is a σ-proper Fredholm map (being a composite of such maps) and so its regular value set V_i' is residual in W_i' (cf. [1]). It follows that the regular value set of the composite

$$\varphi^{-1}(W_i \times W_i') \xrightarrow{\varphi} E \xrightarrow{p_i} E_i'$$

which equals $V_i' \cup (E_i' - W_i')$, is residual in E_i'. The sets $p_i^{-1}(V_i' \cup (E_i' - W_i'))$ are residual in E, hence their intersection V is dense in E. If $q \in V$, then $p_i(q)$ is a regular value of $(p_i \circ \varphi)|\varphi^{-1}(W_i \times W_i')$ and so the origin is a regular value of $(p_i \circ (\varphi - q))|\varphi^{-1}(W_i \times W_i')$ for each i. Therefore, the translate $\varphi - q$ is transverse regular to each E_i on $\cap\varphi^{-1}(W_i \times W_i')$ $= \varphi^{-1}(\cap W_i \times W_i')$.

Remark. Since $\varphi(A)$ is bounded away from 0, for q sufficiently close to 0, $\varphi - q$ also maps A away from 0 and so does $\varphi - tq$ for $0 \leq t \leq 1$. Hence, for small q, φ is homotopic to $\varphi - q$ by a smooth compact one-dimensional homotopy $(X,A)\times I \to (E,E-0)$.

If φ is transverse regular on $X-A$ to a finite dimensional subspace $E^n \subset E$, then it can be shown (cf. [2]) that $(\varphi^{-1}(E^n), \varphi^{-1}(E^n) \cap A)$ is orientable and that an orientation of c_f induces an orientation of $(\varphi^{-1}(E^n), \varphi^{-1}(E^n) \cap A)$. Define $(X^n, A^n) = (\varphi^{-1}(E^n), \varphi^{-1}(E^n) \cap A)$ and let $\varphi^n : (X^n, A^n) \to (E^n, E^n - 0)$ be the map defined by φ. If $y \in E^n$ is a regular value of φ^n close to the origin, then y is a regular value of φ and $\varphi^{-1}(y) = (\varphi^n)^{-1}(y)$. Since (X^n, A^n) inherits its orientation from (X,A), $\mathrm{sgn}(d\varphi^n)_x = \mathrm{sgn}(d\varphi)_x$ for each $x \in \varphi^{-1}(y)$ Thus, $\deg(\varphi, A, 0) = \deg(\varphi^n, A^n, 0)$. However, $\deg(\varphi^n, A^n, 0)$ can be computed by well known homological methods discussed below.

Let $B \subset E$ be an open ball about 0 with $\varphi^{-1}(B) = X-A$ (this is no loss of generality as pointed out earlier) and let $\gamma^n \in \check{H}_c^n(E^n - B)$ be a generator (Čech cohomology with compact supports and coefficients $\mathbb{Z}$). Then, up to sign, $\deg(\varphi^n, A^n, 0)$ is the value on γ^n of the composite homomorphism

$$H_c^n(E^n, E^n - B) \xrightarrow{\varphi^{n*}} H_c^n(X^n, A^n) \approx H_o(X^n - A^n) \overset{\epsilon}{\to} \mathbb{Z}$$

In particular γ^n can be chosen so that the homological degree has the correct sign. If $E^m \subset E^n$ are finite dimensional subspaces of E to which φ is transverse regular on $X-A$, there is a commutative diagram

$$\begin{array}{ccccccc}
\check{H}_c^n(E^n, E^n-B) & \xrightarrow{\varphi^{n*}} & \check{H}_c^n(X^n, A^n) & \approx & H_o(X^n - A^n) & \overset{\epsilon}{\to} & \mathbb{Z} \\
\approx \uparrow & & \uparrow & & \uparrow & & \| \\
\check{H}_c^m(E^m, E^m-B) & \xrightarrow{\varphi^{m*}} & \check{H}_c^m(X^m, A^m) & \approx & H_o(X^m - A^m) & \overset{\epsilon}{\to} & \mathbb{Z}
\end{array}$$

where $H_c^m(E^m, E^m - B) \to H_c^n(E^n, E^n - B)$ is the suspension or the Thom isomorphism of the normal bundle of E^m in E^n sending γ^m to γ^n, and $H_c^m(X^m, A^m) \to H_c^n(X^n, A^n)$ is the induced Thom map.

Let $f : (X,A) \to (E, E-0)$ be a compact finite dimensional (but not necessarily smooth) perturbation of φ, say $f = \varphi + K$. Let $E^m \subset E$ be any finite dimensional subspace of E containing $\operatorname{im} K$. Let $q \in E$ be small and such that $\varphi_q = \varphi - q$ is transverse regular to E^m. Then f is compactly homotopic to $f_q = f - q$ (see the Remark after Lemma 2.1). Since $\operatorname{im} K \subset E^m$, it follows that

$$(f_q^{-1}(E^m) \, , \, f_q^{-1}(E^m) \cap A) = (\varphi_q^{-1}(E^m) \, , \, \varphi_q^{-1}(E^m) \cap A) \, .$$

We define $(X_q^m, A_q^m) = (\varphi_q^{-1}(E^m) \, , \, \varphi_q^{-1}(E^m) \cap A)$. Then we have the homomorphisms

$$H_c^m(E^m, E^m - B) \xrightarrow{f_q^{m*}} H_c^m(X_q^m, A_q^m) \overset{D}{\approx} H_o(X_q^m - A_q^m) \overset{\epsilon}{\to} \mathbb{Z} \, .$$

<u>Lemma 2.2.</u> If E^m, E^n both contain $\operatorname{im} K$ and $f_q, f_{q'}$ are small translations such that φ_q is transverse regular to E^m and $\varphi_{q'}$ is transverse regular to E^n, then

$$\epsilon D\ f_q^{m*}(\gamma^m) = \epsilon D\ f_{q'}^{n*}(\gamma^n)\ .$$

Proof. First assume $E^n = E^m$. There is a compact finite dimensional homotopy $G : (X,A)\times I \to (E,E-0)$ from φ_q to $\varphi_{q'}$ which is a Fredholm map of index 1 on $(X-A)\times I$. Let $\lambda : I \to I$ be a smooth map, strictly positive on $(0,1)$ and such that it and all its derivatives equal 0 on the boundary $\{0,1\}$. Set η equal to the composite $X\times I \xrightarrow{pr} I \xrightarrow{\lambda} I$. Then $G' = \frac{1}{\eta} G$ is Fredholm of index 1 on $(X-A)\times(0,1)$. By Lemma 2.1 there is a point $r \in E$ close to the origin such that $\frac{1}{\eta} G-r$ is transversal regular to E^m $(X-A)\times(0,1)$. It follows that $G - \eta r$ is transverse regular to E^m on $(X-A)\times I$ and homotopic to G by the smooth compact finite dimensional homotopy $G-t\eta r$, $0 \leq t \leq 1$. Pulling back E^m to $X\times I$ by $G_{\eta r} = G - \eta r$ gives a relative manifold (Y^{m+1},C^{m+1}) with boundary $(X_q^m,A_q^m)\times 0 \cup (X_{q'}^m,A_{q'}^m)\times 1$ (in particular, $Y^{m+1}-C^{m+1}$ is a manifold with boundary $(X_q^m-A_q^m)\times 0 \cup (X_{q'}^m-A_{q'}^m)\times 1)$, and there is a diagram of homomorphisms

$$\check{H}_c^m(E^m,E^m-B) \xrightarrow{G_{\eta r}^m} H_c^m(Y^{m+1},C^{m+1}) \overset{D}{\approx} H_1(Y^{m+1}-C^{m+1},(X_q^m-A_q^m)\times 0\cup(X_{q'}^m-A_{q'}^m)\times 1)$$

$$\Big\downarrow \partial$$

$$H_0((X_q^m-A_q^m)\times 0 \cup (X_{q'}^m-A_{q'}^m)\times 1)\ .$$

Since $\epsilon\cdot\partial = 0$, we see that $\epsilon\,\partial\,D\,G_{\eta r}^{m*}(\gamma^m) = 0$. However, it is easy to verify that if $i_0 : X_q^m - A_q^m \to (X_q^m-A_q^m)\times 0$ and $i_1 : X_{q'}^m - A_{q'}^m \to (X_{q'}^m - A_{q'}^m)\times 1$ are the obvious maps, then

$$\partial\ D\ G_{\eta r}^{m*}(\gamma^m) = \pm\,(i_{1*}\ Df_{q'}^{m*}\ (\gamma^m) - i_{0*}\ Df_q^{m*}(\gamma^m))\ .$$ Therefore,

$\epsilon i_{1*}Df_{q'}^{m*}(\gamma^m) = \epsilon i_{0*}\ Df_q^{m*}(\gamma^m)$. Since $\epsilon i_{0*} = \epsilon$ and $\epsilon i_{1*} = \epsilon$, this gives the result in this case.

If $E^n \neq E^m$, we may as well suppose $E^m \subset E^n$. If φ_q is not also transverse regular to E^n replace q by an element, also denoted q , for which this is true. According to the first part of the proof this does not change the value of $\epsilon\, Df_q^{m*}(\gamma^m)$. In this case the result follows from commutativity of the diagram

$$\begin{array}{ccccccc}
H_c^n(E^n, E^n - B) & \xrightarrow{f_q^{n*}} & H_c^n(X_q^n, A_q^n) & \overset{D}{\approx} & H_0(X_q^n - A_q^n) & \overset{\epsilon}{\to} & \mathbb{Z} \\
\approx\uparrow & & \uparrow & & \uparrow & & \| \\
H_c^m(E^m, E^m - B) & \xrightarrow{f_q^{m*}} & H_c^m(X_q^m, A_q^m) & \overset{D}{\approx} & H_0(X_q^m - A_q^m) & \overset{\epsilon}{\to} & \mathbb{Z} .
\end{array}$$

This completes the proof of lemma 2.2.

The degree of f with respect to the origin, denoted by deg(f,A,0) , is defined by

$$\deg(f,A,0) = \deg(f_q^m, A_q^m, 0) .$$

By lemma 2.2 this degree is well defined. It is easy to see that it does not depend on the particular choice of smooth representative φ in the perturbation class. The degree function so defined is an extension of the ordinary one for smooth Fredholm maps of index 0 and of the Leray-Schauder degree (of not necessarily smooth maps) and satisfies conditions analogous to the Leray-Schauder degree [5, p.86].

References

[1] Elworthy, K.D. and Tromba, A.J., Degree theory on Banach manifolds, Proc. Chicago Sym. on Nonlinear Math., Chicago, 1968.

[2] Elworthy, D.K. and Tromba, A.J., Differential structures and Fredholm maps, Proc. Symp. on Global Analysis, Berkeley, 1968.

[3] Geba, K. and Granas, A., Algebraic topology in normed linear spaces I, II, Bull. Acad. Polon. Sci. Ser. Math. 13 (1965) pp. 287-290 and pp. 341-346.

[4] Leray, J. and Schauder, J., Topologie et equations fonctionelles, Ann. Ecole Norm. Sup. 51 (1934) pp. 45-78.

[5] Schwartz, J.T., Nonlinear functional analysis, Notes on mathematics and its applications, Gordon and Breach science publishers, New York 1969.

[6] Smale, S., An infinite dimensional version of Sard's theorem, Amer. Jour. Math., 87 (1965), pp. 861-866.

A FIXED THEOREM FOR FINITE DIFFEOMORPHISM GROUPS GENERATED BY REFLECTIONS

by

Wu-chung Hsiang and Wu-yi Hsiang (1)

Introduction.

Let G be a compact connected Lie group and let T be a fixed maximal torus of G. Via the adjoint representation, the Weyl group of G, W(G) = N(T)/T acts on the tangent space $\mathfrak{T}$ of T at the identity. The action of W(G) on $\mathfrak{T}$ is a group generated by the (ordinary) reflections. This is a fundamental fact in Lie theory. In fact, it is well-recognized that the groups generated by reflections in euclidean space (the Coxeter system [1] [6]) are very interesting and very important. The notion of reflection can be easily generalized to a connected oriented manifold as follows [5] [6]. A diffeomorphism r of a connected oriented manifold M is a (diffeomorphism) reflection if

(i) r reverses the orientation and $r^2 = id$,

(ii) the set $\{x \in M | r(x) \neq x\}$ is disconnected.

We shall call the codimension 1 submanifold $\{x | x \in M, r(x) = x\}$ the hyperplane (possibly disconnected) associated to the reflection r and it is denoted by H_r. It is easy to show that $M-H_r$ has exactly two components [5]. It was proved in [5]

---

(1) The first author was partially supported by the N.S.F. Grant GP-9452, the second author was an Alfred P. Sloan Fellow and was also partially supported by N.S.F. Grant GP-8623.

that a group W of diffeomorphisms acting properly [2] on a simply connected manifold M and generated by a finite set of reflections is a Coxeter system [3] (or a Coxeter group). Therefore, W may actually act on a euclidean space as a group generated by the ordinary reflections. In the course of the proof, a fundamental domain was implicitly constructed.

In this note, we study the homology structure of the finite diffeomorphism groups generated by reflections of a manifold. The main result is the existence of the fixed point set if the number of generators of W and the homology of the manifold M satisfy some conditions. Roughly speaking, if M is Q-acyclic, then the action has a fixed point. We shall first construct the fundamental domain explicitly following the usual construction of the (closed) Weyl chamber. The reason why we repeat this construction is because the explicit description of the fundamental domain is an essential step in the proof of the main result of the paper.

If the manifold M is **Z**-acyclic, then the fundamental domain is a '**Z**-homology simplex'. Because of the existence of the fixed point set, it is immediate that W is a Coxeter system. We shall then estimate the minimal number of the generators of W if M is either a Q-acyclic manifold or a Q-homology sphere. We shall further show that if M is a Q-homology sphere, then W is a Coxeter system.

As a rough summary, the results indicate that finite group of diffeomorphisms generated by (differentiable) reflections on a Q-acyclic manifold or a Q-homology sphere behave very much like usual groups of (ordinary) reflections on a Euclidean space or a sphere, respectively. On the other hand, we shall also show by examples that there are many finite groups of diffeomorphisms generated by reflections on a Euclidean space or a sphere different from the standard ones.

---

[2] This means that the map $\psi : W \times M \to M \times M$ defined by $\psi(w, x) = (wx, x)$ is a proper map.

[3] A Coxeter system is a group W with a subset S satisfying the conditions: (a) each $s \in S$ is of order 2, (b) S is a set of generators of W, (c) if p_{st} denotes the order of st in W, then W is presented by the set of generators S and the relations $(st)^{p_{st}} = 1$ for $s, t \in S$ and $p_{st} < \infty$.

Of course, our real motivation of writing this note is for the applications in [2III]. As an illustration, we take out an example from [2III] showing how to apply the results of this paper to determine the orbit structure of a compact connected Lie group G acting differentiably on a Q-acyclic manifold (e.g., a Euclidean space) with the maximal tori as the principal isotropy subgroups. (Cf., [2, II].)

§1. Fundamental Domain.

Let us first give some simple examples of diffeomorphism groups generated by reflections.

Example 1. Let R^n be the ordinary Euclidean space with the inner product (,). Let x be a non-zero vector of R^n. Then the ordinary reflection with respect to the hyperplane perpendicular to x,

$$r(y) = y - \frac{2(x, y)}{(x, x)} x \qquad \text{for } y \in R^n$$

is a reflection in the general sense. Any group W generated by a finite number of the above reflections in the general sense.

Example 2. The unit sphere S^{n-1} of R^n is invariant under the action of W of Example 1. Every reflection of R^n induces a reflection on S^{n-1} in the general sense if $n-1 \geq 1$. Therefore, W acts on S^{n-1} $(n-1 \geq 1)$ as a diffeomorphism group generated by reflections.

Example 3. Let $T = S^1 \times S^1$ be embedded in R^3 in the standard way. The reflections of Example 1 with respect to xy-plane, yz-plane, zx-plane all leave $S^1 \times S^1$ invariant and induce reflections on $S^1 \times S^1$ in the general sense. So, they generate a diffeomorphism group generated by reflections.

Example 4. Let M_0 be a manifold with boundary $\partial M_0 \neq \emptyset$. The double of M_0, $M = M_0 \cup M_0$ has an obvious reflection r by interchanging the summands. So $\mathbf{Z}_2$ is a diffeomorphism group of M generated by reflections.

Now, let W be a given diffeomorphism group generated by reflections $R = \{r_i | i = 1,\ldots,m\}$ and acting properly on a connected manifold M. Write $H_i = F(r_i, M)$, the fixed point set of r_i, and $M_0 = M - \bigcup_{i=1}^{m} H_i$. Let C be a connected component of M_0. We call C an 'open chamber' of the action and $\overline{C}$ (the closure of C in M), the corresponding 'closed chamber'.

<u>Lemma 1</u>. (i) M_0 <u>is invariant under</u> W.

(ii) <u>For</u> $x \in M$, <u>the isotropy subgroup</u> W_x <u>acts transitively on the open chambers whose closures contain</u> x. <u>In particular</u>, $WC = M_0$ <u>and</u> $W\overline{C} = M$.

<u>Proof</u>. (i) If $r_i, r_j \in R$, so is $r_i r_j r_i$. Therefore, $\bigcup_{i=1}^{m} H_i$ is invariant under W, and so is M_0.

(ii) Let C, C' be two open chambers containing x in their closures, and let $x_1 \in C$, $x_2 \in C'$ in a neighborhood of x. Since $\bigcup_{i \neq j} (H_i \cap H_j)$ is of codim 2 in M, we can find a path from x_1 to x_2 missing $\bigcup_{i \neq j} (H_i \cap H_j)$ and all the hyperplanes which do not contain x. Moreover, we may also assume that this path meets any hyperplane (if at all) transeversely and only at a finite number of points. Following this path, we see that there is a chain of open chambers, $C = C_1,\ldots,C_\ell,C_{\ell+1},\ldots,C_{s+1} = C'$ such that C_ℓ and $C_{\ell+1}$ are separated by a hyperplane containing x, H_{i_ℓ} $(1 \leq i_\ell \leq m)$. It is clear that $r_{i_1} \ldots r_{i_1} \in W_x$ and $r_{i_s} \ldots r_{i_1}(C) = C'$.

In fact, we just proved that we can actually move C to C' by a sequence of reflections with respect to the hyperplane passing through x by the way which we explicitly constructed. Clearly, each of reflection is in W_x. Let us now fix an open chamber C. Since $M-H_j$ has exactly two components, C is in one of them. We denote the one containing C by M_j^+ and the other by M_j^-. We have $C = \bigcap_{j=1}^{m} M_j^+$. Let us arrange the indices in such a way that $\{1,\ldots,k\}$ is the smallest subset of $\{1,\ldots,m\}$ such that $C = \bigcap_{j=1}^{k} M_j^+$. Then H_j $(j = 1,\ldots,k)$ are the hyperplanes such that $\overline{C} \cap H_j$ $(j = 1,\ldots,k)$ are of codim 1 in M. It follows from Lemma 1 that $\{r_1,\ldots,r_k\}$ generates W, and we shall call it the <u>fundamental system</u> of generators of W and $\{H_1,\ldots,H_k\}$ the <u>fundamental system of hyperplanes</u> (associated to C).

Lemma 2. Let $\pi = \{r_1,\ldots,r_k\}$ be the fundamental system of generators of W (associated to C). For $i \in \{1,\ldots,k\}$, $\beta \in \{1,\ldots,m\}$ and $x \in \overline{C}$, if $x \notin H_\beta$ and $i \neq \beta$ then $r_i(x) \in M_\beta^+$.

Proof. We can find a point $y \in H_i \cap \overline{C} \cap M_\beta^+$ such that x and y can be connected by a path in $\overline{C}$. It suffices to prove our assertion for y instead of x. Since M_β^+ is open in M, we have a neighborhood $U(y)$ of y in M_β^+ such that $U(y)$ is invariant under r_i and $r_i|U(y)$ is equivalent to an ordinary reflection in a Euclidean space [7]. So $r_i(y) \in M_\beta^+$ and the lemma is proven.

Theorem 1. (i) $W\overline{C} = M$; and for $x, y \in \overline{C}$, $wx = y$ $(w \in W)$ implies $x = y$.

(ii) The isotropy subgroup $W_x (x \in M)$ is generated by the reflections with respect to the hyperplanes passing through x.

We shall call $\overline{C}$ a fundamental domain of W. Before we prove the theorem, let us describe the fundamental domains of some simple examples. Of course, the proof of Theorem 1 is a direct translation of the proof that the (closed) Weyl chamber is a fundamental domain of the Weyl group acting on the Cartan subalgebra. (Cf. Example 5 below.)

Example 5. Let G be a compact connected Lie group and let T be a fixed maximal torus. The Weyl group $W(G)$ acts on the tangent space $\mathfrak{T}$ of T at the identity e as a group of reflections. The (closed) Weyl chamber is the fundamental domain of the action.

Example 6. Let $(Z_2)^{n-1}$ be the group of diagonal matrices of $SO(n)$ with entries ± 1. Let $\mathfrak{S}$ be the space of quadratic forms $\{Q = \Sigma\, \alpha_{ij}x_ix_j \mid i, j = 1,\ldots,n\}$. $SO(n)$ acts on $\mathfrak{S}$ in the usual way. Let $\mathfrak{S}_0$ be the subspace of $\mathfrak{S}$ with trace zero. If we denote the standard representation of $SO(n)$ on $\mathbb{R}^n$ by ψ, the representation of $SO(n)$ on $\mathfrak{S}$ is customarily denoted by $s^2\psi$ and we shall denote the representation of $SO(n)$ on $\mathfrak{S}_0$ by $(s^2\psi-\theta)$. After we give the usual inner product to $\mathfrak{S}, \mathfrak{S}_0$, the representations $s^2\psi$, $(s^2\psi-\theta)$ become orthogonal. The unit sphere of $\mathfrak{S}_0$ under this inner product is invariant under $SO(n)$ and the fixed point set of $(\mathbb{Z}_2)^{n-1}$ consists of the following quadratic forms:

$$S = \{Q = \Sigma\alpha_i x^2 | \Sigma\alpha_i = 0, \Sigma\alpha_i^2 = 1\}$$

which is a sphere again. The $\mathbb{Z}_2$-Weyl group of $SO(n)$, $N((\mathbb{Z}_2)^{n-1})/(\mathbb{Z}_2)^{n-1}$ acts on S as a group of reflections. The fundamental domain consists of the following forms:

$$\Delta = \{Q = \Sigma\alpha_i x_i^2 | \Sigma\alpha_i = 0, \Sigma\alpha_i^2 = 1 \text{ and } \alpha_1 \geq \cdots \geq \alpha_n\}.$$

Note that $(\mathbb{Z}_2)^{n-1}$ is customarily called a maximal $\mathbb{Z}_2$-torus of $SO(n)$.

Example 7. The fundamental domain of Example 4 consists of the points of T^2 with $x \geq 0$, $y \geq 0$ and $z \geq 0$ where (x, y, z) is the coordinates in $\mathbb{R}^3$.

Proof of Theorem 1. $W\overline{C} = M$ is contained in Lemma 1. Since $\{r_1, \ldots, r_k\}$ generates W, we may write $\sigma = r_{i_1} \ldots r_{i_j}$ $(1 \leq i_j \leq k)$ and if ℓ is the minimal number in the expression, we denote it by $\ell(\sigma)$. For $\sigma = id$, we set $\ell(\sigma) = 0$. We claim that for $x, y \in \overline{C}$ and $\sigma \in W$, $\sigma(x) = y$ implies $x = y$. Let us prove our assertion by an induction on $\ell(\sigma)$. If $\ell(\sigma) = 1$, it follows from Lemma 2. Now, assume the assertion for $\ell(\sigma) \leq \ell-1$. Let $\sigma = r_{i_1} \ldots r_{i_\ell}$ $(1 \leq i_j \leq k)$ and $\sigma' = r_{i_2} \ldots r_{i_\ell}$. Suppose that $\ell(\sigma) = \ell$ and there are $x, y \in \overline{C}$ with $\sigma(x) = y$. By the induction hypothesis, we need only to prove the assertion with the assumption that $\sigma'(x) \in M_j^-$ for $1 \leq j \leq k$. Therefore, $r_{i_1}\sigma'(x) = \sigma(x) = y \in r_{i_1}M_j^-$ and $r_{i_1}(y) \in M_j^-$. By Lemma 2, $i_1 = j$. There is $2 \leq s \leq \ell$ such that $r_{i_2} \ldots r_{i_{s-1}}(x) \in M_j^+$ and $r_{i_2} \ldots r_{i_s}(x) \in M_j^-$. Put $T = r_{i_2} \ldots r_{i_{s-1}}$. Then $T(x) \in M_j^+$, $Tr_{i_s}(x) \in M_j^-$. Consequently, $r_{i_s}(x) \in T^{-1}(M_j^-) = M^-_{T^{-1}(r_j)T}$. By Lemma 2, $r_{i_s} = T^{-1}(r_j)T$. In other words, $Tr_{i_s} = r_jT$ and $\sigma = r_j\sigma' = r_{i_2} \ldots r_{i_{s-1}} r_{i_{s+1}} \ldots r_i$. Hence $\ell(\sigma) < \ell$, and it is a contradiction. So, we only have to show that the isotropy subgroup $W_x (x \in M)$ is generated by the reflections with respect to the hyperplanes passing through x. Without loss of generality, we may assume that $x \in \overline{C}$. Let $\psi \in W_x$. $\psi(C)$ is an open chamber containing x in its closure. By Lemma 1 (or rather the remark after Lemma 1), there is an element σ in W_x

generated by the reflections with respect to the hyperplanes passing through x such that $\sigma(C) = \psi(C)$. So $\psi^{-1}\sigma(\overline{C}) = \overline{C}$. It follows that $\psi^{-1}\sigma \in W_y$ for $y \in \overline{C}$. Hence $\psi^{-}\sigma = \mathrm{id}$. This completes the proof of Theorem 1.

§2. A Fixed Point Theorem.

As a special case of Hopf's Theorem (Cf. p. 35, §2, Ch III, Cohomology Operations, Lectures by N. E. Steenrod, Annals of Mathematics Studies 50), we have the following lemma.

Lemma 3. Let X be $\mathbf{Z}$-acyclic (resp. $\mathbf{Z}_p$-acyclic) manifold with or without boundary and Y be a codimension one submanifold (transversal to ∂X) that decompose X into two parts X_1, X_2 such that

$$X_1 \cup X_2 = X,\ X_1 \cap X_2 = Y.$$

Then one of the three subspaces X_1, X_2, Y is $\mathbf{Z}$-acyclic (resp. $\mathbf{Z}_p$-acyclic) will imply the other two subspaces are also $\mathbf{Z}$-acyclic (resp. $\mathbf{Z}_p$-acyclic).

Proof. The case $\partial X \neq \emptyset$ can be easily reduced to the case $\partial X = \emptyset$. One simply observes that $H_q(X) \cong H_q(X-\partial X)$, $H_q(Y) \cong H_q(Y-\partial Y) \cong H_q(Y-\partial X)$, $H_q(X_1) \cong H_q(X_1-\partial X)$, $H_q(X_2) \cong H_q(X_2-\partial X)$. The case $\partial X = \emptyset$ follows easily from Mayer-Vietoris sequence and duality.

Lemma 4. Let W be a finite diffeomorphism group generated by reflections on M and $\overline{C}$ be a fundamental chamber of W, $M^S = \cup H_i$ be the set of all singular points. If

$$\widetilde{H}_q(M; \mathbf{Q}) = 0 \ \text{for} \ 0 \leq q \leq N,$$

then

$$\widetilde{H}_q(\overline{C}, \partial\overline{C}; \mathbf{Q}) = \widetilde{H}_q(M/W, M^S/W; \mathbf{Q}) = 0$$

also for $0 \leq q \leq N$.

Proof. A q-dimensional chain $c \in C_q(M)$ is called an alternating q-chain (with respect to W) if

$$\sigma(c) = \mathrm{sign}(\sigma)\cdot c \quad \text{for all} \quad \sigma \in W.$$

Let $C_q^A(M; \mathbb{Q})$ be the set of all alternating q-chains. It is clear that $C_*^A(M; \mathbb{Q})$ is a sub-complex of the chain complex $C_*(M; \mathbb{Q})$ and the following alternating map

$$A = \frac{1}{w} \Sigma\, \mathrm{sign}(\sigma)\cdot\sigma, \quad (w = \text{order of } W)$$

is a projection of $C_*(M; \mathbb{Q})$ onto $C_*^A(M; \mathbb{Q})$. Hence, the homology of $C_*^A(M; \mathbb{Q})$ is a direct summand of $H_*(M; \mathbb{Q})$. On the other hand, it is not difficult to see that

$$\tilde{H}_*(\overline{C}, \partial\overline{C}; \mathbb{Q}) \cong \tilde{H}_*(M/W, M^S/W; \mathbb{Q})$$

is exactly the homology of $C_*^A(M; \mathbb{Q})$. Therefore, it follows from the assumption that $\tilde{H}_q(M; \mathbb{Q}) = 0$, $0 \leq q \leq N$, we have

$$\tilde{H}_q(\overline{C}, \partial\overline{C}; \mathbb{Q}) = 0$$

for $0 \leq q \leq N$.

Lemma 5. Let M be a $\mathbb{Q}$-acyclic manifold and $H_0, H_1, \ldots, H_k$ be (k+1) codimension one submanifolds such that

$$H_{i_1} \cap H_{i_2} \cap \ldots \cap H_{i_j} = H(i_1, \ldots, i_j);\ 0 \leq i_1 < \ldots < i_j \leq k$$

are all $\mathbb{Q}$-acyclic of codimension j for $j \leq k$. Suppose that

$$H_0 \cap H_1 \cap \ldots \cap H_k = \emptyset.$$

Let M_i^+ be the half space associated to H_i and containing

$$H^{(i)} = H_0 \cap H_1 \cap \ldots \cap \hat{H}_i \cap \ldots \cap H_k.$$

Then $\overline{C} = \bigcap_{i=0}^{k} M_i^+$ is $\mathbb{Q}$-acyclic,

but $H^k(\overline{C}, \partial\overline{C}; \mathbb{Q}) \cong \mathbb{Q}$

<u>where</u>
$$\partial\overline{C} = \bigcup_{i=0}^{k} \{\overline{C} \cap M_i^-\}.$$

<u>Proof.</u> We shall prove by induction on k. The case $k = 1$ is almost trivial. Since

$$\overline{C} = M_0^+ \cap M_1^+,\ M = M_0^+ \cup M_1^+$$

and M, M_0^+, M_1^+ are all Q-acyclic, it follows easily that $\overline{C}$ is also Q-acyclic. On the other hand, $\partial\overline{C} = H_0 + H_1$ is the disjoint union of two Q-acyclic space, it follows from exactness that $H^{'}(\overline{C}, \partial\overline{C}; Q) \cong Q$.

<u>The induction step</u>: Now, we assume that Lemma 5 holds for $k = m$, and proceed to show that it also holds for $k = m + 1$. Let us consider H_{m+1} as M' and $H_i \cap H_{m+1}$ as $H_i^{'}$ for $i = 0,1,\ldots,m$. It follows from the induction assumption, applying to M' and $\{H_i^{'}; i = 0,\ldots,m\}$, that $\overline{C}_{m+1} = H_{m+1} \cap \overline{C}$ is also Q-acyclic. On the other hand, it is an easy consequence of Lemma 3 that

$$\bigcap_{i=0}^{m} M_i^+ = X$$

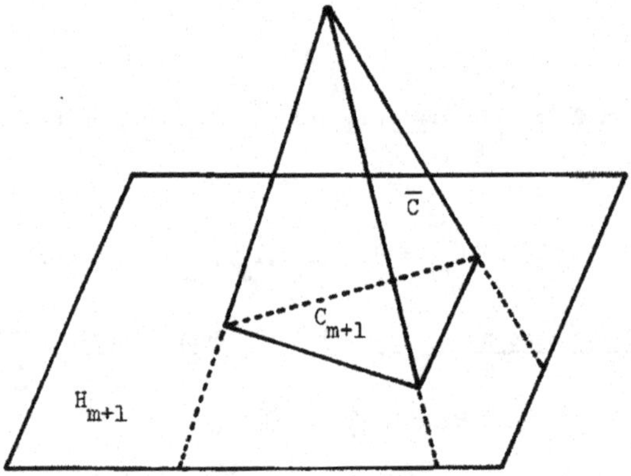

is also Q-acyclic. Hence, again by Lemma 3, $\overline{C}$ is Q-acyclic. It follows easily from induction assumption, applying to suitable intersection of hyperplanes, that all faces of $\overline{C}$ are Q-acyclic. Therefore, it is not difficult to see that

$$\bigcup_{i=0}^{m} \overline{C}_i = \partial\overline{C} - \text{interior}(\overline{C}_{m+1})$$

is also $\mathbb{Q}$-acyclic. Furthermore, we have

$$H^m(\partial\overline{C}, \overline{C}_{m+1}) \simeq H^m(\partial\overline{C})$$

$$\simeq \text{excision}$$

$$H^m(\bigcup_{i=0}^{m} \overline{C}_i, \partial\overline{C}_{m+1}) \simeq H^{m-1}(\partial\overline{C}_{m+1}) \simeq \mathbb{Q}$$

Hence,
$$H^{m+1}(\overline{C}, \partial\overline{C}) \simeq \mathbb{Q}.$$

Theorem 2. Let M be a $\mathbb{Z}$-acyclic (resp. $\mathbb{Z}_p$-acyclic) manifold and W be a finite diffeomorphism group generated by reflections. Then the fixed point set $F(W, M)$ is also $\mathbb{Z}$-acyclic (resp. $\mathbb{Z}_p$-acyclic) and all the faces of its fundamental chamber are also $\mathbb{Z}$-acyclic (resp. $\mathbb{Z}_p$-acyclic).

Proof. We shall prove the above theorem by induction on the number k of fundamental generators of W. Let

$$H_1, \ldots, H_k$$

be the hyperplanes of a fixed fundamental system of generators and let M_i^+, M_i^- be respectively the closed 'positive' and 'negative' half spaces corresponding to H_i, let

$$\overline{C} = \bigcap_{i=1}^{k} M_i^+$$

be the fundamental chamber.

The case $k = 1$. Then $W \simeq \mathbb{Z}_2$ is simply a reflection. Let $X = M$, $Y = H_1$. We have

$$X_1 \subseteq X \to X/W \simeq \overline{C} = X_1,$$

i.e. X_1 is a retraction of X and hence must be $\mathbb{Z}$-acyclic (resp. $\mathbb{Z}_p$-acyclic), and the theorem follows directly from Lemma 3.

The induction step. Now, we assume that Theorem 2 holds for $k \leq n$ and proceed to show that Theorem 2 also holds for $k = (n+1)$.

(A) We claim that $F(W, M) \neq \emptyset$. For otherwise, by our induction assumption and

Lemma 5, we have

$$H^n(\overline{C}, \partial\overline{C}; \mathbb{Q}) \cong \mathbb{Q}$$

which contradicts to Lemma 4. Hence $F(W, M) \neq \emptyset$.

(B) Let

$$H^{(i)+} = H_1 \cap \ldots \cap M_i^+ \cap \ldots \cap H_{n+1},$$

$$H^{(i)-} = H_1 \cap \ldots \cap M_i^- \cap \ldots \cap H_{n+1},$$

and

$$H^{(i)} = H_1 \cap \ldots \cap \hat{H}_i \cap \ldots \cap H_{n+1}.$$

By induction assumption, $H^{(i)}$ is $\mathbb{Z}$-acyclic (resp. $\mathbb{Z}_p$-acyclic). If $H^{(i)+}$ and $H^{(i)-}$ are conjugate under W, then we have

$$\begin{array}{ccccc} H^{(i)+} \subseteq H^{(i)} & \subseteq M & \longrightarrow & M/W = \overline{C} \\ & \searrow & & \cup\!\mid \\ & & & H^{(i)+} \end{array}$$

and hence $H^{(i)+}$ is a retraction of $H^{(i)}$, consequently, $H^{(i)+}$ is also $\mathbb{Z}$-acyclic (resp. $\mathbb{Z}_p$-acyclic). Since

$$H^{(i)} = H^{(i)+} \cup H^{(i)-} \quad \text{and}$$

$$F(W, M) = H^{(i)+} \cap H^{(i)-}$$

is a codimension one submanifold of $H^{(i)}$, it follows from Lemma 3 that $F(W, M)$ is also acyclic.

Next, we consider the case that $H^{(i)+}$ and $H^{(i)-}$ are not conjugate under W. Then $H^{(i)-}$ is conjugate to $H^{(j)+}$ for some $j \neq i$, i.e. $H^{(j)+} \cup H^{(i)+}$ is homeomorphic to $H^{(i)} = H^{(i)-} \cup H^{(i)+}$ which is, by induction assumption, $\mathbb{Z}$-acyclic (resp. $\mathbb{Z}_p$-acyclic). We may first apply Lemma 3 to the case

$$X = H^{(i,j)} = \bigcap_{s \neq i,j} H_s, \quad Y = H^{(j)+} \cup H^{(i)+}$$

$$X_1 = H^{i+j+} = X \cap M_i^+ \cap M_j^+.$$

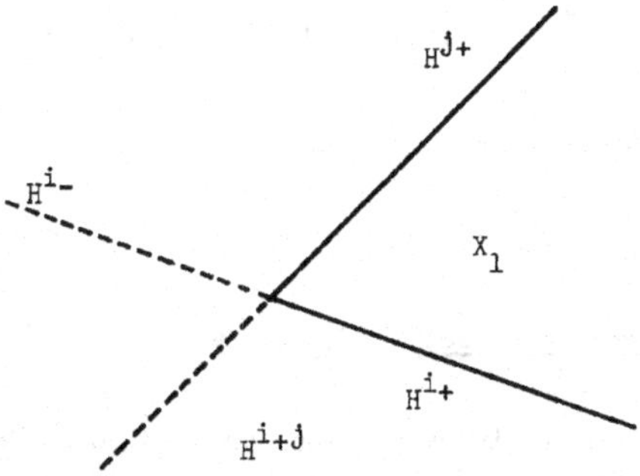

We see that X_1 is $\mathbf{Z}$-acyclic (resp. $\mathbf{Z}_p$-acyclic). Next, we apply Lemma 3 to the case

$$X = H^{i+j} = H^{(i,j)} \cap M_i^+, \quad Y = H^{i+}.$$

Since X_1 is $\mathbf{Z}$-acyclic (resp. $\mathbf{Z}_p$-acyclic), we see that $H^{i+} = Y$ is also $\mathbf{Z}$-acyclic (resp. $\mathbf{Z}_p$-acyclic). Finally, we apply Lemma 3 to the case

$$X = H^i = H^{i+} \cup H^{i-}, \quad X_1 = H^{i+}$$

$$Y = F(W, M) = H^{i+} \cap H^{i-},$$

we get $Y = F(W, M)$ is also $\mathbf{Z}$-acyclic (resp. $\mathbf{Z}_p$-acyclic). From here, it is a straightforward application of Lemma 3 to show that all faces of $\overline{C}$ are also $\mathbf{Z}$-acyclic (resp. $\mathbf{Z}_p$-acyclic). We leave it to the reader. This completes the proof of Theorem 2.

§3. Concluding Remarks.

In fact, we have a somewhat different proof to show the following stronger form of Theorem 2.

Theorem 2'. *Let* W *be a finite diffeomorphism group of* M *generated by* k *fundamental reflections. Suppose that*

$$\tilde{H}_i(M; \mathbb{F}) = 0 \quad \underline{\text{for}} \quad 0 \leq i < k$$

where $\mathbb{F}$ *is a field of coefficients. Then,*

$$\overline{C}(i_1, \ldots, i_s) = \overline{C} \cap H(i_1, \ldots, i_s) \neq \emptyset$$

for $1 \leq i_1 < \ldots < i_s \leq k$ *and*

$$\tilde{H}_j(\overline{C}(i_1, \ldots, i_s); \mathbb{F}) = 0 \quad \underline{\text{for}} \quad 0 \leq j < k - \ell.$$

In particular, the fixed point set $F(W, M) = \emptyset$.

The proof of Theorem 2' is rather long and tedious. We shall not give it in this paper.

The fundamental set of generators $\pi = \{r_1, \ldots, r_k\}$ gives a minimal number of generators of W. Suppose that $\mathbb{F}$ is a fixed field of coefficients. *If* M *is a* $\mathbb{F}$-*acyclic* (or $\mathbb{Z}$-*acyclic*) *manifold of dim* n, *then it follows from Theorem 2 that* $k \leq n$. Let us now prove a similar assertion for a F-homology sphere.

Theorem 3. Let W *be a finite diffeomorphism group of a* F-*homology sphere* M *of dim* n *generated by reflections. Then, the number of the fundamental generators is less than or equal to* (n+1).

Remark. Clearly, this theorem is the best possible because of Example 2.

Proof. Suppose that W has more than (n+1) fundamental generators. By a restriction to a subgroup if necessary, we may assume that W has exactly (n+2) fundamental generators, say $\{r_1, \ldots, r_{n+2}\}$. Let $\{H_1, \ldots, H_{n+2}\}$ be the corresponding fundamental hyperplanes. Any subset of (n+1) fundamental hyperplanes has no common fixed point. For otherwise, we may look at the representation of the isotropy subgroup at such a point and find a finite group generated by ordinary reflections in R^n with (n+1) fundamental generators. This is impossible. Let us now consider the subset $\bigcup_{i=1} \overline{C}(i)$ of $\partial\overline{C}$. By Theorem 2', $\overline{C}(1, \ldots, \hat{i}, \ldots, n+1)$ is non-empty, and hence $\bigcup_{i=1} \overline{C}(i)$ is a closed submanifold of codim 0 in $\partial\overline{C}$.

So $\bigcup_{i=1}^{n+1} \overline{C}(i) = \partial\overline{C}$. Since H_{n+2} is also a fundamental hyperplane. H_{n+2} intersects $\partial\overline{C}$ at a codim 0 submanifold $\overline{C}(n+2)$. This implies that H_{n+1} does not intersect every H_j $(j = 1,\dots,n+1)$ transversely and it is a contradiction. Therefore, W has at most $(n+1)$ fundamental generators.

Let us now recall the canonical Coxeter system associated to a diffeomorphism group of M generated by reflections [5]. Let R be the reflections in W. For $t, t' \in R$, let $p_{tt'}$ denote the order of tt' in W. Let $\overline{W}$ be the group with $t \in R$ as the generators and $(tt')^{p_{tt'}} = 1$ as the relations for $p_{tt'} < \infty$. Clearly, we have the projection $c : \overline{W} \to W$ and we shall call $\overline{W}$ the <u>canonically associated Coxeter system of</u> W. It was proved in [5] that <u>if</u> M <u>is simply connected</u>, <u>then</u> c <u>is an isomorphism</u>. <u>It follows from Theorem 2 immediately that</u> c is an isomorphism if M is $\mathbb{F}$-acyclic and W is finite. In fact, we have the following theorem.

<u>Theorem 4</u>. <u>Let</u> M <u>be either a $\mathbb{F}$-acyclic manifold or a $\mathbb{F}$-homology sphere</u>. <u>Let</u> W <u>be a finite group of diffeomorphisms of</u> M <u>generated by reflections</u>. <u>Then</u>, $\overline{W}$ <u>is isomorphic to</u> W <u>via</u> c.

<u>Proof</u>. We only have to prove our assertion for the $\mathbb{F}$-homology sphere case. If $\dim M = 1$, it is clear. If $\dim M = 2$, M is actually S^2 and it is simply connected. The assertion follows from [5]. So let us assume that $\dim M \geq 3$. By Theorem 2', any three hyperplanes intersect at a common point. Using the local representation at such a point, we see that for $t_1, t_2 \in R$, there is $t_3 \in R$ such that

$$t_1 t_2 t_1 = t_3 \quad \text{in } \overline{W}.$$

Let us now prove our assertion by an induction on the length of the word in the free group $F(R)$ generated by R. Suppose that for $\sigma = t_{i_1} \dots t_{i_{\ell-1}}$ $(\ell-1 \geq 3)$ the image in W is the identity only if it is already the identity in $\overline{W}$. Let $\sigma = t_{i_1} \dots t_{i_\ell}$ be a word whose image in W is the identity. If $\ell \leq n$, we can use Theorem 2' and the local representation again to prove our assertion. If

$\ell = n + 1$ and the hyperplanes $H_{i_1}, \ldots, H_{i_{n+1}}$ correspondent to the reflections $r_{i_1}, \ldots, r_{i_{n+1}}$ do not meet at a common point where r_{i_j} is the image of t_{i_j} in W, then the image of σ in W is not the identity. So for $\ell \geq n + 1$, we can always assume that there are i_j and i_h $(1 \leq j < h \leq \ell)$ such that $t_{i_j} = t_{i_h}$ without loss of generality. Then the image of $\sigma = (t_{i_1} \ldots t_{i_{j-1}})t_{i_j}(t_{i_{j+1}} \ldots t_{i_{h-1}})t_{i_h}(t_{i_{h+1}} \ldots t_{i_\ell})$ in $\overline{W}$ is equal to that of $\sigma' = (t_{i_1} \ldots t_{i_{j-1}})(t_{i_j}t_{i_{j+1}}t_{i_j}) \ldots (t_{i_j}t_{i_{h-1}}t_{i_j})(t_{i_{h+1}} \ldots t_{i_\ell})$ which may be represented by of word of length $\ell - 2$. Then the theorem follows from the induction.

Finally, let us consider some examples.

Example 8. Let us consider Example 5 again, i.e., the Weyl group $W(G)$ acts on $\mathfrak{T}$, the tangent space of a maximal torus T of G. Then the fundamental domain is a closed Weyl chamber $\overline{C}$. $\overline{C}$ is a non-compact contractible manifold with boundary $\partial\overline{C}$. $\partial\overline{C}$ is built up by the walls of the chamber. In fact, for $x \in \partial\overline{C}$, we can find a unique maximal subset $\{H_{i_1}, \ldots, H_{i_\ell}\}$ $(i_1 < \ldots < i_\ell)$ of the fundamental system of hyperplanes $\{H_1, \ldots, H_k\}$ such that $x \in H_{i_1} \cap \ldots \cap H_{i_\ell}$. Let us denote $\overline{C} \cap (H_{i_1} \cap \ldots \cap H_{i_\ell})$ by $\overline{C}(i_1, \ldots, i_\ell)$ for $1 \leq i_j \leq k$, $1 \leq i_1 < \ldots < i_\ell \leq k$. Let $\mathfrak{N} = (N_0, N_1, \ldots, N_\ell)$ be a compact $(\ell+1)$-aid satisfying the following conditions:

(i) N_0 is a contractible manifold of the same dimension as $\overline{C}$, say, of dimension s. Consequently, ∂N_0 is a $\mathbf{Z}$-homology sphere of dimension $(s-1)$.

(ii) $N_i (i = 1, \ldots, \ell)$ is an $(s-i)$-dim $\mathbf{Z}$-acyclic compact manifold with the boundary ∂N_i a $\mathbf{Z}$-homology sphere of $\dim(s-i-1)$.

(iii) N_{i+1} is embedded in ∂N_i as a regular submanifold of dim 0.

We can view $\mathfrak{C} = (\overline{C}, \overline{C}(i_1), \ldots, \overline{C}(i_1, \ldots, i_\ell))$ as an $(\ell+1)$-aid of manifolds (with corners along the boundaries). Let $\mathfrak{D} = (D^s, \ldots, D^{s-\ell})$ be the standard $(\ell+1)$-aid of discs. We can embed $\mathfrak{D}$ in $\mathfrak{C}$ and $\mathfrak{N}$ in the obvious ways and we then identify their images with the orientations reversed. This is the connected sum

$\mathcal{B} = \mathcal{C} \# \mathcal{N}$ of the two $(\ell+1)$-aids. It is easy to construct a retraction $f : \mathcal{B} \to \mathcal{C}$ such that

(i) for $x \in \mathcal{C}$, $f(x) = x$;

(ii) for $x \in N_j - N_{j+1}$,

$$f(x) \in \overline{C}(i_1, \ldots, i_j) - \overline{C}(i_1, \ldots, i_{j+1}).$$

(We understand that $\overline{C}(0) = \overline{C}$.) We can view $\overline{C}$ as the orbit space of $W(G)$ such that a point in $\text{Int}\,\overline{C}(i_1, \ldots, i_t)$ corresponds to the orbit $W(G)/W(r_{i_1}, \ldots, r_{i_t})$ where $W(r_{i_1}, \ldots, r_{i_t})$ denotes the subgroup of $W(G)$ generated by the reflections $r_{i_1}, \ldots, r_{i_t}$. f induces a covering space over $f^{-1}(\text{Int}\,\overline{C}(i_1, \ldots, i_t))$ with $W(G)/W(r_{i_1}, \ldots, r_{i_t})$ as the fibre. The data for piecing the portion of $\mathcal{T}$ over $\text{Int}\,\overline{C}(i_1, \ldots, i_h, \ldots, i_t)$ together with that over $\text{Int}\,\overline{C}(i_1, \ldots, i_t)$ induces a correspondent data for piecing together the induced covering space over $f^{-1}(\text{Int}(\overline{C}(i_1, \ldots, \hat{i}_h, \ldots, i_t))$ with that over $f^{-1}(\text{Int}\,\overline{C}(i_1, \ldots, i_t))$. They satisfy all the necessary compatibility conditions. (Cf. [2I, III].) So there is a manifold M with $W(G)$ acting as a diffeomorphism group generated by reflections. The manifold $B = N_0 \# \overline{C}$ becomes the fundamental domain. <u>Since</u> $W(G)$ <u>acts on</u> $\mathcal{T}$ <u>with non-empty fixed point set</u>, <u>so does</u> $W(G)$ <u>act on</u> M. M is actually diffeomorphic to a Euclidean space [2III]. By this construction, we may have many different finite groups of diffeomorphisms of R^n generated by (differentiable) reflections of R^n. Homologically, they are similar to an ordinary group of reflections, but differentiably they are rather different. Similar construction also holds for the spheres. We leave the details to the readers.

Let us now indicate an example of the applications of the results of this note to differentiable action of compact connected Lie groups [2III].

<u>Example 9</u>. Let G be a compact connected Lie group differentiably acting on a Q-acyclic space M, e.g., a euclidean space. Let T be fixed maximal torus of G. Then the fixed point set of T, $F(T, M)$ is also Q-acyclic. Let us look at the local representation of T at a point $x \in F(T, M)$. This representation is

independent of the choice of x and is called the geometric weight system of the action [3]. Suppose that the geometric weight system is the adjoint representation of G restricted to T. (Cf. [2II].) Now let $W(G)$ act on $F(T, M)$ in the obvious way. Under the present assumption, $W(G)$ acts on $F(T, M)$ as a finite group of diffeomorphisms generated by reflections. Let $\overline{C}$ be the fundamental domain. For $x \in \overline{C}(i_1,\ldots,i_s)$, it is not difficult to see that the isotropy subgroup G_x of the G-action at x contains at least the identity component of the centralizer (in G) of a certain subtorus of corank s in T. By Theorem 2, $W(G)$ has a fixed point in $F(T, M)$. Then this point is also fixed under G. Using the local representation at this point, we see that the orbit structure of this action of G is very similar to the adjoint representation of G which may be identified as the differential at this point.

References

[1] H. S. Coxeter, Regular Polytopes, Methnen & London, (1948).

[2] W. C. Hsiang and W. Y. Hsiang, Differentiable Actions of Compact Connected Lie groups I. Amer. J. Math. 89 (1967), 705-786; II. (to appear in Ann. of Math.); III. (in preparation).

[3] W. Y. Hsiang, On the geometric weight system of differentiable transformation groups on acyclic manifolds, (to appear, memeo. at Univ. of Calif. at Berkeley.)

[4] N. Jacobson, Lie algebra, Interscience publishers, New York (1962).

[5] J. L. Kozul, Lectures on Group of transformations, Tata Inst. of Fund. Research, Bombay, 1965.

[6] ———, Hyperbolic Coxeter Groups, Notes from Univ. of Notre Dame (1967).

[7] D. Montgomery and C. T. Yang, The existence of slices, Ann. of Math. 65 (1957), 108-116.

SPHERICAL FIBRATIONS

BY S.Y. HUSSEINI

Introduction

Suppose that $p : E \to B$ and $p' : E' \to B$ are two fiber spaces with the same fiber S^m, where S^m is the standard Euclidean m-sphere. Recall that a fiber homotopy equivalence

$$\varphi : E \to E'$$

is a fiber-preserving map which covers the identity map $B \to B$ and which is, at the same time, a homotopy equivalence. The object of this article is to answer in part the following problem.

Problem A. Determine when one can replace a given homotopy equivalence

$$f : (E, S^m) \to (E', S^m)$$

by a fiber homotopy equivalence where S^m is identified with the fiber above the basepoint.

This problem is related to the classification part of the second problem, which follows.

Problem B. Given the pair (E, S^m), determine when E is fibered by S^m over a space B, and classify the various possible fibrations.

Problem B is the subject of [1]. There are necessary and sufficient conditions (which generalize the conditions on the existence of Hopf fibrations when E is contractible) on $\Omega(E, S^m)$, the space of based paths in E ending in S^m, which insure that E is fibered by S^m and, therefore, in some sense provide an answer for Problem A. But whereas these conditions are perfectly satisfactory for the purposes of existence of fibrations, it is generally difficult to see how to use them, particularly when the possible fiber homotopy equivalences between E and E' are not necessarily in the homotopy class of f. Our object is to provide a different kind of answer.

It is easy enough to see that some additional conditions are necessary if one hopes for a positive answer. For example, two Hopf fibrations $S^7 \to S^4$ are equivalent if, and only if, the multiplications on S^3 which give rise to them are equivalent (see also [4]).

James and Whitehead study in [3] the problem of the homotopy-type classification of E and E' when B is a sphere. Their study has some bearing on Problem A. It indicates that there are two essentially different cases, depending upon whether or not the fibrations admit sections, and that one cannot hope for a positive answer, even in the easier case when there is a section, unless one stabilizes the problem.

The positive answer we give is a stable result. The fiber homotopy equivalence

$$\varphi : E \to E'$$

is constructed by induction on the skeletons of the base. The given maps f and φ are related in a certain way which can be described roughly as follows: the stable nature of the problem allows us to think of f as effecting a horizontal twist of B by S^m and a vertical twist of S^m by B. (This is precisely the situation if the fibrations are trivial.) In constructing φ one insures that it has the vertical effect of B.

To make the preceding precise is a delicate task and requires use of the RPT-category; the relevant facts and results are summarized in §2. In §3 the problem of the Theorem is given. We conclude with some remarks and examples in §4.

1. Statement of the Main Theorem

It will be assumed tacitly throughout this article (unless otherwise noted) that spaces and maps are in the category of based spaces and basepoint-preserving maps.

Suppose now that

$$(\xi) \qquad p : E \to B$$

is a Hurewicz fibration with fiber F, and assume that B is a simply connected finite CW-complex. By definition, let $E(\Sigma^n \xi)$ be the space obtained from $E \times S^n$, where S^n is the standard Euclidean n-sphere, by collapsing $p^{-1}(x) \vee S^n$ to a point, for all $x \in B$. One can prove easily, either directly or by using the classification theorem of [2], that the map

$$\Sigma^n p : E(\Sigma^n \xi) \to B$$

induced by p is also a Hurewicz fibration. We shall denote it by $\Sigma^n \xi$ and call it the nth suspension of ξ. Two Hurewicz fibrations

$$(\xi) \qquad p : E \to B, \text{ and}$$

$$(\xi') \qquad p' : E' \to B,$$

with the same fiber F and over the simply connected finite CW-complex B are said to be stably fiber homotopically equivalent if, and only if, there is a fiber homotopy equivalence

$$\varphi : E(\Sigma^n \xi) \to E(\Sigma^n \xi'),$$

for some n, covering the identity map $B \to B$.

The main theorem is the following.

THEOREM 1.1. <u>Suppose that</u>

$$(\xi) \qquad p : E \to B, \text{ and}$$

$$(\xi') \qquad p' : E' \to B$$

<u>are the Hurewicz fibrations over the simply connected finite CW-complex</u> B <u>and with the same fiber</u> S^m, <u>where</u> S^m <u>is the standard Euclidean m-dimensional sphere. Then</u> ξ <u>and</u> ξ' <u>are stably fiber homotopically equivalent if, and only if, there is a homotopy equivalence</u>

$$f : (E(\Sigma^n \xi, S^{n+m}) \to (E(\Sigma^n \xi'), S^{n+m}),$$

<u>where</u> $n + m > \dim B$ <u>and</u> S^{n+m} <u>is identified with the fiber over the basepoint.</u>

When B is a sphere, the theorem follows from the classification, up to homotopy type, of the total spaces of sphere bundles over spheres carried out by James and Whitehead in [3]. In general the fiber homotopy equivalences $E \to E'$ obtainable are not homotopic to f but are, however, related to f in certain ways to be described in §3, where the proof is given.

2. <u>The RPT-category</u>

Before we can give the proof, we need a few concepts from the theory of RPT-complexes [1, 2]. Recall that a <u>special complex</u> is a countable CW-complex with a single vertex which we take as a basepoint. An RPT-complex (i.e., a complex of the reduced product type) is a special complex with an associative multiplication

$$A \times A \to A,$$

for which the vertex is a two-sided identity and such that the product of two cells of A is again a cell of A. (See [1, 2] for details.) The purpose of the RPT-theory is to provide suitable combinatorial models for loopspaces and path spaces. For example, if B is a special complex and $\Omega(B)$ is its space of Moore loops, then there is an RPT-complex M and a homomorphism

(2.1) $$\gamma : M \to \Omega(B)$$

which is also a homotopy equivalence, and the indecomposable cells of M are in one-to-one correspondence with the cells of ΩB. Moreover, there is a universal quasi-fibration

$$q : \mathcal{B}M \to B,$$

for M, whose total space $\mathcal{B}M$ is a contractible special complex on which M acts on the right, and a map

$$\mathcal{B}M \to \mathcal{P}B$$

of an M-space to an $\Omega(B)$-space which covers the identity $B \overset{=}{\to} B$ ($\mathcal{P}(B)$ being the space of Moore paths).

These combinatorial models for $\Omega(B)$ and $\mathcal{P}(B)$ allow us to prove a classification theorem [2] for fibrations.

THEOREM 2.1. _Suppose that_

$$p : E \to B$$

is a Hurewicz fibration with fiber F; _and assume that_ E, B, _and_ F _are special complexes and that_ B _is simply connected. Then there is a fiber-preserving map_

$$\varphi' : \mathcal{B}M \times F \to E$$

which induces a fiber homotopy equivalence

$$\varphi : \mathcal{B}M \times_M F \to E$$

which covers the identity map of B _and such that the maps_ $(x, f) \to F$ _defined by the induced action_

$$M \times F \to F$$

give a homomorphism

$$\bar{\gamma} : M \to G(F),$$

where $G(F)$ is the monoid of homotopy equivalences of F with itself. Moreover, if $p : E \to B$ admits a section, then $\bar{\gamma}$ takes M to the submonoid $G_0(F)$ of basepoint-preserving homotopy equivalences.

The following corollary is the starting point of the proof of Theorem 1.1.

COROLLARY 2.2. Suppose that

$$p : E \to B$$

is a Hurewicz fibration as in Theorem 2.1, and let

$$E_r = p^{-1}B_r, \quad r \geq 0,$$

where B_r stands for the r-skeleton of B. Then there are fiber-preserving maps

$$\alpha_i : \partial D^r \times F \to E_r,$$

where $i = 1, \ldots, k_r$, (k_r being the number of cells of $B_r - B_{r-1}$), and a fiber homotopy equivalence

$$E_r \cong E_{r-1} \cup_{\alpha_1} D^r \times F \cup_{\alpha_2} \ldots \cup_{\alpha_{k_r}} D^r \times F$$

which is the identity on E_{r-1}. Moreover, if $p : E \to B$ admits a section, then

$$B_r = B_{r-1} \cup \alpha_1(D^r \times *) \cup \ldots \cup_{\beta_{k_r}} D^r \times *,$$

where $*$ stands for the basepoint of F.

3. The Proof of Theorem 1.1.

In order to prove the theorem, it is enough to consider the following. Suppose that

$$p : E \to B, \quad \text{and}$$
$$p' : E' \to B$$

are two Hurewicz fibrations with fiber S^m over the simply connected finite CW-complex B; and let

$$f : (E, S^m) \to (E', S^m)$$

be a homotopy equivalence where S^m is identified with the fiber above the base-point in B. To prove Theorem 1.1, it is enough to show that there is a fiber homotopy equivalence

$$\varphi : E \to E'$$

covering the identity map of B, provided that $\dim B \leq m - 1$.

Observe that, since $\dim B < m$, we can find sections

$$s : B \to E \quad \text{and} \quad s' : B \to E'$$

for the two fibrations $p : E \to B$ and $p' : E' \to B$. Identify B with the subsets of E and E' given by the sections, and choose cellular structures on E and E' which make the sum $B \vee S^m$ a subcomplex of E and E'. It is clear that the given homotopy equivalence f can be assumed (after a deformation, if necessary) to induce a homotopy equivalence of pairs

$$f : (E, B \vee S^m) \to (E', B \vee S^m)$$

which takes each summand of $B \vee S^m$ into itself. We can also arrange it, after changing $p' : E' \to B$ by a fiber homotopy equivalence if necessary, so that

$$f \mid B \vee S^m = \text{identity}.$$

As in §2, denote the r-skeleton of B by B_r, and let

$$E_r = p^{-1}B_r \quad \text{and} \quad E'_r = p'^{-1}B_r.$$

According to Corollary 2.2, we can assume that

$$E_r = E_{r-1} \cup_{\alpha_1} D^r \times S^m \cup_{\alpha_2} \dots \cup_{\alpha_{k_r}} D^r \times S^m$$

$$E' = E'_{r-1} \cup_{\alpha'_1} D^r \times S^m \cup_{\alpha'_2} \dots \cup_{\alpha'_{k_r}} D^r \times S^m,$$

where

(3.1)
$$\alpha_i : \partial D^r \times S^m \to E_{r-1} \quad \text{and}$$
$$\alpha'_i : \partial D^r \times S^m \to E'_{r-1},$$

for $i = 1, \dots, k_r$ (k_r being the number of cells in $B_r - B_{r-1}$), are suitable fiber maps. Since $p : E \to B$ and $p' : E' \to B$ admit sections, we can assume that the maps α_i and α'_i have been so chosen that

$$\alpha_i(\partial D^r \times \{x_0\}) \subset B_{r-1} \subset E_{r-1} \quad \text{and}$$
$$\alpha'_i(\partial D^r \times \{x_0\}) \subset B_{r-1} \subset E'_{r-1},$$

where x_0 is the basepoint of S^m. Observe that the (r+m)-cells of E_r and E'_r are obtained from the (r+m)-cells of products $D^r \times S^m$. Let

(3.2)
$$\beta_i : \partial D^{r+m} \to E_{r-1} \cup B_r \quad \text{and}$$
$$\beta'_i : \partial D^{r+m} \to E'_{r-1} \cup B_r,$$

for $i = 1, \dots, k_r$, be the attaching maps of the (r+m)-cells.

Consider now the filtration

$$S^m \vee B \subset \dots \subset E_r \cup B \subset \dots \subset E.$$

Suppose that M is the RPT-complex representing ΩE, and let

$$\gamma : M \to \Omega E$$

be the representing homomorphism. The filtration of E induces an ascending filtration

$$\ldots \subset {}^{(r)}M \subset \ldots \subset M$$

of M by sub-RPT-complexes ${}^{(r)}M$ such that

(1) $\gamma | {}^{(r)}M$ is a homotopy equivalence

$${}^{(r)}M \to \Omega(E_r \cup B), \quad \text{and}$$

(2) ${}^{(r)}M$ is generated by ${}^{(r-1)}M$ and the cells $\{e_i^r\}$ in one-to-one correspondence with the cells of $E_r \cup B - E_{r-1} \cup B$, whose characteristic maps

$$\beta_i : (D^{r+m}, \partial D^{r+m}) \to (E_r \cup B, E_{r-1} \cup B)$$

are those of (3.2) above.

We shall construct now a complex useful in measuring the difference between two given homomorphisms of ${}^{(r)}M$ into G, an associative H-space. To start out with, denote by A the RPT-complex of M which corresponds to $\Omega(B)$, and note that ${}^{(r)}M$ is generated by A and the set of cells $\{e^{m+i-1}\}$, with $0 \leq i \leq r$, which are in one-to-one correspondence with the cells of $E_r \cup B - B$. Therefore, according to [1, 2], an element of ${}^{(r)}M$ is an equivalence class of the form

$$(x_1, \ldots, x_n),$$

where each x_j is either an element of A or one of the generating cells $\{e^{m+i-1}\}$. One can easily see that the condition $\dim B < m$ implies that the subset N_r' of ${}^{(r)}M$ consisting of those elements represented by sequences

$$(x_1, \ldots, x_n)$$

such that at most one x_j belongs to a generating cell e^{m+i-1} is actually a subcomplex of ${}^{(r)}M$. Now let $N_r' \cup_A N_r'$ be the complex obtained from two distinct copies of N_r' by joining them along A; and put

$$L_r = \bigcup \partial(D^{i+m-1} \times I)_{j_i},$$

with the disjoint summation ranging over the set of indices (i, j_i), where $0 \leq i \leq r$ and, for each i, $1 \leq j_i \leq k_i$ (k_i being the number of indecomposable cells of dimension i). By definition, let

$$N_r = (N'_r \cup_A N'_r) \cup A \times L_r \times A,$$

where the attaching map

$$A \times \bigcup (D^{i+m-1} \times \partial I)_{j_i} \times A \to N'_r \cup_A N'_r$$

is the one induced by the characteristic map of the cells $D^{i+m-1} \times \{0\}$ and $D^{i+m-1} \times \{1\}$ onto the cells e^{i+m-1} in the left and right copies of N'_r in $N'_r \cup_A N'_r$ respectively. We shall call N_r the <u>difference complex relative to</u> A.

We can now result the proof of the theorem. We shall construct the desired fiber-preserving map

$$\varphi : E \to E'$$

by induction on the skeletons of B. So we put

$$\varphi_0 : E_0 = S^m \to E'_0 = S^m \subset E'$$

equal to the identity map of S^m. Suppose we have been able to define a fiber-preserving map

$$\varphi_r : E_r \to E'_r$$

which covers the identity map $B_r \xrightarrow{=} B_r$ such that the following conditions are satisfied

$(3.3)_r$ $\quad$ $\varphi_r\beta_k$ <u>and</u> $f\beta_k$ <u>are homotopic as maps</u>

$$S^{r+m-1} \to E'_r \cup B,$$

<u>where</u> β_k <u>are the maps defined in</u> (3.2) <u>above.</u>

In order to state the second condition, we need a new concept. Consider therefore the adjoints

$$\tilde{\varphi}_r, \tilde{f} : {}^{(r)}M \to \Omega(E'_r \cup B) \subset \Omega E'.$$

Together they define a map

$$\delta(\tilde{\varphi}_r, \tilde{f}) : N_r/A \to \Omega(E'_r \cup B) \subset \Omega E'$$

by first putting $\delta(\tilde{\varphi}_r, \tilde{f})$ equal to $\tilde{\varphi}_r$ on $D^{m+i-1} \times \{0\}$ and to $\tilde{f}$ on $D^{m+i-1} \times \{1\}$ and then using the homotopy of $(3.3)_r$ to define it on $\partial D^{m+i-1} \times I$. (Cf. [5].) Here N_r/A is the complex obtained from N_r by collapsing A to the vertex.

We shall say that a map

$$h : K \to \Omega E'$$

of the complex K into the loopspace $\Omega E'$ is <u>horizontal</u> if, and only if, the maps

$$\eta h, h : K \to \Omega E'$$

are homotopic. Here η is the composite

$$E' \xrightarrow{\Omega p'} B \xrightarrow{\Omega i'} \Omega E'.$$

The second condition is the following

$(3.4)_r$ <u>The map</u>

$$\delta(\tilde{\varphi}_r, \tilde{f}) : N_r/A \to \Omega E'$$

<u>is horizontal.</u>

We wish to extend the fiber homotopy equivalence φ to the (r+1)-skeleton. For the sake of simplicity, we shall give the proof when $B_{r+1} - B_r$ consists of one cell only, the general case being similar. Then

$$E_{r+1} = E_r \cup_\alpha (D^{r+1} \times S^m) \quad \text{and} \quad E'_{r+1} = E'_r \cup_{\alpha'} D^{r+1} \times S^m,$$

where the attaching maps

$$\alpha : \partial D^{r+1} \times S^m \to E_r \quad \text{and} \quad \alpha' : \partial D^{r+1} \times S^m \to E'_r$$

are those given in $(3.1)_{r+1}$ above. Next choose an RPT-complex $P_{r,m}$ of the form

$$P_{r,m} = e^0 \cup e^{r-1} \cup e^{m-1} \cup e^{r+m-1} \cup \text{higher dimensional cells}$$

to represent the loopspace $\Omega(\partial D^{r+1} \times S^m)$ (see [1, 2]). The maps α and α' define homomorphisms

$$\tilde{\alpha} : P_{r,m} \to {}^{(r)}M \simeq \Omega(E_r \cup B) \quad \text{and} \quad \tilde{\alpha}' : P_{r,m} \to \Omega(E'_r \cup B),$$

which correspond to the adjoints of α and α'. Observe now that the two homomorphisms

$$\tilde{\varphi}_r\tilde{\alpha},\ \tilde{f}\tilde{\alpha} : P_{r,m} \to \Omega(E'_r \cup B)$$

agree on the m-skeleton $e^0 \cup e^{r-1} \cup e^{m-1}$ and hence induce a map

$$d(\tilde{\varphi}_r\tilde{\alpha}, \tilde{f}\tilde{\alpha}) : S^{r+m-1} \to \Omega(E'_r \cup B)$$

as in [5].

Next we note that

$$d(\tilde{\varphi}_r\tilde{\alpha}, \tilde{f}\tilde{\alpha}) = \delta(\tilde{\varphi}_r, \tilde{f}) \circ (\tilde{\alpha} \vee \tilde{\alpha}),$$

where $\tilde{\alpha} \vee \tilde{\alpha}$ is the folding map

$$P_{r,m} \vee P_{r,m} \to N_r$$

induced by the mapping $P_{r,m}$ into N'_r by $\tilde{\alpha}$ [1]. Since $\delta(\tilde{\varphi}_r, \tilde{f})$ is horizontal by induction, we conclude the following.

(3.5) <u>The mapping</u>

$$d(\tilde{\varphi}_r\tilde{\alpha}, \tilde{f}\tilde{\alpha}) : S^{r+m-1} \to \Omega(E'_r \cup B)$$

<u>is horizontal.</u>

Consider next the diagram

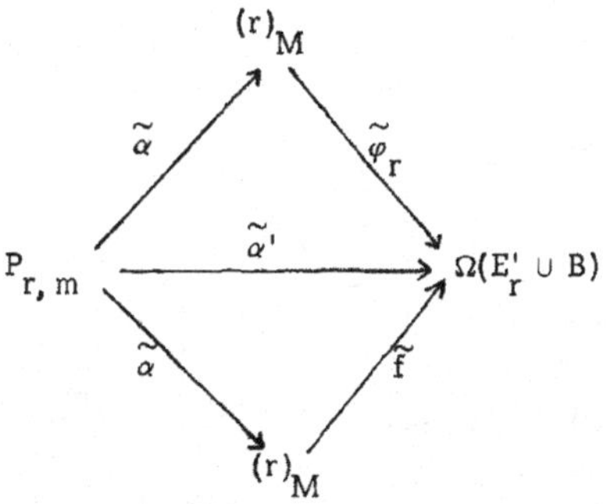

where the triangles are not necessarily commutative. One can easily see that (cf. [5])

$$d(\tilde{\varphi}_r\tilde{\alpha}, \tilde{f}\tilde{\alpha}) = d(\tilde{\varphi}_r\tilde{\alpha}, \tilde{\alpha}') \cdot d(\tilde{\alpha}', \tilde{f}\tilde{\alpha})$$

as maps of S^{m+r-1} into $\Omega(E_r' \cup B)$, or that

(3.6) $$[d(\tilde{\varphi}_r\tilde{\alpha}, \tilde{f}\tilde{\alpha})] = [d(\tilde{\varphi}_r\tilde{\alpha}, \tilde{\alpha}')] + [d(\tilde{\alpha}', \tilde{f}\tilde{\alpha})]$$

as elements of $\pi_{m+r-1}\Omega(E_r' \cup B)$. To compute $[d(\tilde{\varphi}_r\tilde{\alpha}, \tilde{\alpha}')]$, consider the commutative diagram

$$\begin{array}{ccc} P_{r,m} & \xrightarrow{\tilde{\alpha}} & \Omega(E_r \cup B) \\ {\scriptstyle \tilde{\sigma}}\downarrow & & \downarrow{\scriptstyle \tilde{\varphi}_r} \\ P_{r,m} & \xrightarrow{\tilde{\alpha}'} & \Omega(E_r' \cup B) \end{array}$$

where $\tilde{\sigma}$ is the loopspace map induced by the map

$$\alpha'^{-1}\varphi_r\alpha : \partial D^{r+1} \times S^m \to \partial D^{r+1} \times S^m$$

and α'^{-1} is a fiber homotopy inverse to α'. Now let

$$\tau : \partial D^{r+1} \to G(S^m)$$

be the map of ∂D^{r+1} into $G_0(S^m)$, the monoid of basepoint-preserving homotopy equivalences of S^m defined by the equation

$$\alpha'^{-1}\varphi_r\alpha(x, y) = (x, \tau(x)y),$$

for all (x, y) in $\partial D^{r+1} \times S^m$.

LEMMA 3.7. <u>The element</u> $[d(\tilde{\varphi}_r\tilde{\alpha}, \tilde{\alpha}')]$ <u>lies in</u>

$$\mathrm{im}\,(\tilde{i}_* : \pi_{r+m-1}\, S^m \to \pi_{r+m-1}\Omega(E'_r \cup B)),$$

<u>where</u> i <u>is the injection of</u> S^m <u>in</u> $E'_r \cup B$ <u>as the fiber above the basepoint</u>.

<u>Proof.</u> Note that

$$(\tilde{\alpha}')_{\#}[d(\tilde{\sigma}, \mathrm{id})] = [d(\tilde{\alpha}'\tilde{\sigma}, \tilde{\alpha}')] = [d(\tilde{\varphi}_r\tilde{\alpha}, \tilde{\alpha}')],$$

where id stands for the identity map of $P_{r,m}$. But $d(\tilde{\sigma}, \mathrm{id})$ is the adjoint of the map

$$d(\sigma, \mathrm{id}) : S^{r+m} \to \partial D^{r+1} \times S^m.$$

Since the two maps σ and id agree when projected onto $S^r = \partial D^{r+1}$, it follows that the component of $d(\sigma, \mathrm{id})$ which lies in ∂D^{r+1} is trivial. This implies that $d(\tilde{\sigma}, \mathrm{id})$ factors through S^m --which proves the lemma.

LEMMA 3.8. $d(\alpha', f\alpha)$ <u>is null-homotopic as a map of</u> S^{r+m} <u>into</u> $E'_r \cup B$.

<u>Proof.</u> Observe that f induces a homotopy equivalence of pairs

$$(E_{r+1} \cup B, E_r \cup B) \to (E'_{r+1} \cup B, E_r \cup B).$$

But

$$E_{r+1} \cup B = (E_r \cup B) \cup_{\beta} D^{r+m+1} \quad \text{and}$$
$$E'_{r+1} \cup B = (E'_r \cup B) \cup_{\beta'} D^{r+m+1},$$

where β and β' are the attaching maps for the (r+m+1)-cells (see (3.2) above).

Hence the maps

$$\beta f, \beta' : S^{m+r} \to E'_r \cup B$$

are homotopic. After a more or less straightforward computation (see [6]), one sees that

$$d(\beta f, \beta') = d(\alpha f, \alpha')$$

--which proves the lemma.

An immediate consequence of the preceding two lemmas is that

$$[d(\tilde{\varphi}_r \tilde{\alpha}, \tilde{\alpha}')] = 0,$$

since they imply that the element under consideration is both horizontal and lies in the image of ΩS^m. It follows now that we can extend φ_r to a map

$$\varphi'_{r+1} : E_{r+1} \to E'_{r+1}$$

such that $\varphi'_{r+1} | B_{r+1}$ = identity and $\varphi'_{r+1} | E_r$ is the fiber homotopy equivalence $E_r \to E'_r$ obtained by induction. According to [6], or computing directly in the spirit of [3], one can find a fiber homotopy equivalence

$$\varphi''_{r+1} : E_{r+1} \to E'_{r+1}$$

such that

$$\varphi''_{r+1} | E_{r+1} = \varphi_r.$$

The map φ''_{r+1}, however, may not satisfy Condition $(3.4)_r$ of the induction assumption without further modification. So consider the map

$$\delta(\tilde{\varphi}''_{r+1}, \tilde{f}) : N_{r+1}/A \to \Omega E',$$

where N_{r+1}/A is the difference complex and $\delta(\tilde{\varphi}''_{r+1}, \tilde{f})$ is the map induced by the adjoints $\tilde{\varphi}''_{r+1}$ and $\tilde{f}$ of φ''_{r+1} and f respectively. Observe that

$$\delta(\tilde{\varphi}''_{r+1}, \tilde{f}) | (N_r/A) = \delta(\tilde{\varphi}_r, \tilde{f}).$$

The necessary modification is intended to make $\delta(\tilde{\varphi}''_{r+1}, \tilde{f})$ horizontal by choosing φ''_{r+1} appropriately. So let us consider the diagram

$$\begin{array}{ccc} N_{r+1}/A & \xrightarrow{\delta(\tilde{\varphi}_{r+1}, \tilde{f})} & \Omega E' \\ & \searrow^{\delta(\tilde{\varphi}''_{r+1}, \tilde{f})} & \downarrow \eta \\ & & \Omega E' \end{array}$$

where η is the composite

$$\Omega E' \xrightarrow{\Omega p'} \Omega B \xrightarrow{\Omega i'} \Omega E'$$

introduced earlier. We wish to modify φ''_{r+1} so that the preceding diagram becomes homotopy commutative. To begin with, we change $\delta(\tilde{\varphi}''_{r+1}, \tilde{f})$ up to homotopy if necessary, so that the diagram becomes strictly commutative when restricted to N_r/A. Then the difference (in the sense of [4]) of the two maps $\delta(\tilde{\varphi}''_{r+1}, \tilde{f})$ and $\eta \circ \delta(\tilde{\varphi}''_{r\ 1}, \tilde{f})$ is a map

$$h : (N_{r+1}/A)/(N_r/A) \to \Omega E',$$

where $(N_{r+1}/A)/(N_r/A)$ is the complex obtained from N_{r+1}/A by collapsing N_r/A to a point. But

$$(N_{r+1}/A)/(N_r/A) = (A \times (S^{m+r} \vee S^{m+r} \vee S^{m+r}) \times A)/(A \times A),$$

as follows readily from the structure of N_{r+1}. Next we note that h can be factored according to the diagram

$$\begin{array}{ccc} (N_{r+1}/A)/(N_r/A) & \xrightarrow{h} & \Omega E' \\ \downarrow & & \downarrow \| \\ A \times S^{m+r} \times A/(A \times A) & \longrightarrow & \Omega E' , \end{array}$$

where the vertical map on the left is that induced by the folding map

$S^{m+r} \vee S^{m+r} \vee S^{m+r} \to S^{m+r}$ and h is a map of two-sided A-spaces. Therefore, to insure that h is homotopic to 0, it is enough to consider its restriction,

$$h|S^{m+r} : S^{m+r} \to \Omega E'.$$

Now the fact that $p' : E' \to B$ admits a section implies that

$$\pi_{m+r}\Omega E' \cong \pi_{m+r}\Omega S^m \oplus \pi_{m+r}\Omega B.$$

The component of $[h|S^{m+r}]$ in $\pi_{m+r}\Omega S^m$ is represented by a map $h' : S^{m+r} \to \Omega S^m$ which we regard as a map

$$\sigma : S^{r+1} \to G_0(S^m),$$

where $G_0(S^m)$ is the monoid of basepoint-preserving homotopy equivalences of S^m with itself. By definition, let

$$\varphi_{r+1} : E_{r+1} \to E'_{r+1}$$

be the fiber homotopy equivalence satisfying the following:

(i) $\varphi_{r+1}|E_r = \varphi_r$, and

(ii) $\varphi_{r+1}(x, y) = \varphi'_{r+1}(x, \sigma'(x)y)$,

where σ' is the composite

$$D^{r+1} \to S^{r+1} \xrightarrow{\sigma} G_0(S^m) \subset (S^m)^{S^m},$$

the map $D^{r+1} \to S^{r+1}$ being that which sends ∂D^{r+1} to the basepoint. One can now easily check that φ_{r+1} satisfies the two conditions of the induction assumption. This proves the theorem.

4. Remarks.

Suppose that in Problem A the spaces E, E' and B are all smooth closed manifolds and that f is a diffeomorphism. Then, assuming that $\dim B < m$ and $\dim E \geq 5$, one can adjust f up to diffeotopy so that it takes B diffeomorphically onto itself, where B is regarded as a smooth submanifold of E and E' (this is possible since $\dim B < m$). Now, by appealing to the Tubular Neighborhood Theorem, one can change f up to diffeotopy so that it induces a map of closed disk bundles of a tubular neighborhood E_0 of B in E onto a tubular neighborhood E'_0 of B in E'. Note that $E_1 = E - \text{int } E_0$ and $E'_1 = E' - E'_0$ are also closed disk subbundles of E and E'. Hence one can find an orthogonal bundle equivalence $\varphi : E \to E'$ which obviously need not be homotopic to f.

The preceding remark raises the question as to whether or not one can find an equivalence $\varphi : E \to E'$ of the same nature as f. For example, if the given fibrations were Top (S^m)-bundles, Top (S^m) being the group of homeomorphisms of S^m, and f a homeomorphism, would it be possible to find a Top (S^m)-bundle equivalence $E \to E'$?

It is not clear what the answer is, but it seems likely that further conditions on f need to be imposed.

REFERENCES

[1] S.Y. Husseini, "When is a complex fibered by a subcomplex?" Trans. Amer. Math. Soc., 124 (1966), 249-91.

[2] ______, The Topology of the Classical Groups and Related Topics, Gordon and Breach (New York, 1969).

[3] I.M. James and J.H.C. Whitehead, "The homotopy theory of sphere bundles over spheres, I," Proc. London Math. Soc. (3) 4 (1954), 196-218.

[4] I.M. James, "On H-spaces and their homotopy groups," Quart. J. Math. Oxford (2), 11 (1960), 161-79.

[5] ______, "On spaces with a multiplication", Pacific J. Math., 7 (1957), 1083-1100.

[6] T. Kyrouz, Thesis (University of Wisconsin, 1967).

ON THE DECOMPOSABILITY OF FIBRE SPACES

I.M. James

1. Introduction

Let $p: E \to B$ be a fibration with fibre F. We describe E as *decomposable* if E and $B \times F$ have the same homotopy type. This is the case, of course, if the fibration is trivial, in the sense of fibre homotopy type. Another condition is pointed out in I.M. James and J.H.C. Whitehead [8]. Suppose that we have a retraction $g: E \to F$. Then $h = (p,g): E \to B \times F$ is a weak homotopy equivalence. Thus E is decomposable, if the spaces are CW-complexes, and moreover the fibration admits a cross-section. I do not know whether every decomposable fibration admits a cross-section, but the following result shows this to be the case, under certain conditions, when the base is a sphere.

We say that a group A satisfies the *endomorphism condition* if there exist no endomorphisms of A onto itself other than automorphisms. This is the case, for example, if A is finitely generated. Now suppose that $B = S^m$, where $m \geqslant 2$, and that $\pi_{m-1}(F)$ satisfies the endomorphism condition. If E is decomposable then $\pi_{m-1}(E) \approx \pi_{m-1}(F)$ and hence

$$u_*: \pi_{m-1}(F) \approx \pi_{m-1}(E),$$

where $u: F \subset E$. Hence the transgression

$$\Delta: \pi_m(S^m) \to \pi_{m-1}(F)$$

is trivial, in the homotopy exact sequence, and so the fibration admits a cross-section. A similar result is given in I.M. James and J.H.C. Whitehead [8].

2. The brace product

We return to the general case. Consider the homotopy exact sequence

$$\dots \to \pi_r(F) \underset{u_*}{\to} \pi_r(E) \underset{p_*}{\to} \pi_r(B) \to \dots$$

Suppose that we have a cross-section $f: B \to E$. Then $p_* f_* = 1$; hence u_* is injective and p_* is surjective. While the higher homotopy groups of E and $B \times F$ are isomorphic there does not, in general, exist an isomorphism which respects the Whitehead product. To elucidate this point, consider the Whitehead product $[f_*\alpha, u_*\beta] \in \pi_{i+j-1}(E)$, where $\alpha \in \pi_i(B)$, $\beta \in \pi_j(F)$. Since $p_* u_* \beta = 0$ we have $p_*[f_*\alpha, u_*\beta] = 0$, by naturality. Hence by exactness, there exists a (unique) element $\{\alpha, \beta\} \in \pi_{i+j-1}(F)$ such that

$$u_*\{\alpha,\beta\} = [f_*\alpha, u_*\beta]. \tag{2.1}$$

Note that this brace product, as we shall call it, depends on the choice of section f.

For example, consider the case when B and F are spheres, say $B = S^m$ and $F = S^q$. Choose an element $\xi \in \pi_{m-1}O_q$, where O_q denotes the orthogonal group, and take E to be the corresponding q-sphere bundle over S^m with its canonical cross-section. Then

(2.2) $$\{\iota_m,\iota_q\} = \pm J\xi,$$

where ι_r generates $\pi_r(S^r)$ $(r = m,q)$ and J is given by the Hopf construction.

Now suppose, in the general case, that we have a retraction $g: E \to F$ such that gf is nul-homotopic. Then $\{\alpha,\beta\} = g_*u_*\{\alpha,\beta\} = g_*[f_*\alpha,u_*\beta] = 0$, by naturality, since $g_*f_* = 0$. Thus the brace product is an obstruction to the existence of a retraction. Similarly it can be shown that certain brace products are obstructions to decomposability. For example take $B = S^m$ and suppose that E admits a cross-section, as is the case if $\pi_{m-1}(F)$ satisfies the endomorphism condition. Further suppose that $\pi_m(F)$ is finite. Let $q < m$.

If E is decomposable then, under these conditions, it follows easily that

(2.3) $$\{\alpha,\iota_m\} = 0 \qquad (\alpha \in \pi_q(F)).$$

Before applying this, in §4, we briefly recall one of the basic methods for obtaining information about Whitehead products from cohomology.

3. The Hopf construction

Let K be a complex and let $\xi \in \pi_m(K)$ be an element such that

$$\xi^*: H^m(K) \to H^m(S^m)$$

is non-trivial, in mod 2 cohomology. Choose an element $\sigma \in H^m(K)$

such that $\xi^*\sigma$ generates $H^m(S^m)$. Suppose that the Whitehead square $[\xi,\xi] \in \pi_{2m-1}(K)$ is trivial. Then there exists a map

$$f: S^m \times S^m \to K$$

of type (ξ,ξ), in the usual sense. We perform the Hopf construction and obtain from f a map $f': S^{2m+1} \to SK$, where S denotes the suspension functor. Let $K' = e^{2m+2} \cup SK$ denote the mapping cone of f'. Suppose that

$$f^*: H^{2m}(K) \to H^{2m}(S^m \times S^m)$$

is trivial. Then it follows (see E. Thomas [10]) that

$$k^*: H^{2m+2}(K',SK) \to H^{2m+2}(K')$$

is non-trivial, where k denotes the inclusion. Let $\tau \in H^{2m+2}(K')$ be the image of the generator and let $\sigma' \in H^{m+2}(K')$ be the element such that $j^*\sigma' = S^*\sigma$, where $j: SK \subset K'$. By (1.4) of E. Thomas [10] we have $\sigma' \cup \sigma' = \tau$, in the cohomology ring $H^*(K')$, and so

$$Sq^{m+1}: H^{m+1}(K') \to H^{2m+2}(K')$$

is non-trivial. Conclusions can be drawn from this, depending on the cohomology of K, by using the relations of Adem [3].

This procedure is well-known (see J. Adem [3]) when $K = S^m$. However, the following example does not appear to have been published hitherto, although a different proof of the same result is given in M. Gilmore [4].

Proposition (3.1). *Let* m *be odd. Suppose that the Whitehead square* $[1_m,1_m]$ *can be halved in* $\pi_{2m-1}(S^m)$. *Then* m+1 *is a power of two.*

We have $\pi_{2m-1}(S^m) = S_*\pi_{2m-2}(S^{m-1})$, since m is odd, and hence $2\pi_{2m-1}(S^m) = c_*\pi_{2m-1}(S^m)$, where $c\colon S^m \to S^m$ is a map of degree 2. Take $K = e^{m+1} \cup S^m$ to be the mapping cone of c, and take ξ to be the generator of $\pi_m(K)$. Then $[\xi,\xi] = \xi_*[1_m,1_m] = 0$, since $[1_m,1_m] \in c_*\pi_{2m-1}(S^m)$. Following the above procedure we construct a complex

$$K' = e^{2m+2} \cup e^{m+1} \cup S^m.$$

Since the cohomological conditions are satisfied we obtain that

$$Sq^{m+1}\colon H^{m+1}(K') \to H^{2m+2}(K')$$

is non-zero. Hence it follows from the Adem relations between the Steenrod squares that m+1 is a power of two, as asserted.

4. The Stiefel manifold

Consider the Stiefel manifold $V_{n,k}$ of orthonormal k-frames in euclidean n-space, where $1 \leqslant k \leqslant n$. Let $p\colon V_{n,k} \to S^{n-1}$ be defined by taking the last vector of each k-frame. Then p is a fibration and the fibre can be identified with $V_{n-1,k-1}$. We recall (see J.F. Adams [2]) that there exists a cross-section if and only if

(4.1) $$n \equiv 0 \bmod 2^{\varphi(k)},$$

where $\varphi(k)$ denotes the number of integers s in the range $0 < s < k$ such that $s \equiv 0,1,2$ or $4 \bmod 8$. If $V_{n,k}$ is decomposable then (4.1) is a necessary condition, as noted in §1. The main purpose of this lecture is to prove

Theorem (4.2). If $V_{n,k}$ is decomposable then $n = 2^r$ for some integer $r \geqslant \varphi(k)$.

In case k is even we further prove

Theorem (4.3). If $V_{n,k}$ is decomposable, where k is even, then $n = 2,4$ or 8.

It is tempting to conjecture that (4.3) remains true when k is odd, i.e. that $V_{n,k}$ is not decomposable unless $V_{n,k}$ is trivial as a fibre bundle. The first case which has not been settled is that of $V_{16,3}$. It can be shown (see I.M. James [7]) by the method used to prove (4.2) that the complex Stiefel manifold $W_{n,k}$ is indecomposable unless $n = k = 2$, and that the quaternionic Stiefel manifold is never decomposable. These results are relevant to the classification of Hopf homogeneous spaces (see W-Y. Hsiang and J.C. Su [5]).

We fibre O_n over $V_{n,k}$ with fibre O_{n-k}, in the usual way, and consider the transgression operator

$$\Delta: \pi_{n-1}(V_{n,k}) \to \pi_{n-2}(O_{n-k})$$

in the associated homotopy exact sequence. Write $\eta = \Delta\xi$, where ξ denotes the class of a cross-section $f: S^{n-1} \to V_{n,k}$. By (4.3) of I.M. James [6] we have

(4.4) $$\pm S_*^{k-1} J\eta = [\iota_{n-1}, \iota_{n-1}],$$

where J is given by the Hopf construction. Moreover it follows from (2.2) and naturality that

(4.5) $$\pm \alpha_* J\eta = \{\alpha, \iota_{n-1}\},$$

where α generates $\pi_{n-k}(V_{n-1,k-1})$. With these relations in hand the proof of (4.2) proceeds as follows.

Suppose that $V_{n,k}$ is decomposable, where $n \geqslant k \geqslant 2$. If $n = 2^{\varphi(k)}$ or $2^{\varphi(k)+1}$ we have nothing to prove. Suppose, then, that $n \geqslant 3.2^{\varphi(k)}$, and hence $n > 2k$. Then $\pi_{n-1}(V_{n-1,k-1})$ is finite and so it follows from (2.3) and (4.5) that $\alpha_* J\eta = 0$. Consider the mapping cone $L = e^{2n-k-1} \cup S^{n-k}$ of $J\eta$. Since $\alpha_* J\eta = 0$ there exists a map

$$l: L \to V_{n-1,k-1}$$

such that $l|S^{n-k}$ represents α. Recall that $H^*(V_{n-1,k-1})$ is generated, as a ring, by elements whose dimensions do not exceed n-2. It follows at once that

$$l^* H^{2n-k-1}(V_{n-1,k-1}) = 0. \tag{4.6}$$

Now suspend k-1 times and consider the map

$$S^{k-1}l: S^{k-1}L \to S^{k-1}V_{n-1,k-1}.$$

Here $S^{k-1}L = e^{2n-2} \cup S^{n-1}$, where the (2n-2)-cell is attached by a map of class $S_*^{k-1}J\eta$. By (4.4) we can regard $S^{k-1}L$ as obtained from the product $S^{n-1} \times S^{n-1}$ by identifying axes, in the usual way. Write $f = S^{k-1}l \circ \psi$, where

$$\psi: S^{n-1} \times S^{n-1} \to S^{k-1}L$$

is the identification map. By performing the Hopf construction on f we obtain a map $f': S^{2n-1} \to S^k L$. Consider the mapping cone $K' = e^{2n} \cup S^k L$ of f'. It follows from (4.6) that $f^* H^{2n-2}(S^{k-1}V_{n-1,k-1}) = 0$ and hence, as explained in §3, that

$$Sq^n: H^n(K') \to H^{2n}(K')$$

is non-trivial. Now recall that $H^r(K') = 0$, for $n+k-1 \leq r \leq 2n$. Hence it follows from the Adem relations that the interval $[n-k+2,n]$ contains a power of two. But the interval $[n-k+2,n-1]$ contains no power of two, since (4.1) is satisfied. Hence n is a power of two, as asserted.

To prove (4.3) we first recall the following result, which is an immediate consequence of the main theorem of Adams [1] on secondary cohomology operations.

Proposition (4.7). *Let* $n = 2^r$, *where* $r \geq 4$. *Suppose, for some value of* m, *we have a finite complex* K" *such that* $H^{m+q}(K'') = 0$ *for* $1 < q < 2^{r-1} + 2^{r-3}$ *and such that*

$$Sq^1 : H^{m+n-1}(K'') \to H^{m+n}(K'')$$

is trivial. Then

$$Sq^n : H^m(K'') \to H^{m+n}(K'')$$

is trivial.

Suppose that $V_{n,k}$ is decomposable, where k is even. Then $n = 2^r$, by (4.2), where $r \geq \varphi(k)$. We construct the complex K', as before, and note that

$$Sq^1 : H^n(K') \to H^{n+1}(K')$$

is trivial, since k is even, while

$$Sq^n : H^n(K') \to H^{2n}(K')$$

is non-trivial, as we have seen. If $r \geq 4$ the hypotheses of (4.7) are satisfied, with appropriate m, by a dual complex of K', and so we have a contradiction. Therefore $r \leq 3$, which proves (4.3).

5. Appendix

Of course we can extend the scope of this investigation by regarding $V_{n,k}$ as a fibre space over $V_{n,l}$, with fibre $V_{n-l,k-l}$, where $1 < l < k < n$. It is easy to show, by comparison of the Steenrod squares, that this fibration is indecomposable unless first $l = k-1$ and secondly $n \equiv k \mod 4$. But if $V_{n,k}$ has the same homotopy type as $S^{n-k} \times V_{n,k-1}$, with n-k even, then S^{n-k} is a retract of $V_{n,k}$, hence of $V_{n-k+2,2}$, and therefore $n-k+2 = 4$ or 8, as shown in I.M. James and J.H.C. Whitehead [9]. Thus we obtain a contradiction and conclude that none of these fibrations of $V_{n,k}$ is decomposable. A similar result holds for complex and quaternionic Stiefel manifolds.

REFERENCES

[1] J.F. Adams, On the non-existence of elements with Hopf invariant one, Ann. of Math., 72, 20-104 (1960).

[2] J.F. Adams, Vector fields on spheres, Ann. of Math., 75, 603-632 (1962).

[3] J. Adem, The iteration of the Steenrod squares in algebraic topology, Proc. Nat. Acad. Sci., 38, 720-726 (1952).

[4] M. Gilmore, Some Whitehead products on odd spheres, Proc. Amer. Math. Soc., 20, 375-377 (1969).

[5] Wu-Yi Hsiang and J.C. Su, On the classification of transitive effective actions on Stiefel manifolds, Trans. Amer. Math. Soc. 130, 322-336 (1968).

[6] I.M. James, On the iterated suspension, Quart. J. of Math. Oxford (2), 5, 1-10 (1954).

[7] I.M. James, On the homotopy type of Stiefel manifolds (to appear).

[8] I.M. James and J.H.C. Whitehead, Note on factor spaces, Proc. London Math. Soc. (3), 4, 129-137 (1954).

[9] I.M. James and J.H.C. Whitehead, The homotopy theory of sphere-bundles over spheres I, ibid. 196-218.

[10] E. Thomas, on functional cup-products and the transgression operator, Archiv der Mathematik, 12, 435-444 (1961).

Oxford University Mathematical Institute.

THE MILGRAM BAR CONSTRUCTION AS A TENSOR PRODUCT OF FUNCTORS

BY

SAUNDERS MAC LANE

1. Introduction. The classifying space for a topological group can be constructed "algebraically" or "geometrically". For graded differential modules, Eilenberg-Mac Lane [7] used a strictly algebraic bar construction, and Cartan [1] pointed out that this bar construction could be handled effectively by algebraic analogues of classifying space and fiber bundle techniques. Several authors retranslated these algebraic devices back into geometry (Dold-Lashof [4], Rothenberg-Steenrod [14]. Finally in [13] Milgram described for a topological monoid (= associative H-space with identity) a classifying space which had direct ties to the corresponding algebraic bar construction. Steenrod in [15] has developed further properties of Milgram's classifying space. Here we shall show how these properties can be conveniently reformulated using certain recently developed categorical techniques, notably the construction of the tensor product of two functors. The possibility of this reformulation was recognized by both Allen Clark [2] and the author; the author has also much profited from conversations with Milgram on this topic.

Our treatment will begin with a description of some functorial methods. We then indicate that these methods may illuminate the geometrical facts by giving a simple proof of the Hausdorff character of the classifying space, and by using the triangulation of prisms in proving that this construction preserves products.

2. Coends. A category $\mathfrak{C}$ is cocomplete if every functor $T: J \to \mathfrak{C}$ from a small category J to $\mathfrak{C}$ has a colimit $\varinjlim T$ in $\mathfrak{C}$. We recall that such a colimit is an object $A = \varinjlim T$ in $\mathfrak{C}$, regarded as a constant functor $A: J \to \mathfrak{C}$, together with a natural transformation $\lambda: T \to A$ which is universal among natural transformations from T to constant functors. If J is discrete (= all arrows in T are identities = J is a set), a colimit is just a coproduct $\coprod Ti$ of the objects for $i \in J$. If J is the category with just two non-identity arrows $\rightrightarrows$, a colimit is a coequalizer. A category $\mathfrak{C}$ which has all small coproducts and all coequalizers (of pairs $\rightrightarrows$) is necessarily cocomplete.

Let $H(j,k)$ be a functor of two variables $j, k \in J$, contravariant in the first variable and covariant in the second; we write $H: J^{op} \times J \to \mathfrak{C}$, where J^{op} denotes the category "opposite" or "dual" to J. It has recently turned out to be very useful to define a supernatural transformation $\beta: H \to B$, with B an object of $\mathfrak{C}$, to be a family of arrows $\beta_j: H(j,j) \to B$ of $\mathfrak{C}$, one for each object j, such that the square

$$\begin{array}{ccc} H(k,j) & \xrightarrow{H(1,t)} & H(k,k) \\ {\scriptstyle H(t,1)}\downarrow & & \downarrow{\scriptstyle \beta_k} \\ H(j,j) & \xrightarrow[\beta_j]{} & B \end{array} \tag{1}$$

commutes for every arrow $t: j \to k$ of J. (In [12], Mac Lane called β a diagonal spread; Eilenberg-Kelly [6] called β an extraordinary natural transformation and considered more general cases.) There is a familiar example of the notion dual to supernatural. If $*$ denotes the one point set and J is any category, the family of arrows $\beta_j: * \to \hom(j,j)$ which send the one point to the various identity arrows $j \to j$ define a dually supernatural transformation.

A coend of a functor $H: J^{op} \times J \to \mathfrak{C}$ is an object E of $\mathfrak{C}$ together with a supernatural $\kappa: H \to E$ which is universal among supernatural transformations from H. Thus κ is supernatural, and if $\beta: H \to B$ is any supernatural transformation, there is a unique arrow $f: E \to B$ with $\beta_j = f\kappa_j$ for all j. As for any universal, a coend of H is unique, up to isomorphism, if it exists. Following Yoneda [18] we use an integral notation for the coend

$$B = \int H = \int^j H(j, j) = \text{the coend of } H: J^{op} \times J \to \mathcal{C}.$$

As the square (1) indicates, a coend is a special sort of colimit, taken not over J but over a suitable category constructed from J (and indeed this construction goes back to Kan). Put differently, if J is small and $\mathcal{C}$ cocomplete, every functor H has a coend, which may be calculated as the coequalizer of a pair of arrows h_1 and h_2,

(2) $$\coprod_j H(k, j) \underset{h_2}{\overset{h_1}{\rightrightarrows}} \coprod_j H(j, j) \longrightarrow \int^j H(j, j),$$

where h_1 and h_2 are defined on the components of the coproduct by $H(1, t)$ and $H(t, 1)$. Here the first coproduct is taken over all arrows t of J. It is clearly sufficient to take this coproduct over arrows t which (with the identities) generate J under composition.

Though a coend is just a special kind of colimit, it appears to deserve a special name and notation because Day and Kelly [3] and Benabou (unpublished) have shown that "coend" does not reduce to colimit for relative categories (categories with hom-sets in some "closed" category like the category of abelian groups). Starting there, Dubuc in [5] has shown the further effective use of the concept of coend in treating closed categories and categories relative to closed categories.

3. Tensor Products of Functors. Let $\mathcal{X}$ be a cocomplete category with a functor $\otimes: \mathcal{X} \times \mathcal{X} \to \mathcal{X}$; often this "tensor product" will be associative up to a coherent natural isomorphism and will have an identity, up to coherent isomorphism, but we do not require these assumptions. Then two functors $F: J^{op} \to \mathcal{X}$ and $G: J \to \mathcal{X}$ of opposite variance have a tensor product which is an object

(3) $$F \otimes_J G = \int^j (F_j) \otimes (G_j)$$

of $\mathcal{L}$. In view of the description given above for the coend, this object may also be described as the coproduct $\coprod_j (F_j) \otimes (G_j)$ modulo suitable identification (one for each arrow $j \to k$ of J). If J is a ring, regarded as a preadditive category with one object, while $\mathcal{L}$ is the category $\mathcal{Ab}$ of abelian groups with the usual tensor product $\otimes = \otimes_Z$, then a contravariant additive functor $F: J^{op} \to \mathcal{L}$ is just a right J-module, while a covariant additive $G: J \to \mathcal{L}$ is a left J-module. The tensor product $F \otimes_J G$ of these functors is then exactly the usual tensor product of modules: There is just one object j in J, so the coproduct $\coprod_j$ is just the abelian group $F \otimes G$ (generated by the usual elements $f \otimes g$ for $f \in F$, $g \in G$); the arrows $t: j \to j$ are just the elements of the ring J, and the identifications are the usual ones $ft \otimes g = f \otimes tg$.

The interested reader may verify that many of the formal identities for tensor products of modules carry over to tensor products of functors.

4. Compactly Generated Spaces. In our case $\mathcal{L}$ will be the category $\mathcal{CG}$ of compactly generated Hausdorff topological spaces. This category has been carefully examined by Steenrod in [15]. Recall that a Hausdorff space Y is compactly generated when each subset that intersects every compact set of Y in a closed set is itself a closed set, and that a morphism between compactly generated spaces $Y \to Y'$ is just a continuous map between the associated topological spaces. This last property states that $\mathcal{CG}$ is a full subcategory of $\mathcal{Haus}$, the category of Hausdorff spaces and continuous maps between them; we write the inclusion functor as $i: \mathcal{CG} \to \mathcal{Haus}$. Steenrod in [15] constructs to each Hausdorff space X a compactly generated space $k(X)$ and a continuous map $\varepsilon_X: ikX \to X$ (in fact, i is the identity function) which is universal from compactly generated spaces to X

i.e., that every continuous $f\colon iY \to X$ with compactly generated Y factors uniquely as $f = \varepsilon_X f'$ (in fact f' is the same function as f). This statement (without the rest of Steenrod's Theorem 3.2) proves that k is a functor $\mathbf{Haus} \to \mathcal{CH}$ right adjoint to the inclusion functor i and hence that $\mathcal{CH}$ is a coreflective subcategory of $\mathbf{Haus}$, with k the coreflection.

Now any right adjoint preserves products, when they exist. The usual cartesian product topology gives a product $X_1 \times_C X_2$ on $\mathbf{Haus}$; hence any two spaces Y_1, Y_2 in $\mathcal{CH}$ have a product there given as $Y_1 \times Y_2 = k(iY_1 \times_C iY_2)$, with the evident projections on Y_1 and Y_2; we note that this categorical product is the usual cartesian product set with a topology which is not always the usual cartesian product topology. With this product, $\times : \mathcal{CH} \times \mathcal{CH} \to \mathcal{CH}$ is a tensor product, in the sense of §3.

5. <u>The Simplicial Category</u>. Regard each finite ordinal number n as the ordered set $\{0, 1, \ldots, n-1\}$ of all smaller ordinals, and let $\boldsymbol{\Delta}$ be the category whose objects are all finite ordinals and whose arrows $f\colon n \to m$ are all weakly monotone (i.e., order-preserving) functions. The arrows of $\boldsymbol{\Delta}$ thus include the usual "face" and "degeneracy" operations d_i and s_i,

$$0 \longrightarrow 1 \rightrightarrows 2 \ \substack{\longrightarrow\\\longrightarrow\\\longrightarrow}\ 3, \ldots, \quad 1 \longleftarrow 2 \leftleftarrows 3 \ \ldots$$

Explicitly, $d_i\colon n \to n+1$ for $i = 0, \ldots, n$ is that monotone injection with image $\{0, \ldots, \hat{i}, \ldots, n\}$, and $s_i\colon n+1 \to n$ is that monotone surjection with $s_i(i) = s_i(i+1)$ for $i = 0, \ldots, n-1$. These arrows satisfy the familiar identities and generate under composition the whole category $\boldsymbol{\Delta}$. If $\boldsymbol{\Delta}^+$ is the full subcategory omitting the object 0, a functor $(\boldsymbol{\Delta}^+)^{op} \to \mathcal{C}$ is thus a simplicial object in the category $\mathcal{C}$ in the usual sense, except that our dimension n is <u>one greater</u> than the usual geometric dimension. In particular, a functor on $(\boldsymbol{\Delta}^+)^{op}$ to the category of sets is a simplicial set (= complete semisimplicial complex), while a functor on

$\mathbf{\Delta}^{op}$ to sets is an augmented simplicial set.

We systematically include 0 as an object of $\mathbf{\Delta}$ because it will enable us to wholly dispense with the familiar but troublesome manipulation of identities on faces and degeneracies.

Now $\mathbf{\Delta}$ can be interpreted topologically as the category of "standard simplices". This amounts to defining the functor $\sigma_o : \mathbf{\Delta} \to \mathcal{CG}$ which sends the ordinal n to the standard (n-1)-dimensional simplex σ_{n-1} (and 0 to the empty set). With Milgram we regard σ_{n-1} as the set of real numbers $t_1, \ldots, t_{n-1}$ with $0 \leq t_1 \leq \ldots \leq t_{n-1} \leq 1$ with the Euclidean topology, while $\sigma_o(d_i) = d_i$ and $\sigma_o(s_i) = s_i$ are the usual face and degeneracy maps

$$
\begin{aligned}
(4) \qquad d_i(t_1, \ldots, t_{n-1}) &= (0, t_1, \ldots, t_{n-1}), & i &= 0, \\
&= (t_1, \ldots, t_i, t_i, \ldots, t_{n-1}), & 0 &< i < n, \\
&= (t_1, \ldots, t_{n-1}, 1), & i &= n,
\end{aligned}
$$

$$
(5) \qquad s_i(t_1, \ldots, t_n) = (t_1, \ldots, \hat{t}_{i+1}, \ldots, t_n), \qquad 0 \leq i < n.
$$

6. The Milgram Bar Construction. By a $\mathcal{CG}$-monoid G we mean a CG-space G with a continuous multiplication $G \times G \to G$ (written as a product, $(x, y) \longmapsto xy$) which is associative and has a (two-sided) unit e; thus G is just like an associative H-space with unit except that the topology of G must be $\mathcal{CG}$ and the multiplication must be continuous in the compactly generated topology on the product $G \times G$. A left action of G is a pair (ν, A) where A is in $\mathcal{CG}$ and $\nu : G \times A \to A$ in $\mathcal{CG}$ is associative (in the usual way) and makes the unit e act as the identity. All possible left actions (ν, A) of G, with the evident morphisms, from a category $\mathbf{Mod}_G$, and $(\nu, A) \mapsto A$ is a faithful functor to $\mathcal{CG}$. Moreover, each power $G^n = G \times \ldots \times G$ carries an evident left action of G (multiply in the left-most factor). In particular, the one point space $G^o = *$

carries a (trivial) left-action.

For each G we can now define a functor

$$G^{\bullet}: \Delta^{op} \longrightarrow \mathcal{M}od_G$$

which sends n to the n-fold product G^n with the left-action just described, and $d_i: n \to n+1$, $s_i: n+1 \to n$ to the $\mathcal{CG}$-maps $d^i: G^{n+1} \to G^n$ and $s^i: G^n \to G^{n+1}$ defined on $(n+1)$-tuples $(x_o, \ldots, x_n) \in G^{n+1}$ by

$$(6) \qquad d^i(x_o, \ldots, x_n) = (x_o, \ldots, x_{i-1}, x_i x_{i+1}, \ldots, x_n), \quad 0 \leq i < n$$
$$= (x_o, \ldots, x_{n-1}), \qquad i = n$$

$$(7) \qquad s^i(x_o, \ldots, x_{n-1}) = (x_o, \ldots, x_i, e, x_{i+1}, \ldots, x_{n-1}), \quad 0 \leq i < n.$$

These maps all respect the action of G (on the first factor x_o); they satisfy (the duals of) the face and degeneracy relations on d_i and s_i, so do define the requisite contravariant functor. These maps d_i and s_i are, of course, the familiar ones used to describe the standard resolution used in defining the cohomology of a group G, and these identities are well known from that case.

For the cohomology of groups one can also "divide out" the left-action of G by applying the functor $\mathbb{Z} \otimes_G -$ to this resolution (with trivial right action of the group G on $\mathbb{Z}$). In the present case there is a similar functor $* \times_G -$, where the one point set $*$ has trivial right-action by G. Applied to $G^{\bullet}$, it gives a functor

$$G^{\#}: \Delta^{op} \longrightarrow \mathcal{CG}$$

which we may describe directly as follows: $G^{\#}$ sends both 0 and 1 to the one point space $*$, and each $n > 1$ to the $(n-1)$-fold product space G^{n-1} with face and degeneracy maps $d^i: G^n \to G^{n-1}$, $s^i: G^{n-1} \to G^n$ defined by

$$(8) \qquad d^i(x_1, \ldots, x_n) = (x_2, \ldots, x_n), \qquad i = 0,$$
$$= (x_1, \ldots, x_i x_{i+1}, \ldots, x_n), \qquad 0 < i < n,$$
$$= (x_1, \ldots, x_{n-1}), \qquad i = n.$$

(9) $\quad s^i(x_1,\ldots,x_{n-1}) = (x_1,\ldots,x_i,e,x_{i+1},\ldots,x_{n-1})\ ,\quad 0 \le i < n\ ;$

these formulas apply for $n \geq 1$; for $n = 0$, $d^o : * \to *$ is the only possible function.

Both functors $G^\bullet$ and $G^\#$ can be defined conceptually (no d_i, s_i) using the fact that the monoid $2 \to 1 \leftarrow 0$ in $\mathbf{\Delta}$ is universal.

If we regard the first functor $G^\bullet$ as a functor to $\mathcal{CG}$ (compose with the forgetful functor $\mathcal{M}od_G \to \mathcal{CG}$), there is an evident natural transformation

$$p: G^\bullet \to G^\#,$$

with $p_n(x_o, x_1, \ldots, x_n) = (x_o, \ldots, x_n)$; this amounts to "dividing out" the left-action of G.

The Milgram bar construction now assigns to each $\mathcal{CG}$-monoid G two topological spaces, a "total space" E(G) and a "base space" B(G),

(10) $$E(G) = G^\bullet \times_{\mathbf{\Delta}} \sigma_o = \int^n G^n \times \sigma_{n-1}\ ,$$

(11) $$B(G) = G^\# \times_{\mathbf{\Delta}} \sigma_\bullet = \int^n G^{n-1} \times \sigma_{n-1}\ ,$$

moreover, the natural transformation p defined above gives a continuous map $p_G : E(G) \to B(G)$.

In order that the coends used above should exist, we must work in a cocomplete category. So let us calculate the coends above in $\mathcal{T}op$, and so regard E(G) and B(G) for the moment as topological spaces.

This construction is identical, up to language, to that given by Milgram in [13]. For example, the description of E(G) as a coend shows that E(G) is the disjoint union $\coprod_n (G^n \times \sigma_{n-1})$ with certain identifications. Since $\sigma_o(0)$ is the empty set, this is a union for $n > 0$; this amounts to saying that the tensor products could just as well have been taken over the category $\mathbf{\Delta}^+$. The identifications are

(12) $$(d^i\xi, \tau) = (\xi, d_i\tau) \qquad \xi \in G^{n+1}\ ,\quad \tau \in \sigma_{n-1}\ ,$$

(13) $$(s^i\xi, \tau) = (\xi, s_i\tau) \qquad \xi \in G^n\ ,\quad \tau \in \sigma_n\ ;$$

these, with the explicit formulas above for d's and s's , are exactly Milgram's identifications.

We next show that E(G) is a Hausdorff space in two maps -- by direct examination (§7) and by categorical devices (§8).

7. The Separation Axiom Explicitly.

THEOREM. For each $\mathcal{CG}$-monoid G, the spaces E(G) and B(G) are compactly generated.

Proof. The space E(G) is defined by an identificatiom map

$$\coprod_n G^n \times \sigma_{n-1} \longrightarrow E(G).$$

Now by hypothesis each product $G^n \times \sigma_{n-1}$ is completely generated, hence so is the coproduct (disjoint union) of these spaces. Hence it will be enough to show that E(G) is Hausdorff, since Steenrod ([15], Lemma 2.6) proves that an identification map (= a proclusion) $X \rightarrow Y$ with X in $\mathcal{CG}$ and Y Hausdorff has Y in $\mathcal{CG}$.

Now write a point of $G^n \times \sigma_{n-1}$ as $(x_o, x_1, \ldots, x_{n-1}; t_1, \ldots, t_{n-1})$ where the $x_i \in G$ and the t_i are real numbers with $0 \leq t_1 \leq \ldots \leq t_{n-1}$. The identifications above may be used to (possibly) shorten the presentation of the corresponding point of E(G): If $t_1 = 0$, then $(t_1, \ldots, t_{n-1}) = d_o(t_2, \ldots, t_{n-1})$, so

$$(x_o, x_1, \ldots, x_{n-1}; 0, t_2, \ldots, t_{n-1}) = (x_o x_1, x_2, \ldots, x_{n-1}; t_2, \ldots, t_{n-1}).$$

Similarly, if $t_i = t_{i+1}$ we may replace x_i and x_{i+1} by their product and drop t_{i+1}, while if $t_{n-1} = 1$ we may drop both x_{n-1} and t_{n-1}. Again, if one of $x_1, \ldots, x_{n-1}$ is the unit e of G, we may shorten the presentation. As a result, every point of E(G) is equal to a point in normal form $\eta = (y_o, y_1, \ldots, y_{m-1}; s_1, \ldots, s_1, \ldots, x_{m-1})$, with $y_1 \neq e, \ldots, y_{m-1} \neq e$ and $0 < s_1 < \ldots < s_{m-1} < 1$. As Steenrod proves ([16], Corollary 5.4) the normal form is unique. This can also be observed by giving a direct process for getting the normal form in one step: Take $y_o = x_o x_1 \cdots x_i$ where i is 0 or the last index with $t_i = 0$, take $y_1 = x_{i+1} \cdots x_j$ and $s_1 = t_{i+1} = \cdots = t_j$, where $t_j < t_{j+1}$ unless this $y_1 = e$, in which case we drop this y_1 and s_1 and take y_1 to be the product of the next (and largest) string of x's for which the corresponding string of t's is equal, and continue in this way.

This presentation allows us to define certain elementary neighborhoods of a point η in normal form $\eta = (y_o, y_1, \ldots, y_{m-1}; s_1, \ldots, s_{m-1})$. In G, choose open neighborhoods

$$U_o, U_1, \ldots, U_{m-1}, U_m \quad \text{of} \quad y_o, y_1, \ldots, y_{m-1}, \text{ and } e ,$$

respectively, with U_k and U_m disjoint, for every $k \neq m$; in the unit interval, choose disjoint open neighborhoods

$$K_o, K_1, \ldots, K_{m-1}, K_m \quad \text{of} \quad 0, s_1, \ldots, s_{m-1}, \text{ and } 1 .$$

Then for each n define an open set $V_n \subset G^n \times \sigma_{n-1}$, as follows: A point $\xi = (x_o, x_1, \ldots, x_{n-1}; t_1, \ldots, t_{n-1})$ lies in V_n if (i) each t_i lies in some K_j; (ii) there is at least one t_i in each of $K_1, \ldots, K_{m-1}$ (and hence $n \geq m$); (iii) if $t_1, \ldots, t_j$ is the list (possibly empty) of all the t's in K_o, then the product $x x_1 \ldots x_j$ is in U_o; (iv) if $t_i, \ldots, t_\ell$ is the list of all the t's in K_k, then the product $x_i \ldots x_\ell$ is in U_k, for each $k = 1, \ldots, m$. Since the product map $\mu : G \times G \to G$ is continuous, this description does give an open set V_n in $G^n \times \sigma_{n-1}$. Moreover, any identification $\xi = \xi'$ in the tensor product clearly carries a ξ in V_n onto a ξ' in $V_{n'}$; in particular, the provision that $U_k \cap U_m = \emptyset$ for $k \neq m$ insures that no identification can drop all the t's lying in K_k. Hence the union of all these V_n for various n provides an open set in $\coprod G^n \times \sigma_{n-1}$ which defines an open set, call it V, in E(G). This V is the desired "elementary" neighborhood of the given point η in normal form.

Now take two distinct points η and η', both in normal form. We can then separate them by elementary neighborhoods V and V' of the above special form.

Thus, if $\eta' = (y'_o, \ldots, y'_{n-1}; s'_1, \ldots, s'_{n-1})$ is such that there is some s'_j equal to none of the $s_1, \ldots, s_{m-1}$ in η, we choose the K_i above so that none of them contains s', and then choose a suitable neighborhood K'_j of s'_j; an easy argument shows $V \cap V' = \emptyset$. On the other hand if $\eta' = (y'_o, \ldots, y'_{m-1}; s_1, \ldots, s_m)$ with the same s_i as for η, we can appeal to the fact that G is Hausdorff to choose strings of neighborhoods U_i, U'_i so that $V \cap V' = \emptyset$. This completes the proof of the theorem for E(G). The proof for B(G) is analogous.

In the treatment by Steenrod [16], the Hausdorff property is obtained by making the additional assumption that G and its unit e form an NDR (a neighborhood deformation retract; see [15]). The above argument avoids the assumption, essentially by showing that the identifications made in forming E(G) involve only very "nicely" situated face and degeneracy maps on the standard simplices.

8. <u>The Separation Axiom Categorically.</u> One may also construct E(G) and B(G) directly in the category $\mathcal{CG}$ of compactly generated spaces in virtue of the following result:

THEOREM. The category $\mathcal{CG}$ is cocomplete.

<u>Proof.</u> Since the coproduct of compactly generated spaces is again such, it will suffice to construct coequalizers in $\mathcal{CG}$. The construction depends on the adjoints to inclusion functors i and i':

$$\mathcal{CG} \underset{i}{\overset{k}{\rightleftarrows}} \mathcal{H}aus \underset{h}{\overset{i'}{\rightleftarrows}} \mathcal{T}op ;$$

here k is the right adjoint used by Steenrod, which turns each Hausdorff space X into a compactly generated space kX (same points, more closed sets), while h is the left adjoint to i', which sends each topological space Y to its "largest Hausdorff quotient" h(Y). The existence of h follows readily from Freyd's adjoint functor theorem [8]. Since h is a left adjoint, it preserves all colimits which exist, and in particular preserves coequalizers.

The coequalizer of a pair of maps $X \rightrightarrows Y$ in **CH** may now be calculated as follows. First take the coequalizer in **Top** and apply the left adjoint h to get a coequalizer $p: Y \to C$ in **Haus**. Form kC.

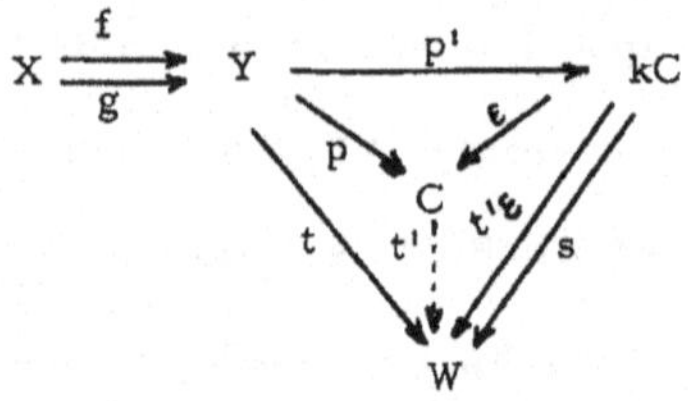

Since $\varepsilon: kC \to C$ is universal, there is a unique $p': Y \to kC$ with $\varepsilon p' = p$; we claim that p' is the desired coequalizer in **CH**. Clearly $p'f = p'g$. If $t: Y \to W$ is any arrow in **CH** with $tf = tg$, then $t = t'p$ for some t', because p is a coequalizer in **Haus** and hence $t = (t'\varepsilon)p'$. If also $t = sp'$ for some other arrow s in **CH**, then $t'\varepsilon = s$ on the level of sets (because p' is surjective); hence $t'\varepsilon$ is unique with $(t'\varepsilon)p' = t$, and $p': Y \to kC$ is the desired coequalizer. This proves the theorem.

For each **CH**-monoid G we can now calculate the spaces $E(G)$ and $B(G)$ directly as coends in **CH**. This will give all requisite properties of the classifying space $B(G)$ directly, without recourse to the detailed separation argument of the last paragraph. That argument now serves only to show that, given the **CH** spaces $G^n \times \sigma_{n-1}$, we get the <u>same</u> coends

$$E(G) = \int^n G^n \times \sigma_{n-1} \ , \quad B(G) = \int^n G^{n-1} \times \sigma_{n-1}$$

calculated in **Top** or in **CH** -- and this information does not seem essential.

9. <u>Acyclicity and other Properties</u>. The essential property of the space $E(G)$ is its acyclicity, which may be established by constructing a suitable contraction of $E(G)$ to a point. This contraction is in fact just the tensor product of two familiar contractions: The contraction h which carries the face $d_0\sigma_n$ of the

n-simplex to the opposite vertex, and the standard contraction h' of the (algebraic) bar construction. To state this in detail, we use the translation functor $T: \Delta \rightarrow \Delta$ which carries each ordinal n to $n+1$ and each monotone $f: n \rightarrow m$ to $Tf: n+1 \rightarrow m+1$; here $(Tf)(0) = 0$ and $(Tf)(i) = 1 + fi$ for $i \neq 0$. Now, if t is any real number, set $\beta(t) = t$ or 1 according as $t \leq 1$ or $t > 1$. Then for I the unit interval define for each ordinal n a continuous map $h_n: I \times \sigma_{n-1} \rightarrow \sigma_n$ by (for $s \in I$, $0 \leq t_1 \leq \ldots \leq t_{n-1} \leq 1$)

$$h_n(s, t_1, \ldots, t_{n-1}) = (s, \beta(s + t_1), \ldots, \beta(s + t_n)) .$$

Then $h_n(0, --): \sigma_{n-1} \rightarrow \sigma_n$ is the face operator d_o and $h_n(1, --)$ maps σ_{n-1} onto one point -- the point $(1, 1, \ldots, 1)$; geometrically, h_n is the obvious linear map deforming $d_o \sigma_n = \sigma_{n-1}$ through σ_n to the opposite vertex. Moreover, one verifies readily that h is in fact a natural transformation of functors

$$h: I \times \sigma_\bullet \rightarrow \sigma_\bullet T, \qquad h_n(0, n) = d_o .$$

(This can be checked readily by replacing the $t_1, \ldots, t_{n-1}$ by the usual barycentric coordinates in a simplex.)

On the other hand, for a $\mathcal{CH}$-monoid G, there is a natural transformation

$$h': G^\bullet \rightarrow GT ,$$

defined for elements $x_o, \ldots, x_{n-1}$ of G by

$$h'_n(x_o, \ldots, x_{n-1}) = (e, x_o, \ldots, x_{n-1}) \in G^{n+1} .$$

This is not a G-map, but a continuous map $G^n \rightarrow G^{n+1}$, and is indeed just the standard contracting homotopy of the algebraic bar construction; one verifies that h' is natural and that $d_o h'_n$ is the identity. In order to combine these two we first construct the diagram

$$\begin{array}{ccc}
G^{Tn} \times \sigma_{T(n-1)} & \longrightarrow & \int^n G^{Tn} \times \sigma_{T(n-1)} \\
\| & & \downarrow g \\
G^{n+1} \times \sigma_n & \longrightarrow & \int^n G^n \times \sigma_{n-1}
\end{array}$$

where the horizontal maps are the (universal) supernatural transformations for the

the two coends displayed. By universality, there exists then a (continuous) map g, as displayed, vertically, on the right.

Since h' and h are both natural transformations, we have a continuous map $\chi = \int^n h' \times h$ as in the diagram

$$\begin{array}{ccccc} I \times E(G) & & \xrightarrow{\quad \bar{h} \quad} & & E(G) \\ \| & & & & \| \\ \int^n G^n \times (I \times \sigma_{n-1}) & \xrightarrow{\chi} & \int^n G^{Tn} \times \sigma_{T(n-1)} & \xrightarrow{g} & \int^n G^n \times \sigma_{n-1} . \end{array}$$

The composite $\bar{h}$ is the desired contraction of E(G); $\bar{h}(1, --)$ maps E(G) to a point, while $\bar{h}(0, --)$ is the identity in virtue of the two equations $h(0, --) = d_o$, and $d_o h' =$ identity, as noted above.

This argument shows that Milgram's contraction map, like his classifying space, is a tensor product.

Other properties of E(G) and B(G), as previously established by Milgram and Steenrod, can now be put in this language. For example, one may prove readily that the projection $E(G) \rightarrow B(G)$ is an identification map (a proclusion in the terminology of Steenrod); indeed, it is the identification map obtained by "dividing out" the action of G.

10. <u>The Product Theorem.</u> Steenrod, going beyond Milgram, proved in [16] a product formula

THEOREM. For two $\mathfrak{CH}$-monoids G and H there is a homeomorphism

$$E(G) \times E(H) \cong E(G \times H)$$

which is also a homomorphism for the action of $G \times H$ on both sides.

We now show how this follows readily from the tensor product formalism. First, the product $G \times H$ in $\mathfrak{CH}$ has two projections $p: G \times H \rightarrow G$ and $q: G \times H \rightarrow H$ which give a diagram

$$E(G) \longleftarrow E(G \times H) \longrightarrow E(H). \tag{14}$$

LEMMA. In sets, the diagram (14) is a product diagram.

Proof. Represent an element of $E(G)$ by a pair $(\xi, u) \in G^n \times \sigma_{n-1}$, where $\xi = (x_o, \ldots, x_{n-1})$ and $u = (t_1, \ldots, t_{n-1}) \in \sigma_{n-1}$. The identification rules above, by inserting more units e in ξ, allow us to replace u by any longer string involving at least the same t's. Therefore points $(\xi, u) \in E(G)$ and $(\eta, v) \in E(H)$ can be written in the form $(\xi, u) \equiv (\xi', w)$ and $(\eta, v) \equiv (\eta', w)$ with the same string w. This shows that every pair of points in $E(G)$ and $E(H)$ comes by projections from a single point $((\xi', \eta'), w)$ in $E(G \times H)$. By putting the latter point into normal form (see §7 above) its uniqueness readily follows. This proves that we have a product in (14).

Now return to the theorem. Since $E(G) \times E(H)$ and $E(G \times H)$ are both products (as sets) of $E(G)$ and $E(H)$ there are unique functions

$$E(G) \times E(H) \underset{\psi}{\overset{\varphi}{\rightleftarrows}} E(G \times H)$$

which commute with the projections to $E(G)$ and to $E(H)$. Since $E(G) \times E(H)$ is a product in $\mathcal{CH}$, the function ψ is continuous. Since both composites $\psi\varphi$ and $\varphi\psi$ are identities, it remains only to show φ continuous. By a standard result, it will suffice to prove continuity on $G^{n+1} \times \sigma_n \times H^{m+1} \times \sigma_m$, regarded as embedded in $E(G) \times E(H)$. Now $\sigma_n \times \sigma_m$ is a prism (product of simplices) and has a standard triangulation into simplices, with one simplex of dimension n+m for each shuffle κ of n letters through m letters (see Mac Lane [12], page 243). Specifically, the point $(u, v) \in \sigma_n \times \sigma_m$, with $u = (t_1, \ldots, t_n)$ and $v = (s_1, \ldots, s_m)$, belongs to the shuffle κ if κ puts the real numbers $t_1, \ldots, t_n, s_1, \ldots, s_m$ into non-decreasing order $w = (r_1, \ldots, r_{n+m})$. Moreover, this shuffle κ can then be represented by the monotonic surjections $f: n+m+1 \to n+1$ and $g: n+m+1 \to n+1$ in the category $\mathbf{\Delta}$ so that, for the maps $\sigma_f: \sigma_{n+m} \to \sigma_n$ and $\sigma_g: \sigma_{n+m} \to \sigma_n$ corresponding to f and g under the functor $\sigma_\bullet$ are

$$\sigma_f(r_1, \dots, r_{n+m}) = (t_1, \dots, t_n), \ \sigma_g(r_1, \dots, r_{n+m}) = (s_1, \dots, s_m).$$

But this means that

$$(\xi, u) = (\xi, \sigma_f w) = (G^f \xi, w) ,$$
$$(\eta, v) = (\eta, \sigma_g w) = (G^g \eta, w) .$$

Therefore, by the lemma above, the map φ when restricted to this particular (n+m)-simplex is

$$(\xi, u), (\eta, v) \longmapsto (G^g \xi \times G^g \eta, w) ,$$

and this map is clearly continuous. Since φ is continuous on each simplex $G^{n+1} \times H^{n+1} \times (\sigma_{n+m})$ of maximum dimension, in the subdivision it is continuous. Q.E.D.

The shuffles here used are exactly those involved in the usual Eilenberg-Zibler theorem. The proof of this theorem (for simplicial complexes) uses only the formal properties of the shuffles, and not the triangulation which they provide for the prism $\sigma_n \times \sigma_m$. The present proof really uses that triangulation. Moreover, Steenrod's proof of this result also depends essentially on the same triangulation.

Steenrod also proves that $E(G) \to B(G)$ is a quasifibration, as well as other results when G is a CW-complex. We have not tried to present these in functorial form.

Bibliography

[1] Cartan, H. Sur les groupes d'Eilenberg-Mac Lane $H(\pi, n)$: I, II, Proc. Nat. Acad. Sci. USA 40 (1954), 467-471, 704-707.

[2] Clark, Allen, Categorical Constructions in Topology, (Preprint, 1969).

[3] Day, B. J. and Kelly, G. M. Enriched Functor Categories, Reports of the Midwest Category Seminar III, Berlin, Heidelberg, New York (Springer) 1969, pp. 178-191.

[4] Dold, A. and Lashof, R. Principal Quasifibrations and Fiber Homotopy Equivalence of Bundles, Ill. J. Math 3 (1959), 285-305.

[5] Dubuc, E. Kan Extensions in Enriched Category Theory. Forthcoming in Springer Lecture Notes Series.

[6] Eilenberg, S. and Kelly, G. M. A Generalization of the Functorial Calculus, Journal of Algebra 1 (1964), 397-402.

[7] Eilenberg, S. and Mac Lane, S. On the Groups $H(\pi, n)$, I, II. Ann. of Math. 58 (1953), 55-106 and 60 (1954), 49-139.

[8] Freyd, P. Abelian Categories, An Introduction to the Theory of Functors, New York (Harper and Row), 1963.

[9] Kan, D. M. Adjoint Functors. Trans. Amer. Math. Soc. 87 (1958), 294-329.

[10] __________, Functors Involving C.S.S. Complexes, Trans. Amer. Math. Soc. 87 (1958), 330-346.

[11] Mac Lane, Saunders. Categorical Algebra, Bull. Amer. Math. Soc. 71 (1965), 40-106.

[12] __________, Homology, Heidelberg and New York (Springer), 1963.

[13] Milgram, R.J. The Bar Construction and Abelian H-Spaces. Ill. J. Math. 11 (1967), 242-250.

[14] Rothenberg, M. and Steenrod, N.E. The Cohomology of Classifying Spaces and H-Spaces. Bull. Amer. Math. Soc. 71 (1965), 872-875.

[15] Steenrod, N.E. A Convenient Category of Topological Spaces, Mich. Math. J. 14 (1967), 132-152.

[16] __________, Milgram's Classifying Space of a Topological Group. Topology, 7 (1968), 319-368.

[17] Ulmer, F. Representable Functors with Values in Arbitrary Categories. Journal of Algebra 8 (1968), 96-129.

[18] Yoneda, N. On Ext and Exact Sequences. Jour. Fac. Sci., Univ. Tokyo 8 (1960), 507-576.

Added in Proof:

Stasheff in his theses (Homotopy Associativity of H-Spaces I & II, Trans. AMS 108 (1963)) points out that his construction of $XP(\infty)$ reduces to

$$\cup \Delta^n \times X^n / \sim$$

if X is a monoid and is homeomorphic to a reduced version of the Dold Lashof B_X (p. 289-290).

In addition, Stasheff and Milgram point out the correspondence

$$\overline{B}(C_*(X)) \to C_*(B_X) \ ,$$

which for Milgram is an isomorphism (using cellular theory in the rare cases X has a cellular multiplication) and in general is a homotopy equivalence.

A GENERAL ALGEBRAIC APPROACH TO STEENROD OPERATIONS

by

J. Peter May

1. Introduction. Since the introduction of the Steenrod operations in the cohomology of topological spaces, it has become clear that similar operations exist in a variety of other situations. For example, there are Steenrod operations in the cohomology of simplicial restricted Lie algebras, in the cohomology of cocommutative Hopf algebras, and in the homology of infinite loop spaces (where they were introduced mod 2 by Araki and Kudo [3] and mod p, $p > 2$, by Dyer and Lashof [6]).

The purpose of this expository paper is to develop a general algebraic setting in which all such operations can be studied simultaneously. This approach allows a single proof, applicable to all of the above examples, of the basic properties of the operations, including the Adem relations. In contrast to categorical treatments of Steenrod operations, the elegant proofs developed by Steenrod [25-30] actually simplify somewhat in our algebraic setting. Further, even the most general existing categorical study of Steenrod operations, that of Epstein [7], cannot be applied to iterated loop spaces.

We emphasize that this is an expository paper. Although a number of new results and new proofs of old results are scattered throughout, the only real claim to originality lies in the basic context. We have chosen to give complete proofs of all results since a large number of minor simplifications in the arguments allows a substantial simplification of the theory as a whole. We have also included a number of topological results which should be well-known but appear not to be in the literature. In particular, in section 10, we give a quick complete calculation of the mod p cohomology Bockstein spectral sequence of $K(\pi, n)$'s and show that Serre's simple proof [23] of the axiomatization of the mod 2 Steenrod operations applies with only slight modifications to the case $p > 2$.

The general theory is presented in the first five sections. Most of the proofs in sections 1, 2, and 4 are based on those of Steenrod [25-30], and those of section 3 are simplifications of arguments of Dyer and Lashof [6]. Via acyclic models and a lemma due to Dold [5], the theory is applied to several simplicial categories and to topological spaces in sections 7 and 8. The standard properties of the Steenrod operations in spaces, except $P^0 = 1$, drop out of the algebraic theory, and $P^0 = 1$ is shown to follow from these properties. In contrast, we prove that $P^0 = 0$ on the cohomology of simplicial restricted Lie algebras. The theory is applied to the cohomology of cocommutative Hopf algebras in section 11; the operations here are important in the study of the cohomology of the Steenrod algebra [13, 18]. The present analysis arose out of work on iterated loop spaces, but this application will appear elsewhere. The material of sections 6 and 9, which is peripheral to the study of Steenrod operations, is presented here with a view towards this application.

1. Algebraic preliminaries; equivariant homology

Let Λ be a commutative (ungraded) ring. By a Λ-complex, we understand a Z-graded differential Λ-module, graded by subscripts, with differential of degree minus one. We say that a Λ-complex K is positive if $K_q = 0$ for $q < 0$ and negative if $K_q = 0$ for $q > 0$. We use Z-graded complexes in order that our theory can be applied equally well to homology and to cohomology. The exposition will be geared to homology, where the notation is slightly simpler, and the notations appropriate for cohomology will be given in section 5. We give some elementary homological lemmas in this section; these extract the slight amount of information about the homology of groups that is needed for the development of the Steenrod operations.

If π is a group, we let $\Lambda\pi$ denote its group ring over Λ. We shall generally speak of $\Lambda\pi$-morphisms rather than π-equivariant Λ-morphisms, and we shall speak of π-morphisms when Λ is understood. Let Σ_r denote the symmetric group on r letters, and let $\pi \subset \Sigma_r$. For $q \in Z$, let $\Lambda(q)$ denote the $\Lambda\pi$-module which is Λ as a Λ-module and has the $\Lambda\pi$-action determined by $\sigma\lambda = (-1)^{qs(\sigma)}\lambda$, where $(-1)^{s(\sigma)}$ is the sign of $\sigma \in \pi$. If M is a $\Lambda\pi$-module, let M(q) denote the $\Lambda\pi$-module $M \otimes \Lambda(q)$ with the diagonal action $\sigma(m \otimes \lambda) = \sigma m \otimes \sigma\lambda$ (where $\otimes = \otimes_{\Lambda}$). If K is a Λ-complex, let $K^r = K \otimes \ldots \otimes K$, r factors K. Via permutation of factors, with the standard sign convention, K^r becomes a $\Lambda\pi$-complex for $\pi \subset \Sigma_r$, and $K^r(q)$ is defined.

Let I denote the Λ-free Λ-complex which has two basis elements e_o and e_1 of degree zero, one basis element e of degree one, and differential $d(e) = e_1 - e_o$. If Γ is a Hopf algebra over Λ and I is given the trivial Γ-module structure, $\gamma a = \varepsilon(\gamma)a$ for $\gamma \in \Gamma$ and $a \in I$, then the notion of a Γ-homotopy $h: f \simeq g$, where $f, g: K \rightarrow L$ are morphisms of Γ-complexes, is equivalent to the notion of a Γ-morphism $H: I \otimes K \rightarrow L$ such that $H(e_1 \otimes k) = f(k)$ and $H(e_o \otimes k) = g(k)$, where

$I \otimes K$ is given the diagonal Γ-action. In fact, H determines h by $h(k) = H(e \otimes k)$ and conversely.

With these notations, we have the following lemma. In all parts, $\Lambda\pi$ acts diagonally on tensor products.

Lemma 1.1. Let $\pi \subset \Sigma_r$ and let V be a positive $\Lambda\pi$-free complex.

(i) There exists a $\Lambda\pi$-morphism $h: I \otimes V \longrightarrow V \otimes I^r$ such that $h(e_o \otimes v) = v \otimes e_o^r$ and $h(e_1 \otimes v) = v \otimes e_1^r$ for all $v \in V$.

(ii) If $f, g: K \longrightarrow L$ are Λ-homotopic morphisms of Λ-complexes, then $1 \otimes f^r, 1 \otimes g^r: V \otimes K^r \longrightarrow V \otimes L^r$ are $\Lambda\pi$-homotopic morphisms of $\Lambda\pi$-complexes.

(iii) If Λ is a field and K is a Λ-complex, then K is Λ-homotopy equivalent to $H(K)$ and $V \otimes K^r$ is $\Lambda\pi$-homotopy equivalent to $V \otimes H(K)^r$.

(iv) Let $v \in V$ satisfy $d(v \otimes 1) = 0$ in $V \otimes_\pi \Lambda$; let K be a Λ-complex and let $a, b \in K_q$ be homologous cycles. Then $v \otimes a^r$ and $v \otimes b^r$ are homologous cycles of $V \otimes_\pi K^r(q)$.

Proof. (i) Let $\varepsilon: I \longrightarrow \Lambda$ be the augmentation $\varepsilon(e_o) = 1 = \varepsilon(e_1)$, and let $J = \mathrm{Ker}(\varepsilon^r)$, $\varepsilon^r: I^r \longrightarrow \Lambda^r = \Lambda$. Define $k: V \longrightarrow V \otimes J$ by $k(v) = v \otimes (e_1^r - e_o^r)$. Since $H(J) = 0$, $H(V \otimes J) = 0$. Define a $\Lambda\pi$-homotopy $s: V \longrightarrow V \otimes J$ from k to the zero map by induction on degree as follows. Let $s_{-1} = 0$; given $s_{i-1}: V_{i-1} \longrightarrow (V \otimes J)_i$, we find easily that $d_i(k_i - s_{i-1}d_i) = 0$. Let $\{x_j\}$ be a $\Lambda\pi$-basis for V_i; for $x \in \{x_j\}$, choose $s_i(x)$ such that $d_{i+1}s_i(x) = k_i(x) - s_{i-1}d_i(x)$, and extend s_i to all of V_i by π-equivariance. The desired $\Lambda\pi$-morphism h is obtained by letting $h(e \otimes v) = s(v)$ for $v \in V$.

(ii) Let $t: I \otimes K \longrightarrow L$ determine a Λ-homotopy from f to g. Then the following composite is a $\Lambda\pi$-morphism which determines a $\Lambda\pi$-homotopy from $1 \otimes f^r$ to $1 \otimes g^r$:

$$I \otimes V \otimes K^r \xrightarrow{h \otimes 1} V \otimes I^r \otimes K^r \xrightarrow{1 \otimes u} V \otimes (I \otimes K)^r \xrightarrow{1 \otimes t^r} V \otimes L^r,$$

where $u: I^r \otimes K^r \longrightarrow (I \otimes K)^r$ is the evident shuffling isomorphism.

(iii) Define $f: H(K) \longrightarrow K$ by sending each element of a basis for $H(K)$ to a chosen representative cycle. $K \cong \operatorname{Im} f \oplus \operatorname{Coker} f$ as a Λ-complex and $\operatorname{Coker} f$ is acyclic and therefore contractible since Λ is a field. The first half follows and implies the second half by (ii).

(iv) Define a morphism of Λ-complexes $f: I \longrightarrow K$, of degree q, by $f(e_1) = a$, $f(e_0) = b$, and $f(e) = (-1)^q c$, where $d(c) = a-b$ in K (the sign ensures that $df(e) = (-1)^q fd(e)$). Let $F: I \otimes V \longrightarrow V \otimes K^r(q)$ be the composite $I \otimes V \xrightarrow{h} V \otimes I^r \xrightarrow{1 \otimes f^r} V \otimes K^r(q)$. A check of signs shows that f^r is a $\Lambda\pi$-morphism, hence that F is a morphism of $\Lambda\pi$-complexes of degree qr. By (i), we find that

$$F(e_i \otimes v) = (1 \otimes f^r)(v \otimes e_i^r) = (-1)^{qr \deg v} v \otimes f(e_i)^r, \quad i = 0 \text{ or } 1.$$

Since π operates trivially on I and $d(v \otimes 1) = 0$ in $V \otimes_\pi \Lambda$, we have that $d(e \otimes v) = (e_1 - e_0) \otimes v$ in $I \otimes_\pi V$. Thus, in $V \otimes_\pi K^r(q)$,

$$dF(e \otimes v) = (-1)^{qr} F(e_1 \otimes v - e_0 \otimes v) = (-1)^{qr(\deg v + 1)}(v \otimes a^r - v \otimes b^r),$$

and this proves the result.

We now consider the cyclic group π of prime order p. We recall the definition of the standard $\Lambda\pi$-free resolution $W = W(p, \Lambda)$ of Λ.

<u>Definition 1.2.</u> Let π be the cyclic group of prime order p with generator α. Let W_i be $\Lambda\pi$-free on one generator e_i, $i \geq 0$. Let $N = 1 + \alpha + \ldots + \alpha^{p-1}$ and $T = \alpha - 1$ in $\Lambda\pi$. Define a differential d, augmentation ε, and coproduct ψ on W by the formulas

(1) $$d(e_{2i+1}) = Te_{2i} \quad \text{and} \quad d(e_{2i}) = Ne_{2i-1}\,; \quad \varepsilon(\alpha^j e_0) = 1\,;$$

$$\psi(e_{2i+1}) = \sum_{j+k=i} e_{2j} \otimes e_{2k+1} + \sum_{j+k=i} e_{2j+1} \otimes \alpha e_{2k} \quad \text{and}$$

$$\psi(e_{2i}) = \sum_{j+k=i} e_{2j} \otimes e_{2k} + \sum_{j+k=i-1} \; \sum_{0 \leq r < s < p} \alpha^r e_{2j+1} \otimes \alpha^s e_{2k+1}.$$

Then W is a differential $\Lambda\pi$-coalgebra and a $\Lambda\pi$-free resolution of Λ. When necessary for clarity, we shall write $W(p, \Lambda)$ for W. Of course,

$W(p,\Lambda) = W(p,Z)\otimes\Lambda$. The structure of $W(p, Z_p)$ shows that $H_*(\pi; Z_p) = H(W(p, Z_p)\otimes_\pi Z_p)$ is given, with its Bockstein operation β and coproduct ψ, by the formula

(2) $H_*(\pi; Z_p)$ has Z_p-basis $\{e_i \mid i \geq 0\}$ such that $\beta(e_{2i}) = e_{2i-1}$ and

$$\psi(e_i) = \sum_{j+k=i} e_j \otimes e_k \text{ if } p = 2 \text{ or } i \text{ is odd}, \quad \psi(e_{2i}) = \sum_{j+k=i} e_{2j}\otimes e_{2k} \text{ if } p > 2.$$

We embed π in Σ_p by $\alpha(1,\ldots,p) = (p, 1,\ldots, p-1)$, where Σ_p acts on $\{1,\ldots,p\}$. We then have the following lemma.

Lemma 1.3. Let $W^{(n)} = \sum_{i\leq n} W_i$ be the n-skeleton of $W = W(p, Z_p)$. Let G be any set of left coset representatives for π in Σ_p. Let K be a Z_p-module with totally ordered basis $\{x_j \mid j \in J\}$. Let $A \subset K^p$ have basis $\{x_j^p \mid j \in J\}$ and let $B \subset K^p$ have basis $\{x_{\gamma(j_p)} \mid \gamma \in G,\ j_1 \leq \ldots \leq j_p,\ j_1 < j_p\}$. Then

$$H(W^{(n)}\otimes_\pi K^p) = (\bigoplus_{i=0}^{n} e_i\otimes A) \oplus (e_o\otimes B) \oplus (\operatorname{Ker} d_n \otimes B), \quad d_n : W_n \longrightarrow W_{n-1}.$$

Proof. It is easy to see that K^p is isomorphic as a $Z_p\pi$-module to $A \oplus (Z_p\pi \otimes B)$, where π acts trivially on A and acts on $Z_p\pi\otimes B$ by its left action on $Z_p\pi$. Since $H(W^{(n)}\otimes_\pi A) = H(W^{(n)}\otimes_\pi Z_p)\otimes A$ and $H(W^{(n)}\otimes_\pi Z_p\pi\otimes B) = H(W^{(n)})\otimes B$, the result follows.

Recall that if π is any subgroup of Σ and if $\gamma \in N(\pi)$, the normalizer of π in Σ, then conjugation by γ defines a homomorphism $\gamma_*: H(\pi; M) \longrightarrow H_*(\pi; M)$ for any $\Lambda\Sigma$-module M. γ_* is the map induced on homology from $\gamma_\# \otimes \gamma : W\otimes_\pi M \longrightarrow W\otimes_\pi M$ where W is any $\Lambda\pi$-free resolution of Λ and $\gamma_\# : W \longrightarrow W$ is any morphism of Λ-complexes such that $\gamma_\#(\sigma w) = \gamma\sigma\gamma^{-1}\gamma_\#(w)$ for $\sigma \in \pi$ and $w \in W$. (It is easy to verify that $\gamma_\#$ exists and that γ_* is independent of the choice of W and of $\gamma_\#$.) Clearly $\gamma_* = 1$ if $\gamma \in \pi$ since we may then define $\gamma_\#(w) = \gamma w$ so that

$$(\gamma_\# \otimes_\pi \gamma)(w\otimes m) = \gamma w\otimes\gamma m = w\otimes m, \quad w \in W \text{ and } m \in M.$$

If $\pi \subset \rho \subset \Sigma$ and $\gamma \in N(\pi) \cap N(\rho)$, then the following diagram commutes:

$$\begin{array}{ccc} H_*(\pi; M) & \xrightarrow{j_*} & H_*(\rho; M) \\ \downarrow \gamma_* & & \downarrow \gamma_* \\ H_*(\pi; M) & \xrightarrow{j_*} & H_*(\rho; M) \end{array}$$

In fact, if W is any $\Lambda\rho$-free resolution of Λ, then W is also a $\Lambda\pi$-free resolution of Λ, and the above diagram results from the observation that j_* is induced from $W \otimes_\pi M \longrightarrow W \otimes_\rho M$. In particular, $j_* = j_*\gamma_*$ if $\gamma \in \rho$.

Lemma 1.4. Let π be cyclic of prime order $p > 2$ and let $q \in Z$. Consider $j_*: H_*(\pi; Z_p(q)) \longrightarrow H_*(\Sigma_p; Z_p(q))$. Then

(i) If q is even, $j_*(e_i) = 0$ unless $i = 2t(p-1)$ or $i = 2t(p-1) - 1$.

(ii) If q is odd, $j_*(e_i) = 0$ unless $i = (2t+1)(p-1)$ or $i = (2t+1)(p-1) - 1$.

Proof. Let k generate the multiplicative subgroup of Z_p, $k^{p-1} = 1$. Let Σ_p operate on Z_p and define $\gamma \in \Sigma_p$ by $\gamma(i) = ki$. Then $\gamma\alpha\gamma^{-1} = \alpha^k$ and γ is an odd permutation in $N(\pi)$. Define $\gamma_\# : W \longrightarrow W$ by

$$\gamma_\#(e_{2i}) = k^i e_{2i}; \quad \gamma_\#(e_{2i+1}) = k^i \sum_{j=0}^{k-1} \alpha^j e_{2i+1}; \quad \gamma_\#(\sigma e_i) = \gamma\sigma\gamma^{-1}\gamma_\#(e_i), \quad \sigma \in \pi$$

Then $\gamma_\# d = d\gamma_\#$ and $\gamma_\# \otimes \gamma$ induces the conjugation γ_* on $H_*(\pi; Z_p(q))$. Since $\gamma \in \Sigma_p$, $j_*\gamma_* = j_*$, hence $j_*(e_i - \gamma_* e_i) = 0$ for all i. γ operates by $(-1)^q$ on $Z_p(q)$ and therefore

$$\gamma_*(e_{2i}) = (-1)^q k^i e_{2i} \quad \text{and} \quad \gamma_*(e_{2i+1}) = (-1)^q k^{i+1} e_{2i+1}.$$

Thus $j_*(e_{2i}) = 0$ unless $1 - (-1)^q k^i \equiv 0 \bmod p$ and $j_*(e_{2i+1}) = 0$ unless $1 - (-1)^q k^{i+1} \equiv 0 \bmod p$. Clearly $k^i \equiv 1 \bmod p$ if and only if $i = t(p-1)$ for some t and $k^i \equiv -1 \bmod p$ if and only if $2i = (2t+1)(p-1)$ for some t. The result follows easily.

2. The definition and elementary properties of the operations

We now define a large algebraic category $\mathcal{C}(p,n)$ on which the Steenrod operations will be defined. Steenrod operations will be obtained for particular categories of interest by obtaining functors to $\mathcal{C}(p,n)$. The interest of the integer n in the following definition lies solely in the applications to iterated loop spaces. For all other known applications, only the case $n = \infty$ is relevant.

Definitions 2.1. Let Λ be a commutative ring, let r be an integer, and let π be a subgroup of Σ_r. Let W be a $\Lambda\pi$-free resolution of Λ, let V be a $\Lambda\Sigma_r$-free resolution of Λ, and let $j: W \longrightarrow V$ be a morphism of $\Lambda\pi$-complexes over Λ. Assume that $W_0 = \Lambda\pi$ with generator e_0. Let $0 \leq n \leq \infty$ and let $W^{(n)}$ and $V^{(n)}$ denote the n-skeletons of W and V. Define a category $\mathcal{C}(\pi, n, \Lambda)$ as follows. The objects of $\mathcal{C}(\pi, n, \Lambda)$ are pairs (K, θ), where K is a homotopy associative differential Λ-algebra and $\theta: W^{(n)} \otimes K^r \longrightarrow K$ is a morphism of $\Lambda\pi$-complexes such that

(i) The restriction of θ to $e_0 \otimes K^r$ is Λ-homotopic to the iterated product $K^r \longrightarrow K$, associated in some fixed order, and

(ii) θ is $\Lambda\pi$-homotopic to a composite $W^{(n)} \otimes K^r \xrightarrow{j \otimes 1} V^{(n)} \otimes K^r \xrightarrow{\phi} K$, where ϕ is a morphism of $\Lambda\Sigma_r$-complexes.

A morphism $f: (K, \theta) \longrightarrow (K', \theta')$ in $\mathcal{C}(\pi, n, \Lambda)$ is a morphism of Λ-complexes $f: K \longrightarrow K'$ such that the diagram

$$\begin{array}{ccc} W^{(n)} \otimes K^r & \xrightarrow{\quad\theta\quad} & K \\ \Big\downarrow{\scriptstyle 1 \otimes f^r} & & \Big\downarrow{\scriptstyle f} \\ W^{(n)} \otimes (K')^r & \xrightarrow{\quad\theta'\quad} & K' \end{array}$$

is $\Lambda\pi$-homotopy commutative. A morphism f is said to be perfect if $\theta'(1 \otimes f^r) = f\theta$, with no homotopy required, and $\mathcal{P}(\pi, n, \Lambda)$ denotes the subcategory of $\mathcal{C}(\pi, n, \Lambda)$ having the same objects (K, θ) and all perfect morphisms between them. Λ is itself an object of $\mathcal{C}(\pi, n, \Lambda)$, with

$\theta = \varepsilon \otimes 1: W^{(n)} \otimes \Lambda^r \longrightarrow \Lambda^r = \Lambda$, and an object $(K, \theta) \in \mathcal{C}(\pi, n, \Lambda)$ is said to be unital if K has a two-sided homotopy identity e such that $\eta: \Lambda \longrightarrow K$, $\eta(1) = e$, is a morphism in $\mathcal{C}(\pi, n, \Lambda)$. The tensor product of objects (K, θ) and (L, θ') in $\mathcal{C}(\pi, n, \Lambda)$ is the pair $(K \otimes L, \tilde{\theta})$, where $\tilde{\theta}$ is the composite

$$W^{(n)} \otimes (K \otimes L)^r \xrightarrow{\psi \otimes U} W^{(n)} \otimes W^{(n)} \otimes K^r \otimes L^r \xrightarrow{1 \otimes T \otimes 1} W^{(n)} \otimes K^r \otimes W^{(n)} \otimes L^r$$
$$\xrightarrow{\theta \otimes \theta'} K \otimes L$$

Here U is the evident shuffling isomorphism, $T(x \otimes y) = (-1)^{\deg x \deg y}\, y \otimes x$, and $\psi: W \longrightarrow W \otimes W$ is any fixed $\Lambda\pi$-morphism over Λ; conditions (i) and (ii) are clearly satisfied by the pair $(K \otimes L, \tilde{\theta})$. An object $(K, \theta) \in \mathcal{C}(\pi, n, \Lambda)$ is said to be a Cartan object if the product $K \otimes K \longrightarrow K$ is a morphism in $\mathcal{C}(\pi, n, \Lambda)$. When π is cyclic of prime order p, we agree to choose W to be the explicit resolution $W(p, \Lambda)$ of Definition 1.2, and we abbreviate $\mathcal{C}(\pi, n, Z_p)$ to $\mathcal{C}(p, n)$ and $\mathcal{P}(\pi, n, Z_p)$ to $\mathcal{P}(p, n)$. An object $(K, \theta) \in \mathcal{C}(p, n)$ is said to be reduced mod p if (K, θ) is obtained by reduction mod p from an object $(\tilde{K}, \tilde{\theta}) \in \mathcal{C}(\pi, n, Z)$ such that $\tilde{K}$ is a flat Z-module.

We can now define the Steenrod operations in the homology $H(K)$ of an object $(K, \theta) \in \mathcal{C}(p, n)$. Observe that if $x \in H(K)$ and $0 \leq i \leq n$, then $e_i \otimes x^p$ is a well-defined element of $H(W^{(n)} \otimes_\pi K^p) \cong H(W^{(n)} \otimes_\pi H(K)^p)$ by Lemmas 1.1 and 1.3; here (iv) of Lemma 1.1 applies, since $\pi \subset \Sigma_p$ contains only even permutations, and shows that $e_i \otimes x^p$ is represented by $e_i \otimes a^p \in W^{(n)} \otimes_\pi K^p$ for any representative cycle a of x.

Definitions 2.2. Let $(K, \theta) \in \mathcal{C}(p, n)$ and let $x \in H_q(K)$. For $0 \leq i \leq n$, define $D_i(x) \in H_{pq+i}(K)$ by $D_i(x) = \theta_*(e_i \otimes x^p)$, $\theta_*: H(W^{(n)} \otimes_\pi K^p) \longrightarrow H(K)$. If $p = 2$, define $P_s: H_q(K) \longrightarrow H_{q+s}(K)$ for $s \leq q+n$ by the formula

(i) $\qquad P_s(x) = 0$ if $s < q$; $P_s(x) = D_{s-q}(x)$ if $s \geq q$.

If $p > 2$, define $P_s: H_q(K) \longrightarrow H_{q+2s(p-1)}(K)$ for $2s(p-1) \leq q(p-1)+n$ and define

$\beta P_s : H_q(K) \longrightarrow H_{q+2s(p-1)-1}(K)$ for $2s(p-1) \leq q(p-1)+n+1$ by the formulas

(ii) $P_s(x) = 0$ if $2s < q$; $P_s(x) = (-1)^s \nu(q) D_{(2s-q)(p-1)}(x)$ if $2s \geq q$ and

$\beta P_s(s) = 0$ if $2s \leq q$; $\beta P_s(x) = (-1)^s \nu(q) D_{(2s-q)(p-1)-1}(x)$ if $2s > q$,

where $\nu(2j+\varepsilon) = (-1)^j (m!)^{\varepsilon}$, j any integer, $\varepsilon = 0$ or 1, $m = \frac{1}{2}(p-1)$,

or, equivalently since $(m!)^2 \equiv (-1)^{m+1} \mod p$, $\nu(q) = (-1)^{q(q-1)m/2} (m!)^q$

Observe that, if $n = \infty$, the P_s and, if $p > 2$, the βP_s are defined for all integers s and that βP_s is a single symbol which is not a priori related to any Bockstein operation. The P_s and βP_s are appropriately defined for applications to homology; as shown in section 5, the appropriate formulation for cohomology is obtained by a simple change of notation.

The following proposition contains most of the elementary properties of the D_i, P_s, and βP_s. In particular, if $p > 2$, it shows that the P_s and βP_s account for all non-trivial operations D_i and that βP_s is the composition of P_s and the Bockstein β provided that (K, θ) is reduced mod p.

Proposition 2.3. Let $(K, \theta) \in \mathcal{C}(p, n)$ and consider $D_i : H_q(K) \longrightarrow H_{pq+i}(K)$.

(i) If $f: K \longrightarrow K'$ is a morphism in $\mathcal{C}(p, n)$, then $D_i f_* = f_* D_i$.

(ii) If $i < n$, then D_i is a homomorphism.

(iii) D_o is the p-th power operation in the algebra $H(K)$ and if (K, θ) is unital, then $D_i(e) = 0$ for $i \neq 0$, where $e \in H_o(K)$ is the identity.

(iv) If $p > 2$ and $i < n$, then $D_i = 0$ unless either

(a) q is even and $i = 2t(p-1)$ or $i = 2t(p-1) - 1$ for some t, or

(b) q is odd and $i = (2t+1)(p-1)$ or $i = (2t+1)(p-1) - 1$ for some t.

(v) If (K, θ) is reduced mod p and β is the Bockstein, then

(a) $\beta D_{2i} = D_{2i-1}$ if either $p > 2$ or q is even, and $2i < n$

(b) $\beta D_{2i+1} = D_{2i}$ if $p = 2$ and q is odd, and $2i+1 < n$.

Proof. Part (i) is immediate from the definitions and from Lemma 1.1 (iv), and part (iii) is immediate from the definitions.

(ii) Let $a, b \in K_q$ be cycles and define $\Delta(a,b) = (a+b)^p - a^p - b^p \in K^p$. $\Delta(a,b)$ is a sum of monomials involving both a's and b's, and π permutes such monomials freely. Let $c \in K^p$ be a sum of monomials whose permutations under π give each monomial of $\Delta(a,b)$ exactly once. Then $\Delta(a,b) = Nc$. If i is odd, then $d(e_{i+1} \otimes c) = e_i \otimes Nc$ and if i is even, then $d(T^{p-2}e_{i+1} \otimes c) = e_i \otimes Nc$ in $W^{(n)} \otimes_\pi K^p$, $i < n$, since $T^{p-1} = N$ in $Z_p\pi$. Thus $e_i \otimes \Delta(a,b)$ is a boundary and therefore D_i is a homomorphism, $i < n$.

(iv) In the notations of Definition 2.1, we have that θ is homotopic to a composite $W^{(n)} \otimes_\pi K^p \xrightarrow{j \otimes 1} V^{(n)} \otimes_\pi K^p \rightarrow V^{(n)} \otimes_{\Sigma_p} K^p \xrightarrow{\phi} K$. Since nothing is changed by tensoring with two copies of $Z_p(q)$, this composite can equally well be written as

$$W^{(n)}(q) \otimes_\pi K^p(q) \xrightarrow{j \otimes 1} V^{(n)}(q) \otimes_{\Sigma_p} K^p(q) \xrightarrow{\phi} K.$$

Let $a \in K_q$ be a cycle. Then, bythe definition of $K^p(q)$, a^p is a basis for a trivial Σ_p-subcomplex of $K^p(q)$. Therefore, if $j(e_i) = d(f)$ in $V^{(n)} \otimes_{\Sigma_p} Z_p(q) = V^{(n)}(q) \otimes_{\Sigma_p} Z_p$, then $d(f \otimes a^p) = j(e_i) \otimes a^p$ in $V^{(n)}(q) \otimes_{\Sigma_p} K^p(q)$. For $i < n$, j induces $j_*: H_i(\pi; Z_p(q)) \longrightarrow H_i(\Sigma_p; Z_p(q))$, and the desired conclusion now follows immediately from Lemma 1.4.

(v) Let (K, θ) be the mod p reduction of $(\tilde{K}, \tilde{\theta})$. Let $a \in \tilde{K}_q$ satisfy $d(a) = pb$. An easy calculation demonstrates that, in $\tilde{K}^p$,

$$d(a^p) = pNba^{p-1} \quad \text{if } p > 2 \text{ or } q \text{ is even;}$$

$$d(a^2) = 2Tab \quad \text{if } p = 2 \text{ and } q \text{ is odd.}$$

In the former case, if $2i < n$, then, in $W(p, Z)^{(n)} \otimes_\pi \tilde{K}^p$,

$$\begin{aligned} d(e_{2i} \otimes a^p) &= e_{2i-1} \otimes Na^p + pe_{2i} \otimes Nba^{p-1} \\ &\equiv p[e_{2i-1} \otimes a^p + d(T^{p-2}e_{2i+1} \otimes ba^{p-1})] \bmod p^2, \end{aligned}$$

since $T^{p-1} \equiv N \bmod p$. In the latter case, if $2i+1 < n$, then

$$d(e_{2i+1} \otimes a^2) = e_{2i} \otimes Ta^2 - 2e_{2i+1} \otimes Tab \equiv 2[e_{2i} \otimes a^2 - d(e_{2i+2} \otimes ab)] \bmod 4.$$

Thus, in $H(W(p, Z_p)^{(n)} \otimes_\pi K^p)$, if $\bar{a}$ is the mod p reduction of a, then $\beta\{e_{2i} \otimes \bar{a}^p\} = \{e_{2i-1} \otimes \bar{a}^p\}$ in case (a) and $\beta\{e_{2i+1} \otimes \bar{a}^2\} = \{e_{2i} \otimes \bar{a}^2\}$ in case (b). Since θ is the mod p reduction of the map $\theta: W(p, Z)^{(n)} \otimes_\pi K^p \longrightarrow K$, $\beta\theta_* = \theta_*\beta$, and the result follows.

Of course, (i) and (ii) imply that the P_s and βP_s are natural homomorphisms (except, if $n < \infty$, for the last operation). A check of constants gives the following corollary of part (iii).

Corollary 2.4. Let $(K, \theta) \in \mathcal{C}(p, n)$. Then $P_q(x) = x^p$ if $p = 2$ and $x \in H_q(K)$ or if $p > 2$ and $x \in H_{2q}(K)$. If (K, θ) is unital, then $P_s(e) = 0$ for $s \neq 0$.

The implications of (iv) and (v) for the P_s and βP_s are clear if $p > 2$. If $p = 2$, we have the following corollary of (v).

Corollary 2.5. If $(K, \theta) \in \mathcal{C}(2, \infty)$ is reduced mod 2, then $\beta P_{s+1} = sP_s$.

The following result is the external Cartan formula.

Proposition 2.6. Let (K, θ) and (L, θ') be objects of $\mathcal{C}(p, n)$. Let $x \in H_q(K)$ and $y \in H_r(L)$. Consider $x \otimes y \in H(K) \otimes H(L) = H(K \otimes L)$.

(i) If $p = 2$, then $D_i(x \otimes y) = \sum_{j+k=i} D_j(x) \otimes D_k(y)$ for $i \leq n$.

(ii) If $p > 2$, then $D_{2i}(x \otimes y) = (-1)^{mqr} \sum_{j+k=i} D_{2j}(x) \otimes D_{2k}(y)$ for $2i \leq n$, and

$$D_{2i+1}(x \otimes y) = (-1)^{mqr} \sum_{j+k=i} (D_{2j+1}(x) \otimes D_{2k}(y) + (-1)^q D_{2j}(x) \otimes D_{2k+1}(y)) \quad \text{for } 2i+1 \leq n.$$

Proof. By Lemmas 1.1 and 1.3, we may work in $W^{(n)} \otimes_\pi [H(K) \otimes H(L)]^p$. Since π operates trivially on $(x \otimes y)^p$, we may compute $\theta_*(e_i \otimes (x \otimes y)^p)$ by means of the induced coproduct on $W^{(n)} \otimes_\pi Z_p$, as given in (2) of Definition 1.2. The result follows by direct calculation from

$$\theta_*(e_i \otimes (x \otimes y)^p) = (\theta_* \otimes \theta'_*)(1 \otimes T \otimes 1)(\psi \otimes U)(e_i \otimes (x \otimes y)^p).$$

A trivial verification of constants, together with part (iv) of Proposition 2.3, yields the following corollary.

Corollary 2.7. Let (K, θ) and (L, θ') be objects of $\mathcal{C}(p, n)$. Let $x \in H_q(K)$ and $y \in H_r(L)$. Then $P_s(x \otimes y) = \sum_{i+j=s} P_i(x) \otimes P_j(y)$ and, if $p > 2$, $\beta P_{s+1}(x \otimes y) = \sum_{i+j=s} (\beta P_{i+1}(x) \otimes P_j(y) + (-1)^q P_i(x) \otimes \beta P_{j+1}(y))$.

Of course, if (K, θ) is a Cartan object in $\mathcal{C}(p, n)$, then the corollary and the naturality of the operations imply that the P_s and, if $p > 2$, the βP_s on $H(K)$ satisfy the internal Cartan formulas

(1)
$$P_s(xy) = \sum_{i+j=s} P_i(x) P_j(y) \quad \text{and}$$
$$\beta P_{s+1}(xy) = \sum_{i+j=s} (\beta P_{i+1}(x) P_j(y) + (-1)^{\deg x} P_i(x) \beta P_{j+1}(y)) .$$

3. Chain level operations, suspension, and spectral sequences

In this section, we define chain level Steenrod operations and use them to prove that the homology operations commute with suspension. The chain level operations can also be used to study the behavior of Steenrod operations in spectral sequences and, in particular, we shall prove a general version of the Kudo transgression theorem.

Theorem 3.1. Let $(K, \theta) \in \mathcal{C}(p, \infty)$. Then there exist functions $P_s : K_q \longrightarrow K_{q+s}$ if $p = 2$ and $P_s : K_q \longrightarrow K_{q+2s(p-1)}$ and $\beta P_s : K_q \longrightarrow K_{q+2s(p-1)-1}$ if $p > 2$ which satisfy the following properties.

(i) $dP_s = P_s d$ and $d\beta P_s = -\beta P_s d$

(ii) If a is a cycle which represents $x \in H(K)$, then $P_s(a)$ and $\beta P_s(a)$ are cycles which represent $P_s(x)$ and $\beta P_s(x)$.

(iii) If $f : (K, \theta) \longrightarrow (K', \theta')$ is a morphism in $\mathcal{P}(p, \infty)$, so that $f\theta = \theta'(1 \otimes f^p)$, then $fP_s = P_s f$ and $f\beta P_s = \beta P_s f$.

Proof. Let $a \in K_q$ and write $b = d(a) \in K_{q-1}$. In the case $p = 2$, define

(1) $P_s(a) = \theta(c)$, where $c = e_{s-q+1} \otimes b \otimes a + e_{s-q} \otimes a \otimes a \in W \otimes_\pi K^2$.

The verification of (i), (ii), and (iii) is trivial. Thus assume that $p > 2$. Let (a, b) denote the subcomplex of K with basis a and b, so that $(a, b)^p \subset K^p$. Define $s : (a, b) \longrightarrow (a, b)$, of degree one, by $s(a) = 0$ and $s(b) = a$. Then $ds + sd = 1$ on (a, b). Let $S = 1^{p-1} \otimes s$ on $(a, b)^p$. Then $dS + Sd = 1$ on $(a, b)^p$ and S is given explicitly by $S(ea) = 0$ and $S(eb) = (-1)^{\deg e} ea$ for $e \in (a, b)^{p-1}$. Define $t_i \in (a, b)^p$ for $0 \leq i \leq p$ by the inductive formula

(2) $t_o = b^p$; $t_1 = b^{p-1}a$; $t_{2k} = S(\alpha^{-1} t_{2k-1} - t_{2k-1})$; $t_{2k+1} = S(Nt_{2k})$.

Since $dS + Sd = 1$, an easy calculation demonstrates that

(3) $d(t_1) = t_o$; $d(t_{2k}) = (\alpha^{-1} - 1)t_{2k-1}$ and $d(t_{2k+1}) = Nt_{2k}$, $1 \leq k \leq m$.

A straightforward induction, which uses the explicit formula for S, yields

(4) $t_{2k} = \sum_I (-1)^{kq}(k-1)!\, b^{i_1}a^2 b^{i_2}a^2 \cdots b^{i_k}a^2$, $1 \le k \le m$, summed over all

k-tuples $I = (i_1, \ldots, i_k)$ such that $\sum_j i_j = p-2k$; and

(5) $t_{2k+1} = \sum_I (-1)^{kq}\, k!\, b^{i_1}a^2 \ldots b^{i_k}a^2 b^{i_{k+1}}a$, $0 \le k \le m$, summed over all

$(k+1)$-tuples $I = (i_1, \ldots, i_{k+1})$ such that $\sum_j i_j = p-1-2k$.

In particular, $t_p = t_{2m+1} = (-1)^{mq} m!\, a^p$ (since each $i_j = 0$). Now let $j = (2s-q+1)(p-1)$ and define chains c and c' in $W \otimes_\pi K^p$ by the following formulas (where, by convention, $e_i = 0$ if $i < 0$):

(6) $c = \sum_{k=0}^{m} (-1)^k e_{j-2k} \otimes t_{2k+1} - \sum_{k=1}^{m} (-1)^k e_{j+1-2k} \otimes (\alpha^{-1} - 1)^{p-2} t_{2k}$;

(7) $c' = \sum_{k=0}^{m} (-1)^k e_{j-1-2k} \otimes t_{2k+1} + \sum_{k=1}^{m} (-1)^k e_{j-2k} \otimes t_{2k}$.

Then an easy computation, which uses Definition 1.2 and (3), gives

(8) $d(c) = e_j \otimes b^p$ and $d(c') = -e_{j-1} \otimes b^p$ $(j = (2s-q+1)(p-1))$

In calculating $d(c)$, the salient observations are that $Nt_p = 0$, that $\alpha e_i \otimes t = e_i \otimes \alpha^{-1} t$ for $t \in K^p$ by the very definition of a tensor product, and that $(\alpha^{-1} - 1)^{p-1} = N$ in $Z_p\pi$. Finally, define

(9) $P_s(a) = (-1)^s \nu(q-1)\theta(c)$ and $\beta P_s(a) = (-1)^s \nu(q-1)\theta(c')$.

If a is a cycle, $b = 0$, then $t_i = 0$ for $i < p$ and $t_p = (-1)^{mq} m!\, a^p$, hence

(10) $c = (-1)^{m(q+1)} m!\, e_{(2s-q)(p-1)} \otimes a^p$ and $c' = (-1)^{m(q+1)} m!\, e_{(2s-q)(p-1)-1} \otimes a^p$.

It is easy to verify that $\nu(q) = (-1)^{m(q+1)} m!\, \nu(q-1)$ and now (ii) is obvious from (9) and (10) and (i) follows from (8), (9), and (10), applied to the cycle $b \in K_{q-1}$. Part (iii) is immediate from (9).

The remaining results of this section are corollaries of the theorem and its proof. We first define and study a very general notion of suspension.

<u>Definition.</u> Let $f: K' \longrightarrow K$ and $g: K \longrightarrow K''$ be morphisms of Λ-complexes such that $gf = 0$. Define σ: Ker $f_* \longrightarrow$ Coker g_* by the formula $\sigma\{b'\} = \{g(a)\}$, where b' represents $\{b'\} \in$ Ker f_* and $d(a) = f(b')$ in K. It is trivial to verify that σ is well-defined, and we call σ the suspension.

We can now prove that the P_s commute with suspension. We remark that if $n = \infty$, the hypotheses of the next theorem simplify to the requirement that f and g be morphisms in $\mathcal{P}(p, \infty)$ such that $gf = 0$. For $n < \infty$, the stated hypotheses arise in practice in the study of iterated loop spaces.

<u>Theorem 3.3.</u> Let $(K', \theta') \in \mathcal{C}(p, n+1)$ and let $(K'', \theta'') \in \mathcal{C}(p, n)$. Let K be a Z_p-complex and let $f: K' \longrightarrow K$ and $g: K \longrightarrow K''$ be morphisms of complexes such that $gf = 0$. Define a subcomplex $\tilde{K}$ of $W^{(n+1)} \otimes K^p$ by

$$\tilde{K} = W^{(n+1)} \otimes f(K')^p + \overline{W}^{(n+1)} \otimes f(K')^{p-1} \otimes K + W^{(n)} \otimes K^p,$$

where $\overline{W}^{(n+1)} = W^{(n)} \oplus Z_p e_{n+1}$ (that is, $e_{n+1} \in \overline{W}^{(n+1)}$ but $\alpha^i e_{n+1} \notin \overline{W}^{(n+1)}$ for $i \leq i < p$). Suppose given a π-morphism $\theta: \tilde{K} \longrightarrow K$ (where, by convention, π does not act on $e_{n+1} \otimes f(K')^{p-1} \otimes K$) such that the following diagram is commutative:

$$\begin{array}{ccccc} W^{(n+1)} \otimes (K')^p & \xrightarrow{1 \otimes f^p} & \tilde{K} & \xrightarrow{1 \otimes g^p} & W^{(n)} \otimes (K'')^p \\ \downarrow \theta' & & \downarrow \theta & & \downarrow \theta'' \\ K' & \xrightarrow{f} & K & \xrightarrow{g} & K'' \end{array}$$

(Here $gf = 0$ ensures that $(1 \otimes g^p)(\tilde{K}) \subset W^{(n)} \otimes (K'')^p$.) Observe that Ker f_* is closed under the P_s and βP_s and that there are well-defined induced P_s on Coker g_*. Let $x \in$ Ker f_*. Then $\sigma P_s(x) = P_s \sigma(x)$ and $\sigma \beta P_s(x) = -\beta P_s \sigma(x)$ whenever $P_s(x)$ and $\beta P_s(x)$ are defined.

<u>Proof.</u> Let $\deg(x) = q-1$ and let $b' \in K'$ represent x. Let $b = f(b')$ and let $d(a) = b$ in K, so that $g(a)$ represents $\sigma(x)$. The hypothesis guarantees that if s is such that $P_s(x)$ or $\beta P_s(x)$ is defined, then the chain level operation $P_s(a)$ or $\beta P_s(a)$ constructed in the previous proof is also defined. Of course, this is

clear if $n = \infty$; if $n < \infty$, we need only verify that all elements involved in the definition of $P_s(a)$ or $\beta P_s(a)$ are present in $\tilde{K}$. For example, if $p = 2$, the last operation $P_s(x)$ occurs for $s = q+n$ and then $P_s(a) = \theta(c)$, where $c = e_{n+1} \otimes b \otimes a + e_n \otimes a \otimes a$, and c is indeed in $\tilde{K}$. Now our diagram ensures that $fP_s(b') = P_s f(b')$, hence $fP_s(b') = dP_s(a)$, and that $gP_s(a) = P_s g(a)$. $\sigma P_s(x) = P_s \sigma(x)$ follows from the definition of σ, and the proof that $\sigma\beta P^s(x) = -\beta P_s \sigma(x)$ is equally simple.

Note that if $p > 2$ and all objects are reduced mod p, then $\sigma\beta = -\beta\sigma$, which is consistent with the theorem. The theorem implies that $\sigma(x^p) = 0$ and that $\sigma\beta P_s(x) = 0$ if $p > 2$ and $\deg(x) = 2s-1$; if (K'', θ'') is reduced mod p, the latter statement becomes $\beta\sigma(x)^p = 0$. The operation $\beta P_s(x)$, $\deg(x) = 2s-1$, plays a special role in many applications; the following very useful technical result about about this operation is known as the Kudo transgression theorem. It applies to the Dyer-Lashof operations in the homology Serre spectral sequence of the path-space fibration $\Omega^n X \to P\Omega^{n-1}X \to \Omega^{n-1}X$, to the Steenrod operations in the cohomology Serre spectral sequence of a fibration $F \to E \to B$ (with $K' \to K \to K''$ being $C^*(B) \to C^*(E) \to C^*(F)$, graded by subscripts) and to the spectral sequence of Adams [1, p. 210] for cocommutative Hopf algebras.

Theorem 3.4. Assume, in addition to the hypotheses of Theorem 3.3, that K has an increasing fitration $\{F_i K\}$, that $H_0(K') = Z_p = H_0(K'')$, and that there is a morphism of complexes $\pi: K \otimes f(K') \to K$ such that either

(i) K', K, and K'' are positively graded, $F_i K = 0$ if $i < 0$, $F_i K_i = K$ if $i \geq 0$, $f(K') \subset F_0 K$, $\pi(F_i K \otimes f(K')) \subset F_i K$, and f and g induce isomorphisms $E^2 f: H_j(K') \to E^2_{oj} K$ and $E^2 g: K \to H_i(K'')$ and π induces a morphism $E^2\pi: E^2_{ij} K \otimes E^2_{ok} K \to E^2_{i,j+k} K$ such that the composite morphism $E^2\pi[(E^2 g)^{-1} \otimes E^2 f]: H_i(K'') \otimes H_j(K') \to E^2_{ij} K$ is an isomorphism; or

(ii) K', K, and K'' are negatively graded, $F_i K = K$ if $i \geq 0$, $F_{i-1}K_i = 0$ if $i \leq 0$, $f(K'_i) \subset F_i K$, $\pi(F_i K \otimes f(K'_j)) \subset F_{i+j} K$, and f and g induce isomorphisms $E^2 f: H_i(K') \to E^2_{i,o} K$ and $E^2 g: E^2_{oj} K \to H_j(K'')$ and π induces a morphism $E^2\pi: E^2_{ij} K \otimes E^2_{ko} K \to E^2_{i+k,j} K$ such that the composite morphism $E^2\pi[(E^2 g)^{-1} \otimes E^2 f]: H_j(K'') \otimes H_i(K') \to E^2_{ij} K$ is an isomorphism.

Let τ be the transgression, $\tau = d_t: E^t_{to} K \to E^t_{0,t-1} K$ in (i) and $\tau = d_{1-t}: E^{1-t}_{ot} K \to E^{1-t}_{t-1,o} K$ $(t < 0)$ in (ii). Then τ is the inverse additive relation to σ, and if $y \in H_q(K'')$ transgresses to $x \in H_{q-1}(K')$, then $P_s(y)$ and if $p > 2$, $\beta P_s(y)$ transgresses to $P_s(x)$ and $-\beta P_s(x)$, whenever the operations are defined. Moreover, if $p > 2$ and $q = 2s$, then $y^{p-1} \otimes x$ transgresses to $-\beta P_s(x)$ (that is, $d_{q(p-1)}(y^{p-1} \otimes x) = -\beta P_s(x)$ in case (i) and $d_{1-q(p-1)}(y^{p-1} \otimes x) = -\beta P_s(x)$ in case (ii) provided that

(iii) if $a_j \in F_{i_j} K$, then $\theta(e_k \otimes a_1 \otimes \dots \otimes a_p) \in F_i K$, $i = \sum i_j + k$, and

(iv) The restriction of θ to $e_o \otimes K^{p-1} \otimes f(K')$ induces a morphism $E^2\theta: (E^2_{*o} K)^{p-1} \otimes E^2_{o*} K \to E^2 K$ in (i) and $E^2\theta: (E^2_{o*} K)^{p-1} \otimes E^2_{*o} K \to E^2 K$ in (ii) such that $E^2\theta = E^2\pi[(E^2 g)^{-1}\phi(E^2 g)^{p-1} \otimes 1]$, where $\phi: H(K'')^{p-1} \to H(K'')$ is the iterated product.

Proof. Let $y \in H_q(K'')$. By the definition of the differentials in the spectral sequence of a filtered complex, $\tau(y)$ is defined if and only if y is represented by $g(a)$ for some $a \in K_q$ such that $d(a) = f(b')$ for some cycle $b' \in K'_{q-1}$, and then $x = \tau(y) = \{b'\}$. Thus the first statement follows from the properties of the chains $P_s(a)$ and $\beta P_s(a)$. For the second statement, consider $\beta P_s(a)$, $q = 2s$. Since $d\beta P_s(a) = -\beta P_s f(b')$, a and b' as above, it suffices to prove that $\beta P_s(a)$ represents $y^{p-1} \otimes x$ in $E^2 K$. $\beta P_s(a) = -m!\,\theta(c')$ by (9) of the proof of Theorem 3.1 and the observation that $\nu(q-1) = \nu(2s-1) = (-1)^{s-1} m!$. In the definition (7) of c', the term with $k = m$ in the first sum involves e_{-1} (since

$q = 2s$ implies $j = p-1$) and is therefore zero. The term with $k = m$ in the second sum is $(-1)^m e_0 \otimes t_{p-1}$ where, by (4), $t_{p-1} = (m-1)! \sum_{i=0}^{m-1} a^{2i} ba^{2(m-i)}$, $b = d(a)$. It is easy to see that

$$\sum_{i=0}^{m-1} a^{2i} ba^{2(m-i)} = P(\alpha)a^{p-1}b \ , \quad \text{where} \quad P(\alpha) = \sum_{i=1}^{m} \alpha^{2i-1} \ ,$$

and direct calculation shows that $P(\alpha) = m + Q(\alpha)$, in $Z_p\pi$, where $Q(\alpha) = \sum_{j=1}^{m} j(\alpha^{2j} + \alpha^{2j+1})$. Let $c'' = (-1)^m (m-1)! \, e_1 \otimes Q(\alpha)a^{p-1} \otimes b \in W \otimes K^p$. Then $c' - d(c'') = (-1)^m m! \, e_0 \otimes a^{p-1}b$ plus a linear combination of terms $e_i \otimes g$ such that g has $i+1$ factors b and $p-i-1$ factors a. Condition (iii) ensures that each $\theta(e_i \otimes g)$ has lower filtration than does $\theta(e_0 \otimes a^{p-1}b)$ and condition (iv) ensures that $\theta(e_0 \otimes a^{p-1}b)$ represents $y^{p-1} \otimes x \in E^2K$. Since $\beta P_s(a)$ and $-m! \, \theta(c' - d(c'')) = \theta(e_0 \otimes a^{p-1}b)$ represent the same element of E^2K, the proof is complete.

The following proposition gives a general prescription for the study of Steenrod operations in spectral sequences; it will be useful in the study of the cohomology of the Steenrod algebra in [18]. In the applications, the determination of the function f is often quite difficult and depends on how the given θ was constructed.

<u>Proposition 3.5.</u> Let (K, θ) be an increasingly filtered object of $\mathcal{C}(p, \infty)$ Suppose given a function $f(i, j, k)$ such that

(i) If $a_t \in F_{i_t} K_{i_t + j_t}$, where $\sum i_t = i$ and $\sum j_t = j$, then

$$\theta(e_k \otimes a_1 \otimes \ldots \otimes a_p) \in F_{f(i,j,k)} K_{i+j+k} \ ;$$

(ii) $f(i, j, k) > f(i-r, j+r-1, k+1)$, $r \geq 1$; and

(iii) $f(i, j, k) \geq r + f(i-pr, j+p(r-1), k+p-1)$, $r \geq 1$.

Let $y \in E^r_{ij}K$. Then there exist elements $P_s(y) \in E^t_{k\ell}K$ and, if $p > 2$, $\beta P_s(y) \in E^{t'}_{k'\ell'}$ such that $d^t P_s(y) = P_s d^r(y)$ and $d^{t'} \beta P_s(y) = -\beta P_s d^r(y)$, where

(iv) If $p = 2$, then $k = f(2i, 2j, s-i-j)$, $\ell = i+j+s-k$, and $t = k - f(2i-2r, 2j+2(r-1), s+1-i-j)$,

(v) If $p > 2$, then $k = f(pi, pj, (2s-i-j)(p-1))$, $\ell = i+j+2s(p-1)-k$, and $t = k - f(pi-pr, pj+p(r-1), (2s+1-i-j)(p-1))$.

(vi) If $p > 2$, then $k' = f(pi, pj, (2s-i-j)(p-1)-1)$, $\ell' = i+j+2s(p-1)-1-k$, and $t' = k' - f(pi-pr, pj+p(r-1), (2s+1-i-j)(p-1)-1)$.

Proof. Let $a \in F_iK_{i+j}$ represent y and let $b = d(a) \in F_{i-r}K_{i+j-1}$. Consider the chain $P_s(a)$ constructed in Theorem 3.1. By (ii), all summands of $P_s(a)$ other than that involving $e_n \otimes a^p$ (for the appropriate n) have lower filtration than k and, by (i), $\theta(e_n \otimes a^p)$ has filtration k. Since $dP_s(b)$, where $P_s(b) \in F_{k-t}K$ by (i) and $k-t \geq r$ by (iii), the statement about $P_s(y)$ follows. The proof for $\beta P_s(y)$ is similar.

4. The Adem relations

We here show that the Adem relations are valid for the Steenrod operations in $H(K)$ if $(K, \theta) \in \mathcal{C}(p, \infty)$ satisfies certain hypotheses. The general algebraic context is distinctly advantageous in the proof. We are able to exploit a trick (Lemma 4.3) used by Adem to prove the classical Adem relations, and this trick would not be available in a categorical approach to Steenrod operations since it depends on the usage of objects of $\mathcal{C}(p, \infty)$ which are not present in many categories of interest, such as infinite loop spaces and cocommutative Hopf algebras.

We require some notations and definitions before we can proceed to the proof.

Let Σ_{p^2} act as permutations on $\{(i,j) \mid 1 \leq i \leq p,\ 1 \leq j \leq p\}$. Embed π in Σ_{p^2} by letting $\alpha(i,j) = (i+1, j)$. Define $\alpha_i \in \Sigma_{p^2}$, $1 \leq i \leq p$, by

$\alpha_i(i,j) = (i,j+1)$ and $\alpha_i(k,j) = (k,j)$ for $k \neq i$, and let $\beta = \alpha_1 \cdots \alpha_p$ so that $\beta(i,j) = (i,j+1)$. Then

(1) $\alpha\alpha_i = \alpha_{i+1}\alpha$; $\alpha_i\alpha_j = \alpha_j\alpha_i$; and $\alpha\beta = \beta\alpha$.

Let α_i generate π_i and β generate ν, so that π_i and ν are cyclic of order p. Let $\sigma = \pi\nu$ and let τ be generated by the α_i and α. Then $\sigma \subset \tau$, τ is a p-Sylow subgroup of Σ_{p^2}, and τ is a split extension of $\pi_1 \cdots \pi_p$ by π.

Let $W_1 = W$ and $W_2 = W$ regarded, respectively, as π-free and ν-free resolutions of Z_p. Let ν operate trivially on W_1, let π operate trivially on W_2, and let σ operate diagonally on $W_1 \otimes W_2$. Then $W_1 \otimes W_2$ is a σ-free resolution of Z_p.

If M is any ν-module, let τ operate on M^p by letting α operate by cyclic permutation and by letting α_i operate on the i-th factor M as does β. Let α_i operate trivially on W_1. Then τ operates on W_1 and we let τ operate diagonally on $W_1 \otimes M^p$. In particular, $W_1 \otimes W_2^p$ is then a τ-free resolution of Z_p.

Let K be any Z_p-complex. We let Σ_{p^2} operate on K^{p^2} by permutations with the (i, j)-th factor K being the j-th factor K in the i-th factor K^p of $K^{p^2} = (K^p)^p$. We let ν operate in the standard fashion on $W_2 \otimes K^p$ (β acting as cyclic permutation on K^p). By the previous paragraph, this fixes an operation of τ on $W_1 \otimes (W_2 \otimes K^p)^p$.

Let Y be any Σ_{p^2}-free resolution of Z_p and let $w: W_1 \otimes W_2^p \longrightarrow Y$ be any τ-morphism over Z_p. w exists since Y is acyclic, and any two choices of w are τ-homotopic.

With these notations, we have the following definition.

<u>Definition 4.1.</u> Let $(K,\theta) \in \mathcal{C}(p,n)$. We say that (K,θ) is an Adem object if there exists a Σ_{p^2}-morphism $\xi: Y^{(n)} \otimes K^{p^2} \longrightarrow K$ such that the following diagram is τ-homotopy commutative:

$$\begin{array}{ccccc}
(W_1 \otimes W_2^{p})^{(n)} \otimes K^{p^2} & \xrightarrow{w \otimes 1} & Y^{(n)} \otimes K^{p^2} & & \\
\Big\downarrow {\scriptstyle 1 \otimes U} & & & \searrow^{\xi} & \\
 & & & & K \\
 & & & \nearrow_{\theta} & \\
W_1^{(n)} \otimes (W_2^{(n)} \otimes K^{p})^{p} & \xrightarrow{1 \otimes \theta^{p}} & W_1^{(n)} \otimes K^{p} & &
\end{array}$$

Here U is the evident shuffle map, and is clearly a τ-morphism (Σ_{p^2} acts trivially on K and α_i acts trivially on $W_1^{(n)} \otimes K^p$).

For clarity, we only treat the case $n = \infty$ below. The relations obtained will be valid for operations on $H_q(K)$, (K, θ) an Adem object of $\mathcal{C}(p, n)$, provided that n is sufficiently large relative to q.

We first show that the tensor product of Adem objects is an Adem object and then use this fact to show that any relations valid on $H_{q_i}(K)$ for all Adem objects (K, θ) and suitable q_i will necessarily be valid on $H_q(K)$ for arbitrary q.

Lemma 4.2. If (K, θ) and (L, θ') are Adem objects of $\mathcal{C}(p, \infty)$, then $(K \otimes L, \tilde{\theta})$ is an Adem object of $\mathcal{C}(p, \infty)$.

Proof. $\tilde{\theta}$ is as defined in Definitions 2.1. By hypothesis, we are given ξ and ξ' such that (K, ξ) and (K, η') are objects of $\mathcal{C}(\Sigma_{p^2}, \infty, Z_p)$, hence we may define $\tilde{\xi}$ as in Definitions 2.1 so that $(K \otimes L, \tilde{\xi}) \in \mathcal{C}(\Sigma_{p^2}, \infty, Z_p)$. We must show that the diagram of Definition 4.1, for $K \otimes L$, is τ-homotopy commutative, and this follows easily from a simple chase of a large diagram and the definition of $\tilde{\theta}$ and $\tilde{\xi}$. The crucial point is the observation that since $W_1 \otimes W_2^p$ is τ-free and $Y \otimes Y$ is acyclic, the following diagram is τ-homotopy commutative where V is the evident shuffle and $\psi : Y \longrightarrow Y \otimes Y$ is any given Σ_{p^2}-coproduct:

$$\begin{array}{ccccc} W_1 \otimes W_2^p & \xrightarrow{\psi \otimes \psi^p} & W_1 \otimes W_1 \otimes (W_2 \otimes W_2)^p & \xrightarrow{V} & W_1 \otimes W_2^p \otimes W_1 \otimes W_2^p \\ \Big\downarrow{\scriptstyle w} & & & & \Big\downarrow{\scriptstyle w \otimes w} \\ Y & & \xrightarrow{\qquad\quad \psi \qquad\quad} & & Y \otimes Y \end{array}$$

Let F_p denote the free associative algebra generated by $\{P_s | s \in Z\}$ and, if $p > 2$, $\{\beta P_s | s \in Z\}$. Let $J_p \subset F_p$ denote the two-sided ideal consisting of all elements $a \in F_p$ such that $ax = 0$ for all $x \in H(K)$ and all Adem objects $(K, \theta) \in \mathcal{C}(p, \infty)$. Let $B_p = F_p/J_p$. B_p is a universal Steenrod algebra. Both the classical Steenrod algebra and the Dyer-Lashof algebra [17] are quotients of B_p.

Lemma 4.3. Let $a \in F_p$. Let $\{q_i | i \geq 0\}$ be a strictly decreasing sequence of integers. Suppose that $ax = 0$ for all $x \in H_{q_i}(K)$, $i \geq 0$, and all Adem objects $(K, \theta) \in \mathcal{C}(p, \infty)$. Then $a \in J_p$.

Proof. Let K be an Adem object in $\mathcal{C}(p, \infty)$ and let $x \in H_q(K)$. We must prove that $ax = 0$. Choose $r < 0$ such that $q + r = q_i$ for some i. There exists an Adem object $(L_r, \theta_r) \in \mathcal{C}(p, \infty)$ and a class $y \in H_r(L_r)$ such that $P_o(y) = y$, $P_s(y) = 0$ for $s \neq 0$, and $\beta P_s(y) = 0$ for all s. Such an object can easily be constructed explicitly, but it is quicker to appeal to the results of section 8, which show that the singular cochains of a $(-r)$-sphere, graded by non-positive subscripts, provide such an object. Now $(K \otimes L_r, \tilde{\theta})$ is an Adem object of $\mathcal{C}(p,\infty)$ by the previous lemma. By the external Cartan formula, Corollary 2.7, $a(x \otimes y) = ax \otimes y$. Since $x \otimes y \in H_{q_i}(K \otimes L_r)$, $a(x \otimes y) = 0$ and therefore $ax = 0$, as was to be shown.

The Adem relations will be proven by chosing the diagram of Definition 4.1, and we shall need some information about the homology of σ, τ, and Σ_{p^2}. Let $\phi: W_1 \otimes W_2 \longrightarrow W_1 \otimes W_2^p$ be a σ-morphism over Z_p. Define $\gamma \in \Sigma_{p^2}$ by $\gamma(i, j) = (j, i)$. Observe that $\gamma^2 = 1$ and $\gamma\alpha = \beta\gamma$. For $q \in Z$, conjugation by γ

gives a commutative diagram

$$\begin{array}{ccccc}
H_*(\sigma; Z_p(q)) & \xrightarrow{\phi_*} & H_*(\tau; Z_p(q)) & \xrightarrow{w_*} & H_*(\Sigma_{p^2}; Z_p(q)) \\
\downarrow{\scriptstyle \gamma_*} & & & & \downarrow{\scriptstyle 1} \\
H_*(\sigma; Z_p(q)) & \xrightarrow{\phi_*} & H_*(\tau; Z_p(q)) & \xrightarrow{w_*} & H_*(\Sigma_{p^2}; Z_p(q))
\end{array}$$

Thus $w_*(\phi_* - \phi_* \gamma_*) = 0$. The following lemmas compute γ_* and ϕ_*. Note that $H_*(\tau; Z_p(q)) = H_*(\tau; Z_p)$ since τ contains only even permutations.

Lemma 4.4. γ_* is given on $H_*(\sigma; Z_p(q))$ by $\gamma_*(e_i \otimes e_j) = (-1)^{ij+mq} e_j \otimes e_i$.

Proof. Define $\gamma_\# : W_1 \otimes W_2 \longrightarrow W_1 \otimes W_2$ by the formula

$$\gamma_\# (\alpha^k e_i \otimes \beta^\ell e_j) = (-1)^{ij} \alpha^\ell e_j \otimes \beta^k e_i .$$

Then $d\gamma_\# = \gamma_\# d$ and $\gamma_\#(\mu x) = (\gamma\mu\gamma^{-1})\gamma_\#(x)$ for $\mu \in \sigma$ and $x \in W_1 \otimes W_2$. Thus $\gamma_\# \otimes \gamma: (W_1 \otimes W_2) \otimes_\sigma Z_p(q) \longrightarrow (W_1 \otimes W_2) \otimes_\sigma Z_p(q)$ induces γ_*. Since the sign of γ is $(-1)^m$, $\gamma \cdot 1 = (-1)^{mq}$ in $Z_p(q)$, and the result follows.

Before computing ϕ_*, we fix notations concerning binomial coefficients.

Notations 4.5. Let i and j be integers. Define $(i, j) = (i+j)!/i!j!$ if $i \geq 0$ and $j \geq 0$ $(0! = 1)$ and define $(i, j) = 0$ if $i < 0$ or $j < 0$. Recall that if $i \geq 0$ and $j \geq 0$ have p-adic expansions $i = \sum a_k p^k$ and $j = \sum b_k p^k$, then $(i, j) \equiv \prod_k (a_k, b_k) \bmod p$. Clearly $(a_k, b_k) \not\equiv 0 \bmod p$ if and only if $a_k + b_k < p$, hence $(i, j) \not\equiv 0 \bmod p$ if and only if $\sum (a_k + b_k) p^k$ is the p-adic expansion of $i+j$.

Lemma 4.6. $H_*(\tau; Z_p) = H_*(\pi; H_*(\nu; Z_p)^p)$ and $\phi_*: H_*(\sigma; Z_p) \longrightarrow H_*(\tau; Z_p)$ is given by the following formulas (with sums taken over the integers).

(i) If $p = 2$, $\phi_*(e_r \otimes e_s) = \sum_k (k, s-2k) e_{r+2k-s} \otimes e_{s-k}^2$; and

(ii) If $p > 2$, $\phi_*(e_r \otimes e_s) = \sum_k (-1)^k \nu(s)(k, [s/2] - pk)\, e_{r+(2pk-s)(p-1)} \otimes e^p_{s-2k(p-1)}$

$$-\delta(r)\delta(s-1) \sum_k (-1)^k \nu(s-1)(k, [{}^{s-1}/2] - pk) e_{r+p+(2pk-s)(p-1)} \otimes e^p_{s-2k(p-1)-1},$$

where $\nu(2j+\varepsilon) = (-1)^j (m!)^{\varepsilon}$ and $\delta(2j+\varepsilon) = \varepsilon$, j any integer, $\varepsilon = 0$ or 1 .

Proof. Let $\overline{W}_2 = W_2 \otimes_{\nu} Z_p = H_*(\nu\,; Z_p)$. By the definition of the action of τ on $W_1 \otimes W_2^{\,p}$, we have that $(W_1 \otimes W_2^p) \otimes_{\tau} Z_p = W_1 \otimes_{\pi} \overline{W}_2^{\,p}$ as a Z_p-complex, and the first part follows. Of course, $H_*(\tau; Z_p)$ is now computed by Lemma 1.3. ϕ_* could be computed directly, but it is simpler to use topology. Let $K(Z_p, 1) = E/\nu$, where ν operates properly on the acyclic space E, so that, by [14, IV 11], $C_*(E) = Z_p \nu \otimes C_*(E/\nu)$, with Z_p coefficients. Let $D: E \longrightarrow E^p$ be the iterated diagonal. Then $1 \otimes D: W_1 \otimes C_*(E) \longrightarrow W_1 \otimes C_*(E^p)$ is a σ-morphism, and we shall obtain a σ-morphism $\Phi: W_1 \otimes C_*(E^p) \longrightarrow W_1 \otimes C_*(E)^p$ in Lemma 7.1. Let $d = \Phi(1 \otimes D)$ and let $f: W_2 \longrightarrow C_*(E)$ be a ν-morphism over Z_p. Since $W_1 \otimes W_2$ is σ-free and $W_1 \otimes C_*(E)^p$ is acyclic (and τ-free, with the evident τ-action), the following diagram is σ-homotopy commutative.

$$\begin{array}{ccc} W_1 \otimes W_2 & \xrightarrow{\ \ \phi\ \ } & W_1 \otimes W_2^p \\ \Big\downarrow {\scriptstyle 1 \otimes f} & & \Big\downarrow {\scriptstyle 1 \otimes f^p} \\ W_1 \otimes C_*(E) & \xrightarrow{\ \ d\ \ } & W_1 \otimes C_*(E)^p \end{array}$$

Therefore $\phi_* = d_*: H_*(\sigma; Z_p) \longrightarrow H_*(\tau; Z_p)$, and this can clearly be computed from the quotient map $d: W_1 \otimes_{\pi} C_*(E/\nu) \longrightarrow W_1 \otimes_{\pi} C_*(E/\nu)^p$. We shall prove the following formulas in Proposition 9.1.

(a) If $p = 2$, $d_*(e_r \otimes e_s) = \sum_k e_{r+2k-s} \otimes P_*^{\,k}(e_s)^2$; and

(b) If $p > 2$, $d_*(e_r \otimes e_s) = \sum_k (-1)^k \nu(s) e_{r+(2pk-s)(p-1)} \otimes P_*^{\,k}(e_s)^p$

$$- \delta(r) \sum_k (-1)^k \nu(s-1) e_{r+p+(2pk-s)(p-1)} \otimes P_*^k \beta(e_s)^p .$$

Here the $P_*^{\,k}$ are the duals to the Steenrod operations in $H^*(K(Z_p, 1); Z_p) = H^*(\nu; Z_p)$. The latter operations satisfy $P^o = 1$ and the internal Cartan formula, hence, by (1.2), if w_t is dual to e_t, then

(c) If $p = 2$, $P^k(w_t) = (k, t-k) w_{k+t}$, hence $P_*^k(e_s) = (k, s-2k) e_{s-k}$; and

(d) If $p > 2$, $P^k(w_t) = (k, [t/2] - k)w_{t+2k(p-1)}$, hence

$$P_*^k(e_s) = (k, [s/2] - pk)e_{s-2k(p-1)} .$$

Combining (a) and (c), we obtain (i). Combining (b), (d), and $\beta(e_i) = \delta(i-1)e_{i-1}$, we obtain (ii).

Theorem 4.7. The following relations among the P_s and βP_s are valid on all homology classes of all Adem objects in $\mathcal{C}(p,\infty)$.

(i) If $p = 2$ and $a > 2b$, $P_a P_b = \sum_i (2i-a, a-b-i-1)P_{a+b-i}P_i$

(ii) If $p > 2$ and $a > pb$, $P_a P_b = \sum_i (-1)^{a+i}(pi-a, a-(p-1)b-i-1)P_{a+b-i}P_i$

and $\beta P_a P_b = \sum_i (-1)^{a+i}(pi-a, a-(p-1)b-i-1)\beta P_{a+b-i}P_i$

(iii) If $p > 2$ and $a \geq pb$, $P_a \beta P_b = \sum_i (-1)^{a+i}(pi-a, a-(p-1)b-i)\beta P_{a+b-i}P_i$

$$- \sum_i (-1)^{a+i}(pi-a-1, a-(p-1)b-i)P_{a+b-i}\beta P_i$$

and

$$\beta P_a \beta P_b = - \sum_i (-1)^{a+i}(pi-a-1, a-(p-1)b-i)\beta P_{a+b-i}\beta P_i$$

Proof. Note first that the second relations of (ii) and (iii) are implied by the first for objects which are reduced mod p, but are logically independent in our general setting. Let (K, θ) be an Adem object in $\mathcal{C}(p,\infty)$ and let $x \in H_q(K)$. Definition 4.1 implies that we have a Z_p-homotopy commutative diagram

$$\begin{array}{ccccc}
(W_1 \otimes W_2^p) \otimes_\tau K^{p^2}(q) & \xrightarrow{w \otimes 1} & Y \otimes_{\Sigma_{p^2}} K^{p^2}(q) & \xrightarrow{\xi} & K(q) \\
\downarrow{\scriptstyle 1 \otimes u} & & & & \| \\
W_1 \otimes_\pi (W_2 \otimes K^p)^p(q) & \xrightarrow{1 \otimes \theta^p} & W_1 \otimes_\pi K^p(q) & \xrightarrow{\theta} & K(q)
\end{array}$$

Since x^{p^2} is Σ_{p^2} invariant in $K^{p^2}(q)$, we have, for all r and s,

(a) $\xi_*(w \otimes 1)_*(e_r \otimes e_s^p \otimes x^{p^2}) = \xi_*(w_*(e_r \otimes e_s^p) \otimes x^{p^2})$.

In the other direction, U introduces the sign $(-1)^{mqs}$ and we have

(b) $\theta_*(1\otimes\theta^p)_*(1\otimes U)_*(e_r\otimes e_s^p\otimes x^{p^2}) = (-1)^{mqs}D_rD_s(x)$.

Since $w_*\phi_* = w_*\phi_*\gamma_*$, Lemma 4.4 gives the formula

(c) $w_*\phi_*(e_r\otimes e_s) = (-1)^{rs+mq}w_*\phi_*(e_s\otimes e_r)$.

Combining formulas (a) and (c), we obtain the formula

(d) $\xi_*(w\otimes 1)_*(\phi_*(e_r\otimes e_s)\otimes x^{p^2}) = (-1)^{rs+mq}\xi_*(w\otimes 1)_*(\phi_*(e_s\otimes e_r)\otimes x^{p^2})$.

In view of (b), (d) gives relations on iterated operations, and these relations are explicit since ϕ_* is known. We prove the three parts of the theorem successively In all parts, the statements about binomial coefficients are verified by writing out the p-adic expansions of the relevant integers and appealing to the remarks in Notations 4.5.

(i) By (b) and (d), Lemma 4.6 implies the formula

(e) $\sum_k (k, s-2k)D_{r+2k-s}D_{s-k}(x) = \sum_\ell (\ell, r-2\ell)D_{s+2\ell-r}D_{r-\ell}(x)$.

Formula (e) is valid for all r and s, and we set $r = a-2q$ and $s = b-q$ for our fixed $a > 2b$. If we then change variables to $j = b-k$ and $i = a-q-\ell$ and apply Definition 2.2, we obtain

(f) $\sum_j (b-j, 2j-b-q)P_{a+b-j}P_j(x) = \sum_i (a-q-i, 2i-a)P_{a+b-i}P_i(x)$.

The condition $a > 2b$ guarantees that the same terms do not appear with non-zero coefficients on both sides of (f). Now suppose that $q = b-2^t+1$ for some $t > 0$. Then, if $j \neq b$,

$$(b-j, 2j-b-q) = (b-j, 2^t-1-2(b-j)) = 0.$$

On the right side of (f), $P_{a+b-i}P_i(x) = 0$ unless $2^t-1 = b-q \geq 2i-a$, while if $2^t > 2i-a$, then

$$(a-q-i, 2i-a) = (2i-a, a-b-i-1+2^t) = (2i-a, a-b-i-1).$$

Thus (f) reduces to the desired relation (i) when $q = b-2^t+1$ for some $t > 0$. By Lemma 4.3, it follows that (i) holds for all q.

(ii) Observe that $\beta w_* = w_* \beta$ and that, by the proof of part (v) of Proposition 2.3, the Bockstein operation β in $H_*(\tau; Z_p)$ is given by $\beta(e_k \otimes e_\ell^p) = \beta(e_k) \otimes e_\ell^p$, where $\beta(e_k) = \delta(k-1)e_{k-1}$. Now since (c) holds with ϕ_* replaced by $\beta\phi_*$, so does (d); that is,

(d') $$\xi_*(w \otimes 1)_*(\beta\phi_*(e_r \otimes e_s) \otimes x^{p^2}) = (-1)^{rs+mq} \xi_*(w \otimes 1)_*(\beta\phi_*(e_s \otimes e_r) \otimes x^{p^2}).$$

Replace r and s by 2r and 2s in (d) and (d') and let $\varepsilon = 0$ or 1; then, by (b) and Lemma 4.6, (d) and (d') imply the following formula for $\varepsilon = 0$ and $\varepsilon = 1$, respectively.

(g) $$\sum_k (-1)^k \nu(2s)(k, s-pk) D_{2r+(2pk-2s)(p-1)-\varepsilon} D_{2s-2k(p-1)}(x)$$
$$= \sum_\ell (-1)^{\ell+mq} \nu(2r)(\ell, r-p\ell) D_{2s+(2p\ell-2r)(p-1)-\varepsilon} D_{2r-2\ell(p-1)}(x).$$

In (g), set $r = a(p-1)-pqm$ and $s = b(p-1)-qm$ and change variables to $j = b-k$ and $i = a-mq-\ell$. Let $\beta^0 P_s = P_s$ and $\beta^1 P_s = \beta P_s$, by abuse; then, by Definitions 2.2 and a check of constants, we obtain

(h) $$\sum_j (-1)^{b+j}(b-j, pj-b-mq)\beta^\varepsilon P_{a+b-j} P_j(x)$$
$$= \sum_i (-1)^{a+i}(a-mq-i, pi-a)\beta^\varepsilon P_{a+b-i} P_i(x) .$$

Again, $a > pb$ ensures that the same terms do not appear on both sides of (h). Now suppose that $q = 2b - 2(1+p+\ldots+p^{t-1})$, $t > 0$. Then $(b-j, pj-b-mq) = 0$ unless $j = b$ and, on the other side of (h), $\beta^\varepsilon P_{a+b-i} P_i(x) = 0$ unless $1+\ldots+p^{t-1} = b-q/2 \geq pi-a$, when $p^t > pi-a$ implies

$$(a-mq-i, pi-a) = (pi-a, a-(p-1)b+p^t-1-i) = (pi-a, a-(p-1)b-i-1).$$

Thus (h) reduces to the desired relations (ii) when $q = 2b-2(1+\ldots+p^{t-1})$ for some $t > 0$. By Lemma 4.3, it follows that (ii) holds for all q.

(iii) Replace r and s by 2r and 2s-1 in (d) and (d'); then, by (b) and Lemma 4.6, (d) and (d') imply the following formula for $\varepsilon = 0$ and $\varepsilon = 1$, respectively.

(i) $\sum_k (-1)^{k+mq} \nu(2s-1)(k, s-1-pk) D_{2r+(2pk-2s+1)(p-1)-\varepsilon} D_{2s-1-2k(p-1)}(x)$

$= (1-\varepsilon) \sum_\ell (-1)^{\ell+mq} \nu(2r)(\ell, r-p\ell) D_{2s-1+(2p\ell-2r)(p-1)} D_{2r-2\ell(p-1)}(x)$

$- \sum_\ell (-1)^\ell \nu(2r-1)(\ell, r-1-p\ell) D_{2s+(2p\ell-2r+1)(p-1)-\varepsilon} D_{2r-1-2\ell(p-1)}(x).$

In (i), set $r = a(p-1)-pqm$ and $s = b(p-1)-qm$ and change variables to $j = b-k$ and $i = a-mq-\ell$. By Definitions 2. 2, we obtain

(j) $\sum_j (-1)^{b+j} (b-j, pj-b-mq-1) \beta^\varepsilon P_{a+b-j} \beta P_j(x)$

$= (1-\varepsilon) \sum_i (-1)^{a+i} (a-mq-i, pi-a) \beta P_{a+b-i} P_i(x)$

$- \sum_i (-1)^{a+i} (a-mq-i, pi-a-1) \beta^\varepsilon P_{a+b-i} \beta P_i(x)$.

Again, $a \geq pb$ ensures that the same terms do not occur on both sides of the equation. Now suppose that $q = 2b-2p^t$, $t > 0$. Then $(b-j, pj-b-mq-1) = 0$ unless $j = b$. On the other side of (j), $\beta P_{a+b-i} P_i(x) = 0$ unless $p^t > pi-a$, when

$$(a-mq-i, pi-a) = (pi-a, a-(p-1)b-i+(p-1)p^t) = (pi-a, a-(p-1)b-i),$$

and $\beta^\varepsilon P_{a+b-i} \beta P_i(x) = 0$ unless $p^t > pi-a-1$, when

$$(a-mq-i, pi-a-1) = (pi-a-1, a-(p-1)b-i+(p-1)p^t) = (pi-a-1, a-(p-1)b-i) .$$

Thus (j) reduces to the desired relations (iii) when $q = 2b-2p^t$ for some $t < 0$, and Lemma 4. 3 implies that (iii) holds for all q.

Remark 4. 8. It should be observed, for use in section 9, that the relations (f), (h), and (j) derived in the proof above are valid for arbitrary integers a and b (without the restrictions $a > pb$ or $a \geq pb$). Indeed, these conditions on a and b were only required in order to obtain disjoint non-trivial terms on the two sides of the cited equations.

5. Reindexing for cohomology.

We have geared our discussion to homology, but the reformulation appropriate to cohomology is obtained by a minor and standard change of notation. Thus let K be a Z_p-complex Z-graded by superscripts, with d of degree plus one. If we regrade K by $K_{-q} = K^q$, then the theory of the previous sections applies. Equivalently, we can regrade W by non-positive superscripts and reformulate the theory. Obviously, this in no way changes the proofs. Let $(K, \theta) \in \mathcal{C}(p, \infty)$, with K and W graded by superscripts, and let $x \in H^q(K)$. Then $D_i(x) = \theta_*(e^{-i} \otimes x^p) \in H^{pq-i}(K)$, $i \geq 0$, and we may define $P^s(x) = P_{-s}(x)$ and, if $p > 2$, $\beta P^s(x) = \beta P_{-s}(x)$. Explicitly, $P^s(x)$ and $\beta P^s(x)$ are defined by the formulas

(1) If $p = 2$, $P^s(x) = D_{q-s}(x) \in H^{q+s}(K)$, where $D_i = 0$ for $i < 0$; and

(2) If $p > 2$, $P^s(x) = (-1)^s \nu(-q) D_{(q-2s)(p-1)}(x) \in H^{q+2s(p-1)}(K)$ and

$\beta P^s(x) = (-1)^s \nu(-q) D_{(q-2s)(p-1)-1}(x) \in H^{q+2s(p-1)+1}(K)$, where

$D_i = 0$ for $i < 0$ and if $q = 2j - \varepsilon$, $\varepsilon = 0$ or 1, then $\nu(-q) = (-1)^j (m!)^{\varepsilon}$.

Of course, if $p = 2$, we should write $P^s = Sq^s$ in order to conform to standard notations, but we prefer to retain the notation P^s. In this way, the Cartan formula and Adem relations are formally the same in the cases $p = 2$ and $p > 2$.

The P^s and βP^s are natural homomorphisms and are defined for all integers s. If $(K, \theta) \in \mathcal{C}(p, \infty)$ and $x \in H^q(K)$, then

(3) If $p = 2$, $P^s(x) = 0$ if $s > q$ and $P^q(x) = x^2$; and

(4) If $p > 2$, $P^s(x) = 0$ if $2s > q$, $\beta P^s(x) = 0$ if $2s \geq q$, and $P^s(x) = x^p$ if $2s = q$.

Note that we do not claim that $P^s(x) = 0$ if $s < 0$ or that $P^0 = 1$; these formulas are not true in general. If (K, θ) is unital, then $P^s(e) = 0$ for $s \neq 0$. If (K, θ) is reduced mod p, then

(5) $\beta P^{s-1} = sP^s$ if $p = 2$ and βP^s is the composition of P^s and the Bockstein β if $p > 2$.

The external Cartan formula now reads

(6) $$P^s(x \otimes y) = \sum_{i+j=s} P^i(x) \otimes P^j(y) \text{ and, if } p > 2,$$

$$\beta P^{s+1}(x \otimes y) = \sum_{i+j=s} (\beta P^{i+1}(x) \otimes P^j(y) + (-1)^{\deg x} P^i(x) \otimes \beta P^{j+1}(y)).$$

We have $\sigma P^s = P^s \sigma$ and $\sigma \beta P^s = -\beta P^s \sigma$ and of course the Kudo transgression theorem takes on a more familiar form with grading by superscripts in case (ii). The Adem relations, reformulated in terms of the P^s, take on the form given in the following corollary.

Corollary 5.1. The following relations among the P^s and βP^s are valid on all cohomology classes of all Adem objects in $\mathcal{C}(p, \infty)$

(i) If $p \geq 2$, $a < pb$, and $\varepsilon = 0$ or 1 if $p > 2$, $\varepsilon = 0$ if $p = 2$, then

$$\beta^{\varepsilon} P^a P^b = \sum_i (-1)^{a+i}(a - pi, (p-1)b - a+i-1)\beta^{\varepsilon} P^{a+b-i} P^i$$

(ii) If $p > 2$, $a \leq pb$, and $\varepsilon = 0$ or 1, then

$$\beta^{\varepsilon} P^a \beta P^b = (1 - \varepsilon) \sum_i (-1)^{a+i}(a - pi, (p-1)b - a+i-1)\beta P^{a+b-i} P^i$$
$$- \sum_i (-1)^{a+i}(a-pi-1, (p-1)b-a+i)\beta^{\varepsilon} P^{a+b-i} \beta P^i$$

(where, by abuse of notation, $\beta^0 P^s = P^s$ and $\beta^1 P^s = \beta P^s$).

While the two forms of the Adem relations given in Theorem 4.7 and the corollary are completely equivalent, they work out quite differently in practice. The relations of Theorem 4.7 apply to positive complexes, in homology, with $a, b \geq 0$; but $a, b \geq 0$ in Theorem 4.7 corresponds to $a, b \leq 0$ in the corollary, which is designed for use in cohomology with $a, b \geq 0$. For this reason, the Dyer-Lashof algebra [17], which operates on the homology of infinite loop spaces, is a very different algebraic object than the classical Steenrod algebra.

6. Cup-i products, Browder operations, and higher Bocksteins.

We here discuss $\cup_i$-products and certain homology operations of two variables, which were first studied by Browder in [4]; these operations occur in the presence of a $\cup_n$-product and the absence of a $\cup_{n+1}$-product and are central to the study of the homology of (n+1)-fold loop spaces. We shall also obtain a very useful result, Proposition 6.8, on higher Bockstein operations. In section 10, we shall show that this result suffices to give a complete computation of the mod p cohomology Bockstein spectral sequence of $K(\pi, n)$ for any Abelian group π and any prime p.

Throughout this section, Λ is a commutative ring, π is the cyclic group of order 2 with generator α, and W is the canonical $\Lambda\pi$-free resolution of Λ. Let $\Delta_i = \alpha + (-1)^i \in \Lambda\pi$, so that $d(e_i) = \Delta_i e_{i-1}$ for $i \geq 1$. If $(K, \theta) \in \mathcal{C}(\pi, n, \Lambda)$, then we may assume that the restriction of θ to $e_0 \otimes K \otimes K$ agrees with the given product on K by (i) of Definition 2.1.

Definition 6.1. Let $(K, \theta) \in \mathcal{C}(\pi, n, \Lambda)$ and let $x \in K_q$ and $y \in K_r$. For $0 \leq i \leq n$, define $x \cup_i y = (-1)^{\frac{1}{2}i(i+1)} \theta(e_i \otimes x \otimes y)$. Then $\cup_0$ is the product on K and if $i > 0$, then $\cup_i : K \otimes K \longrightarrow K$ is a chain homotopy of degree i from $\cup_{i-1}$ to $(-1)^{i-1} \cup_{i-1} \cdot \alpha$; that is,

(i) $$d(x \cup_i y) = (-1)^i d(x) \cup_i y + (-1)^{i+q} x \cup_i d(y) + x \cup_{i-1} y + (-1)^{i+qr} y \cup_{i-1} x.$$

If $\Lambda = Z_2$ and $x \in K_q$ is a cycle, then $P_{i+q}\{x\} = D_i\{x\} = \{x \cup_i x\}$, which, in cohomology, was Steenrod's first definition [25] of the squares. We now define the Browder operations for $(K, \theta) \in \mathcal{C}(\pi, n, \Lambda)$.

Definition 6.2. Let $(K, \theta) \in \mathcal{C}(\pi, n, \Lambda)$, $n < \infty$, and let $x \in H_q(K)$ and $y \in H_r(K)$. Observe that if a and b are representative cycles for x and y, then $\Delta_{n+1} e_n \otimes a \otimes b$ is a cycle in $W^{(n)} \otimes K^2$ whose homology class $\Delta_{n+1} e_n \otimes x \otimes y$ depends only on x and y. Define $\lambda_n(x, y) \in H_{q+r+n}(K)$ by

$\lambda_n(x, y) = (-1)^{nq+1}\theta_*(\Delta_{n+1}e_n \otimes x \otimes y)$. Note that we have chosen not to pass to equivariant homology; of course, we can do so, and, in $W^{(n)} \otimes_\pi K^2$,

$$(-1)^{nq+1}\Delta_{n+1}e_n \otimes a \otimes b = (-1)^{nq+n}e_n \otimes a \otimes b - (-1)^{nq+qr}e_n \otimes b \otimes a.$$

Thus $\lambda_n(x, y)$ is represented by $(-1)^{nq+n+\frac{1}{2}n(n+1)}(a \cup_n b - (-1)^{n+qr} b \cup_n a)$.

The following proposition contains many of the elementary properties of the λ_n; its proof is immediate from the definition.

Proposition 6.3. Let $(K, \theta) \in \mathcal{C}(\pi, n, \Lambda)$, $n < \infty$, and consider $\lambda_n : H_q(K) \times H_r(K) \longrightarrow H_{q+r+n}(K)$.

(i) λ_n induces a homomorphism $\lambda_n : H_q(K) \otimes H_r(K) \longrightarrow H_{q+r+n}(K)$

(ii) If $f: K \longrightarrow K'$ is a morphism in $\mathcal{C}(\pi, n, \Lambda)$, then $\lambda_n(f_* \otimes f_*) = f_*\lambda_n$

(iii) If θ is the restriction to $W^{(n)} \otimes K^2$ of $\theta' : W^{(n+1)} \otimes K^2 \longrightarrow K$, then $\lambda_n = 0$

(iv) If $n = 0$, then $\lambda_o(x, y) = xy - (-1)^{qr}yx$

(v) If (K, θ) is unital and the restriction of θ to $W^{(n)} \otimes (e \otimes K + K \otimes e)$ is homotopic to $\varepsilon \otimes \phi$, ϕ the product, then $\lambda_n(x, e) = 0 = \lambda_n(e, y)$.

(vi) $\lambda_n(x, y) = (-1)^{qr+1+n(q+r+1)}\lambda_n(y, x)$ and, if $2 = 0$ in Λ, $\lambda_n(x, x) = 0$

(Note that the first part implies $2\lambda_n(x, x) = 0$ if $n+q$ is even.)

The λ_n satisfy the following analog of the external Cartan formula.

Proposition 6.4. Let (K, θ) and (L, θ') be in $\mathcal{C}(\pi, n, \Lambda)$, $n < \infty$ and Λ a field. Let $x \in H_q(K)$, $x' \in H_r(K)$, $y \in H_s(L)$ and $y' \in H_t(L)$. Then

$$\lambda_n(x \otimes y, x' \otimes y') = (-1)^{r(s+n)}xx' \otimes \lambda_n(y, y') + (-1)^{s(r+t+n)}\lambda_n(x, x') \otimes y'y.$$

Proof. Let a, a', b, b' represent x, x', y, y' respectively. Let $c = (-1)^{n(q+s)+1}\tilde{\theta}(\Delta_{n+1}e_n \otimes a \otimes b \otimes a' \otimes b')$, so that c represents $\lambda_n(x \otimes y, x' \otimes y')$. By (1) of Definition 1.2,

$\psi(e_n) = \sum_{j=0}^{n} e_j \otimes \alpha^j e_{n-j}$ in W, and the definition of $\tilde{\theta}$ shows that

$$c = \sum_{j=0}^{n} (-1)^{rs+n(1+r+s)-j(q+r)} \theta(e_j \otimes a \otimes a') \otimes \theta'(\alpha^j e_{n-j} \otimes b \otimes b')$$

$$- \sum_{j=0}^{n} (-1)^{rs+n(r+s)-j(q+r)} \theta(\alpha e_j \otimes a \otimes a') \otimes \theta'(\alpha^{j+1} e_{n-j} \otimes b \otimes b')$$

Let $e = \sum_{j=1}^{n} (-1)^{rs+n(r+s)-(j+1)(q+r)} \theta(e_j \otimes a \otimes a') \otimes \theta'(\alpha^j e_{n+1-j} \otimes b \otimes b')$. Then a straightforward calculation demonstrates that

$$c + d(e) = (-1)^{r(s+n)+sn+n} \theta(e_0 \otimes a \otimes a') \otimes \theta'(\Delta_{n+1} e_n \otimes b \otimes b')$$

$$+ (-1)^{s(r+n)+qn+n} \theta(\Delta_{n+1} e_n \otimes a \otimes a') \otimes \theta'(\alpha^{n+1} e_0 \otimes y \otimes y') .$$

Since L is homotopy commutative for $n > 0$, $\theta'(\alpha^{n+1} e_0 \otimes b \otimes b')$ represents $(-1)^{st} y'y$ for any n, and the result follows.

We next prove that the λ_n commute with suspension.

<u>Proposition 6.5.</u> Let $(K', \theta') \in \mathcal{C}(\pi, n+1, \Lambda)$ and $(K'', \theta'') \in \mathcal{C}(\pi, n, \Lambda)$. Let K be a Λ-complex and let $f: K' \longrightarrow K$ and $g: K \longrightarrow K''$ be morphisms of complexes such that $gf = 0$. Define

$$\tilde{K} = W^{(n+1)} \otimes f(K')^2 + \overline{W}^{(n+1)} \otimes f(K') \otimes K + W^{(n)} \otimes K^2 ,$$

where $\overline{W}^{(n+1)} = W^{(n)} + \Lambda e_{n+1}$ $(\alpha e_{n+1} \notin \overline{W}^{(n+1)})$. Suppose given a π-morphism $\theta: \tilde{K} \longrightarrow K$ such that the following diagram is commutative:

$$\begin{array}{ccccc} W^{(n+1)} \otimes K' \otimes K' & \xrightarrow{1 \otimes f \otimes f} & \tilde{K} & \xrightarrow{1 \otimes g \otimes g} & W^{(n)} \otimes K'' \otimes K'' \\ \downarrow \theta' & & \downarrow \theta & & \downarrow \theta'' \\ K' & \xrightarrow{f} & K & \xrightarrow{g} & K'' \end{array}$$

Let $x, y \in \operatorname{Ker} f_*$. Then $\sigma\lambda_{n+1}(x, y) = \lambda_n(\sigma x, \sigma y) \in \operatorname{Coker} g_*$.

<u>Proof.</u> Let $a' \in K'_q$ and $b' \in K'_r$ represent x and y respectively. Let $a = f(a')$ and $b = f(b')$ and choose $u \in K_{q+1}$ and $v \in K_{r+1}$ such that $d(u) = a$

and $d(v) = b$. Define $c \in \tilde{K}$ by

$$c = (-1)^q \Delta_{n+1} e_n \otimes u \otimes v - (-1)^{n+q} e_{n+1} \otimes a \otimes v - (-1)^{r(q+1)} e_{n+1} \otimes b \otimes u.$$

Then a straightforward calculation demonstrates that

$$d(c) = e_{n+1} \otimes a \otimes b + (-1)^{n+qr} e_{n+1} \otimes b \otimes a = (-1)^n \Delta_{n+2} e_{n+1} \otimes a \otimes b$$

Thus $(-1)^{(n+1)q+1} f\theta'(\Delta_{n+2} e_{n+1} \otimes a' \otimes b') = (-1)^{(n+1)(q+1)} d\theta(c)$ and

$$(-1)^{(n+1)(q+1)} g\theta(c) = (-1)^{n(q+1)+1} \theta''(\Delta_{n+1} e_n \otimes g(u) \otimes g(v)),$$

by our commutative diagram, and this proves the result.

The analog for the Browder operations of the Adem relations is the following Jacobi identity: let $x \in H_q(K)$, $y \in H_r(L)$, and $z \in H_s(K)$; then, under appropriate hypotheses,

$$(-1)^{(q+n)(s+n)} \lambda_n(x, \lambda_n(y, z)) + (-1)^{(r+n)(q+n)} \lambda_n(y, \lambda_n(z, x)) + (-1)^{(s+n)(r+n)} \lambda_n(z, \lambda_n(x, y)) = 0,$$

and, if $3 = 0$ in Λ and $q+n$ is odd, $\lambda_n(x, \lambda_n(x, x)) = 0$. We omit the proof as an easier geometric argument can be obtained for the homology of $(n+1)$-fold loop spaces. This identity, and the identity of (vi) of Proposition 6.3, lead to a notion of λ_n-algebra which generalizes that of Lie algebra (or λ_o-algebra). There is also a notion of restricted λ_n-algebra which is important for the applications. In the case $\Lambda = Z_2$, the restriction is already present in our algebraic context; it is the last Steenrod operation for an object $(K, \theta) \in \mathcal{C}(2, n)$. The following addendum to Proposition 2.3 gives some properties of this operation that are needed in the study of $(n+1)$-fold loop spaces.

<u>Proposition 6.5.</u> Let $(K, \theta) \in \mathcal{C}(2, n)$. Let $\xi_n = P_{q+n}: H_q(K) \longrightarrow H_{2q+n}(K)$. Then

(i) $\xi_n(x+y) = \xi_n(x) + \xi_n(y) + \lambda_n(x, y)$, and

(ii) $\beta\xi_n(x) = (q+n-1)P_{q+n-1}(x) + \lambda_n(x, \beta x)$ if (K, θ) is reduced mod 2

Proof. For (i), if a and b represent x and y, then, in K^2, $(a+b)^2 = a^2 + b^2 + \Delta_{n+1}ab$, and the error term $\Delta_{n+1}e_n \otimes a \otimes b$ yields the stated deviation from additivity of ξ_n. Part (ii) follows from a glance at the proof of (v) of Proposition 2.3.

We now relate the λ_n to the Bockstein operations on $H(K)$ when $(K, \theta) \in \mathcal{C}(\pi, n, Z_p)$ is reduced mod p. In contrast to the Steenrod operations, the higher Bocksteins are all of interest.

Proposition 6.7. Let $(K, \theta) \in \mathcal{C}(\pi, n, Z_p)$ be reduced mod p. Let $x, y \in H(K)$, $\deg(x) = q$. Assume that $\beta_r(x)$ and $\beta_r(y)$ are defined. Then $\beta_r\lambda_n(x, y)$ is defined and, modulo indeterminacy,

$$\beta_r\lambda_n(x, y) = \lambda_n(\beta_r x, y) + (-1)^{n+q}\lambda_n(x, \beta_r y).$$

Proof. Let $(K, \theta) = (\tilde{K} \otimes Z_p, \tilde{\theta} \otimes Z_p)$. Let $a, b \in K$ be such that their mod p reductions $\bar{a}, \bar{b} \in K$ represent x and y. We may assume that $d(a) = p^r a'$ and $d(b) = p^r b'$; the mod p reduction $\bar{a}'$ and $\bar{b}'$ of a' and b' represent $\beta_r(x)$ and $\beta_r(y)$. In $W^{(n)} \otimes \tilde{K}^2$, $d(\Delta_{n+1}e_n \otimes a \otimes b) = (-1)^n p^r \Delta_{n+1}e_n \otimes (a' \otimes b + (-1)^q a \otimes b')$. By reduction mod p and a check of signs, this implies the result.

Surprisingly, the following fundamental result appears not to be in the literature, although it is presumably well-known. It allows complete calculation of the mod p homology Bockstein spectral sequence of $QX = \varinjlim \Omega^n S^n X$ for any space X and, as we shall show later, the mod p cohomology Bockstein spectral sequence of $K(\pi, n)$. Together with the previous result, it also suffices for the computation of the mod p homology Bockstein spectral sequence of $\Omega^n S^n X$, $n \geq 1$.

Proposition 6.8. Let K be a Z-graded associative differential ring which is flat as a Z-module. Let K have a $\smile_1$-product such that

(a) $d(a \smile_1 b) = -d(a) \smile_1 b - (-1)^{\deg a} a \smile_1 d(b) + ab - (-1)^{\deg a \deg b} ba$

and, for case (ii), such that the Hirsch formula (b) holds

(b) $ab \cup_1 c = (-1)^{\deg a} a(b \cup_1 c) + (-1)^{\deg b \deg c}(a \cup_1 c)b$.

Let β_r denote the r-th mod p Bockstein on $H(K \otimes Z_p)$, $\beta_1 = \beta$. Let $y \in H_{2q}(K \otimes Z_p)$ and assume that $\beta_{r-1}(y)$ is defined, $r \geq 2$. Then $\beta_r(y^p)$ is defined and, modulo indeterminacy,

(i) If $p = 2$ and $r = 2$, $\beta_2(y^2) = \beta(y)y + P_{2q}\beta(y)$

(ii) If $p > 2$ and $r = 2$, $\beta_2(y^p) = \beta(y)y^{p-1} + \sum_{j=1}^{m} j\lambda_1(\beta(y)y^{j-1}, \beta(y)y^{p-j-1})$

(iii) If $p \geq 2$ and $r \geq 3$, $\beta_r(y^p) = \beta_{r-1}(y)y^{p-1}$.

Proof. Let $b \in K_{2q}$ be such that its mod p reduction $\bar{b}$ represents y. We may assume that $d(b) = p^{r-1}a$, and then a is a cycle whose mod p reduction $\bar{a}$ represents $\beta_{r-1}(y)$. Clearly we have

$$d(b^p) = p^{r-1}\sum_{i=1}^{p} b^{i-1}ab^{p-i}, \quad \text{and}$$

$$d(ab^{p-i} \cup_1 b^{i-1}) \equiv ab^{p-1} - b^{i-1}ab^{p-i} \mod p^{r-1}, \quad 2 \leq i \leq p.$$

Therefore $d(b^p + p^{r-1}\sum_{i=2}^{p} ab^{p-i} \cup_1 b^{i-1}) \equiv p^r ab^{p-1} \mod p^{2r-2}$.

If $r \geq 3$, then $2r-2 > r$ and part (iii) follows. Thus let $r = 2$; we must now take into account the terms arising from $d(b) = pa$ in $d(ab^{p-i} \cup_1 b^{i-1})$. If $p = 2$, then

$$d(b^2 + 2a \cup_1 b) = 4ab + 4a \cup_1 a.$$

Since the mod 2 reduction of $a \cup_1 a$ represents $P_{2q}\beta(y)$, this proves (i). Thus assume that $p > 2$. Then

$$d(b^p + p\sum_{i=2}^{p} ab^{p-i} \cup_1 b^{i-1}) = p^2 ab^{p-1} + p^2 c + p^2 c', \quad \text{where}$$

$$c = \sum_{i=2}^{p-1}\sum_{j=1}^{p-i} ab^{j-1}ab^{p-i-j} \cup_1 b^{i-1} \quad \text{and} \quad c' = \sum_{1 \leq j < i \leq p} ab^{p-i} \cup_1 b^{j-1}ab^{i-j-1}.$$

By the Hirsch formula, and a separate reindexing of the two resulting sums, we find that

$$c = \sum_{2 \leq j < i \leq p} [(ab^{i-j-1} \cup_1 b^{j-1})ab^{p-i} - ab^{p-i}(ab^{i-j-1} \cup_1 b^{j-1})].$$

Therefore if $e = \sum_{2 \leq j < i \leq p} ab^{p-i} \cup_1 (ab^{i-j-1} \cup_1 b^{j-1})$, then

$$d(e) \equiv -c + \sum_{2 \leq j < i \leq p} ab^{p-i} \cup_1 (ab^{i-2} - b^{j-1}ab^{i-j-1}) \mod p \,.$$

Comparing $d(e)$ to c', we easily find that

$$d(b^p + p\sum_{i=2}^{p} ab^{p-i} \cup_1 b^{i-1} + p^2 e) \equiv p^2 ab^{p-1} + p^2 \sum_{i=2}^{p} (i-1)ab^{p-i} \cup_1 ab^{i-2}$$

$$\equiv p^2 ab^{p-1} + p^2 \sum_{j=1}^{m} j(ab^{p-j-1} \cup_1 ab^{j-1} - ab^{j-1} \cup_1 ab^{p-j-1}) \mod p^3 \,.$$

By Definition 6.2, this implies (ii) and so completes the proof.

Of course, the terms involving λ_1 in (ii) are zero if K admits a $\cup_2$-product. The general result is needed for second loop spaces. The Hirsch formula is valid for the cochains of a space [8, 16], for the chains of a second loop space [10], and for the dual of the bar construction of a cocommutative Hopf algebra [16]. In connection with this formula, we make the following remarks.

<u>Remarks 6.9.</u> Let K be an associative differential Z_p-algebra, $p > 2$, with a $\cup_1$-product which satisfies the Hirsch formula. Define $< >^p: H_{2s-1}(K) \longrightarrow H_{2sp-2}(K)$ as follows. Let a_1 represent $x \in H_{2s-1}(K)$ and define $a_i = \frac{1}{i} a_{i-1} \cup_1 a_1$ for $2 \leq i < p$. Then $d(a_i) = \sum_{j=1}^{i-1} a_j a_{i-j}$ and $\tilde{a} = \sum_{j=1}^{p-1} a_j a_{p-j}$ is a cycle. A computation demonstrates that if $\{a_i' \mid 1 \leq i < p\}$ is any set of elements of K such that a_1' represents x and $d(a_i') = a_j' a_{i-j}'$ for $2 \leq i < p$, then $\tilde{a}' = \sum_{j=1}^{p-1} a_j' a_{p-j}'$ is homologous to $\tilde{a}$. Thus the class of $\tilde{a}$ depends only on x, and we define $<x>^p = \{\tilde{a}\}$. In the applications, it is often the case that if $(K, \theta) \in \mathcal{C}(p, p-2)$, then $<x>^p = -\beta P_s(x)$. Kraines [11] has proven this result for the cohomology of spaces, where it reads

$<x>^p = -\beta P^s(x)$ for $x \in H^{2s+1}(K)$, and Kochman [10] has proven it for the homology of iterated loop spaces. A general proof within our algebraic context should be possible, but appears to be difficult.

7. The category of simplicial Λ-modules

We here develop some machinery that will allow us to apply the theory of the previous sections to a large simplicial category $\mathcal{D}\mathcal{A}$. We shall specialize to specific categories of interest in the next section. We assume familiarity with the basic definitions of the theory of simplicial objects and of acyclic models (see, e.g., [15, §1,2,28,29]). Let Λ be a commutative ring, and let $\mathcal{A}$, $\mathcal{C}\mathcal{A}$ and $\mathcal{S}\mathcal{A}$ denote the categories of (ungraded) Λ-modules, positively graded Λ-complexes, and simplicial Λ-modules. Let $C: \mathcal{S}\mathcal{A} \longrightarrow \mathcal{C}\mathcal{A}$ be the normalized chain complex functor (for $K \in \mathcal{S}\mathcal{A}$, $C(K)$ is the quotient of K, regarded as a chain complex with $d = \sum (-1)^i d_i$, by the subcomplex generated by the degenerate simplices). Define $H_*(K) = H(C(K))$ and $H^*(K) = H(C^*(K))$, where $C^*(K) = \mathrm{Hom}_\Lambda(C(K), \Lambda)$ is given the differential $\delta(f)(k) = (-1)^{q+1} f(dk)$ for $f \in C^q(K)$ and $k \in C_{q+1}(K)$. The following key lemma is based on ideas of Dold [5].

Lemma 7.1. Let π be a subgroup of Σ_r and let W be a $\Lambda\pi$-free resolution of Λ such that $W_0 = \Lambda\pi$ with $\Lambda\pi$-generator e_0. Let $K_1, \ldots, K_r \in \mathcal{S}\mathcal{A}$; then there exists a morphism of Λ-complexes

$$\Phi: W \otimes C(K_1 \times \ldots \times K_r) \longrightarrow W \otimes C(K_1) \otimes \ldots \otimes C(K_r)$$

which is natural in the K_i and satisfies the following properties:

(i) For $\sigma \in \pi$, the following diagram is commutative:

$$\begin{array}{ccc} W \otimes C(K_1 \times \ldots \times K_r) & \xrightarrow{\Phi} & W \otimes C(K_1) \otimes \ldots \otimes C(K_r) \\ \downarrow \sigma & & \downarrow \sigma \\ W \otimes C(K_{\sigma(1)} \times \ldots \times K_{\sigma(r)}) & \xrightarrow{\Phi} & W \otimes C(K_{\sigma(1)}) \otimes \ldots \otimes C(K_{\sigma(r)}). \end{array}$$

(ii) Φ is the identity homomorphism on $W \otimes C_o(K_1 \times \ldots \times K_r)$.

(iii) $\Phi(e_o \otimes k_1 \otimes \ldots \otimes k_r) = e_o \otimes \xi(k_1 \otimes \ldots \otimes k_r)$ if $k_i \in K_i$ is a j-simplex, where $\xi : C(K_1 \times \ldots \times K_r) \to C(K_1) \otimes \ldots \otimes C(K_r)$ is the Alexander-Whitney map.

(iv) $\Phi(W \otimes C_j(K_1 \times \ldots \times K_r)) \subset \sum_{k \leq rj} W \otimes [C(K_1) \otimes \ldots \otimes C(K_r)]_k$.

Moreover, any two such Φ are naturally equivariantly homotopic

Proof. Since $(K \times L)_j = K_j \otimes L_j$, formulas (ii) and (iii) make sense. Write $A_j = C_j(K_1 \times \ldots \times K_r)$ and $B_j = [C(K_1) \otimes \ldots \otimes C(K_r)]_j$. We construct Φ on $W_i \otimes A_j$ by induction on i and for fixed i by induction on j. Formula (ii) defines Φ for $j = 0$ and all i and formulas (i) and (iii) define Φ for $i = 0$ and all j. Thus let $i \geq 1$ and $j \geq 1$ and assume that Φ is defined for $i' < i$ and for the given i and $j' < j$. Choose a $\Lambda\pi$-basis $\{w_k\}$ for W_i. It suffices to define Φ on $w \otimes x$ for $w \in \{w_k\}$ and $x \in A_j$, since Φ can then be uniquely extended to all of $W_i \otimes A_j$ by (i). Let $\Lambda\Delta[j]$ denote the free simplicial Λ-module generated by the standard simplicial j-simplex [15, p. 14]. Then the functor $w \otimes A_j$ is represented by the r-fold Cartesian product $\Lambda\Delta[j]^r$, and $W \otimes B(\Lambda\Delta[j]^r)$ is acyclic. Therefore $\Phi(w \otimes \Delta_j \otimes \ldots \otimes \Delta_j)$ can be defined by choosing a chain whose boundary is $\Phi d(w \otimes \Delta_j \otimes \ldots \otimes \Delta_j)$, and Φ can be carried over to arbitrary $w \otimes k_1 \otimes \ldots \otimes k_r$ by representability. Now (i), (ii), and (iii) are clearly satisfied and (iv) follows from the fact that $C_k(\Lambda\Delta[j]) = 0$ for $k > j$. The proof that Φ is unique up to natural equivariant homotopy is equally simple.

Remarks 7.2. Define $\Psi: W \otimes C(K_1) \otimes \ldots \otimes C(K_r) \longrightarrow W \otimes C(K_1 \times \ldots \times K_r)$ by $\Psi = 1 \otimes \eta$ where $\eta: C(K_1) \otimes \ldots \otimes C(K_r) \longrightarrow C(K_1 \times \ldots \times K_r)$ is the shuffle map. Since η, unlike ξ, is commutative, Ψ is equivariant. By an easy acyclic models proof, $\Phi\Psi$ and $\Psi\Phi$ are equivariantly homotopic to the respective identity maps.

We shall only be interested in the case $K_1 = \ldots = K_r$; here $\Phi: W \otimes C(K^r) \longrightarrow W \otimes C(K)^r$ is a natural morphism of $\Lambda\pi$-complexes. The general case was required in the proof in order to have the representability of the functors A_j. Starting with objects of the following category $\mathcal{DA}$, we shall use Φ to obtain diagonal approximations and so to pass to the category $\mathcal{P}(\pi, \infty, \Lambda) \subset \mathcal{C}(\pi, \infty, \Lambda)$ defined in Definitions 2.1.

Definitions 7.3. Let $\mathcal{DA}$ denote the following category. The objects of $\mathcal{DA}$ are pairs (K, D) where $K \in \mathcal{SA}$ and $D: K \longrightarrow K \times K$ is a morphism in $\mathcal{SA}$ such that $(D \times 1) = (1 \times D)D$ and $tD = D$, where $t(x \otimes y) = y \otimes x$. The morphisms $f: (K, D) \longrightarrow (K', D')$ in $\mathcal{DA}$ are those morphisms $f: K \longrightarrow K'$ in $\mathcal{SA}$ such that $(f \times f)D = D'f$.

Each $K \in \mathcal{SA}$ admits the natural diagonal $D(k) = k \otimes k$, and $\mathcal{SA}$ is thereby embedded as a full subcategory of $\mathcal{DA}$. However, an object $K \in \mathcal{SA}$ may admit other interesting diagonals. For example, if K is a simplicial cocommutative coassociative Λ-coalgebra, then the coproduct $\psi: K \longrightarrow K \times K$ is a permissible diagonal; that is, $(K, \psi) \in \mathcal{DA}$. The following remarks will be of use in the study of relative and reduced cohomology.

Remarks 7.4. (i) If $L \subset K$ in $\mathcal{SA}$, define $H_*(K, L) = H(C(K/L))$ and $H^*(K, L) = H(C^*(K/L))$. If $(K, D) \in \mathcal{DA}$ and $D(L)$ is contained in $L \times K + K \times L$, then K/L admits the diagonal $\overline{D}$ induced from the composite $K \xrightarrow{D} K \times K \xrightarrow{\pi \times \pi} K/L \times K/L$, where $\pi: K \longrightarrow K/L$ is the projection, and then π is a morphism in $\mathcal{DA}$.

(ii) Let $\tilde{\Lambda} = \Lambda\Delta[0] \in \mathcal{S}\mathcal{a}$; thus $\tilde{\Lambda}_n = \Lambda$ for $n \geq 0$, each d_i and s_i is the identity, and $C(\tilde{\Lambda}) = C_o(\tilde{\Lambda}) = \Lambda$. Give $\tilde{\Lambda}$ the natural diagonal. We say that $(K, D) \in \mathcal{D}\mathcal{a}$ is unital if we are given a monomorphism $\nu : \tilde{\Lambda} \longrightarrow K$ in $\mathcal{D}\mathcal{a}$ and an epimorphism $\varepsilon : K \longrightarrow \tilde{\Lambda}$ in $\mathcal{S}\mathcal{a}$ such that $\varepsilon\nu = 1$ and $(\varepsilon \times 1)D = (1 \times \varepsilon)D$ (where $\tilde{\Lambda} \times K = K = K \times \tilde{\Lambda}$). If (K, D) is unital and $IK = \operatorname{Ker} \varepsilon$, then $K = \nu(\tilde{\Lambda}) \oplus IK$ and for $k \in IK$, $D(k) = k \otimes \nu(1) + \nu(1) \otimes k + \bar{D}(k)$, where $\bar{D}(k) \in IK \times IK$. Clearly, $(IK, \bar{D})$ is isomorphic to $(K/\nu(\tilde{\Lambda}), \bar{D})$ in $\mathcal{D}\mathcal{a}$.

If $(K, D) \in \mathcal{D}\mathcal{a}$, then $C^*(K)$ is an associative differential Λ-algebra, with cup product defined as the composite

(1) $\quad \cup : C^*(K) \otimes C^*(K) \xrightarrow{\alpha} [C(K) \otimes C(K)]^* \xrightarrow{\xi^*} C^*(K \times K) \xrightarrow{D^*} C^*(K).$

Here α is the natural map, $\alpha(x \otimes y)(k \otimes \ell) = (-1)^{\deg y \deg k} x(k)y(\ell)$, and ξ is the Alexander-Whitney map. If (K, D) is unital, then $C^*(K)$ is unital (via ν^*) and augmented (via ε^*).

We now define a functor $\Gamma : \mathcal{D}\mathcal{a} \longrightarrow \mathcal{P}(\pi, \infty, \Lambda)$ and then show how to use Γ to apply our general theory to $H^*(K)$ for $(K, D) \in \mathcal{D}\mathcal{a}$ in the case $\Lambda = Z_p$.

<u>Definitions 7.5.</u> Let $(K, D) \in \mathcal{D}\mathcal{a}$ and write D for the iterated diagonal $K \longrightarrow K^r$. Let $\pi \subset \Sigma_r$ and let W be a $\Lambda\pi$-free resolution of Λ with $W_o = \Lambda\pi$. Define $\Delta : W \otimes C(K) \longrightarrow C(K)^r$ to be the composite

(2) $\quad \Delta : W \otimes C(K) \xrightarrow{1 \otimes D} W \otimes C(K^r) \xrightarrow{\Phi} W \otimes C(K)^r \xrightarrow{\varepsilon \otimes 1} C(K)^r$

Let $\alpha : C^*(K)^r \longrightarrow [C(K)^r]^*$ be the natural map and define a $\Lambda\pi$-morphism $\theta : W \otimes C^*(K)^r \longrightarrow C^*(K)$ by the formula

(3) $\quad \theta(w \otimes x)(k) = (-1)^{\deg w \deg x} \alpha(x)(\Delta(w \otimes k))$, $w \in W$, $x \in C^*(K)^r$, $k \in C(K)$.

Since θ may be defined for $\pi = \Sigma_r$ and then factored through $j : W \longrightarrow V$ as in Definition 2.1, and the resulting composite is $\Lambda\pi$-homotopic to the original θ defined in terms of W, θ satisfies condition (ii) of Definition 2.1. By Lemma 7.1, formula (3) specializes to give

(4) $\quad \theta(e_o \otimes x) = D^* \xi^* \alpha(x)$ for any $x \in C^*(K)^r$, and

(5) $\quad \theta(w \otimes x) = \mathcal{E}(w)D^{*}\alpha(x)$ if $x \in C^{0}(K)^{r}$ and $w \in W$.

By (1) and (4), θ satisfies condition (i) of Definition 2.1. Since θ is natural on morphisms in $\mathcal{DA}$, we thus obtain a contravariant functor $\Gamma\colon \mathcal{DA} \to \mathcal{P}(\pi, \infty, \Lambda)$ by letting $\Gamma(K, D) = (C^{*}(K), \theta)$ on objects and $\Gamma(f) = C^{*}(f)$ on morphisms. By (5), if (K, D) is unital in $\mathcal{DA}$ then $\Gamma(K, D)$ is unital in $\mathcal{P}(\pi, \infty, \Lambda)$. If $\Lambda = Z_p$, π is cyclic of order p, and $(K, D) = (\tilde{K} \otimes Z_p, \tilde{D} \otimes Z_p)$ where $\tilde{K}$ is a Z-free simplicial Z-module, we agree to choose θ for K to be the mod p reduction of θ for $\tilde{K}$; then $\Gamma(K, D)$ is reduced mod p (since $C(\tilde{K})$ is Z-free and therefore $C^{*}(\tilde{K})$ is Z-flat, as required by Definition 2.1).

Observe that, by Definition 6.1, we now have $\cup_i$-products in $C^{*}(K)$ for any $(K, D) \in \mathcal{DA}$. When $\Lambda = Z_p$, the results of Proposition 2.3 will clearly apply to the Steenrod operations P^{s} defined on the cohomology of objects $(K, D) \in \mathcal{DA}$. If (K, D) and (L, D') are objects of $\mathcal{DA}$, then $K \times L$ admits the diagonal $\tilde{D} = (1 \times t \times 1)(D \times D')$; if D and D' are the natural diagonals, then so is $\tilde{D}$. Thus $(C^{*}(K \times L), \theta)$ is defined in $\mathcal{C}(\pi, \infty, \Lambda)$. The following lemma compares $(C^{*}(K \times L), \theta)$ to $(C^{*}(K) \otimes C^{*}(L), \tilde{\theta})$ and will imply the applicability of the external Cartan formula to $H^{*}(K \times L)$ when $\Lambda = Z_p$.

<u>Lemma 7.6.</u> For any objects (K, D) and (L, D') in $\mathcal{DA}$, the following diagram is $\Lambda\pi$-homotopy commutative

$$\begin{array}{ccc}
W \otimes C^{*}(K \times L)^{r} & \xrightarrow{\quad\theta\quad} & C^{*}(K \times L) \\
{\scriptstyle 1 \otimes (\xi^{*})^{r}} \Big\uparrow \Big\downarrow {\scriptstyle 1 \otimes (\eta^{*})^{r}} & & {\scriptstyle \xi^{*}} \Big\uparrow \Big\downarrow {\scriptstyle \eta^{*}} \\
W \otimes [C^{*}(K) \otimes C^{*}(L)]^{r} & \xrightarrow{\quad\tilde{\theta}\quad} & C^{*}(K) \otimes C^{*}(L)
\end{array}$$

That is, η^{*} and ξ^{*} are morphisms in the category $\mathcal{C}(\pi, \infty, \Lambda)$.

<u>Proof.</u> By the definitions of θ and $\tilde{\theta} = (\theta \otimes \theta)(1 \otimes T \otimes 1)(\psi \otimes U)$, it suffices to show that the following diagram is $\Lambda\pi$-homotopy commutative:

$$\begin{array}{ccc}
W \otimes C(K \times L) & \xrightarrow{\Delta = (\varepsilon \otimes 1)\Phi(1 \otimes \tilde{D})} & C(K \times L)^r \\
1\otimes\eta \uparrow \; \downarrow 1 \otimes \xi & & \eta^r \uparrow \; \downarrow \xi^r \\
W \otimes C(K) \otimes C(L) & \xrightarrow{U(\Delta \otimes \Delta)(1 \otimes T \otimes 1)(\psi \otimes 1 \otimes 1)} & [C(K) \otimes C(L)]^r
\end{array}$$

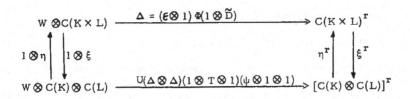

Since $(1 \otimes D \otimes 1 \otimes D')(1 \otimes T \otimes 1)(\psi \otimes 1 \otimes 1) = (1 \otimes T \otimes 1)(\psi \otimes 1 \otimes 1)(1 \otimes D \otimes D')$, if we let $\phi = (\varepsilon \otimes 1)\Phi$ and let $u: K^r \times L^r \to (K \times L)$ be the evident shuffle, so that $\tilde{D} = u(D \times D')$, then this diagram becomes

$$\begin{array}{ccccccc}
W\otimes C(K\times L) & \xrightarrow{1 \otimes D \times D'} & W \otimes C(K^r \times L^r) & \xrightarrow{1 \otimes u} & W \otimes C([K \times L]^r) & \xrightarrow{\phi} & C(K \times L)^r \\
1\otimes\eta \uparrow \; \downarrow 1 \otimes \xi & & 1\otimes\eta \uparrow \; \downarrow 1 \otimes \xi & & & & \eta^r \uparrow \; \downarrow \xi^r \\
W\otimes C(K)\otimes C(L) & \xrightarrow{1\otimes D \otimes D'} & W\otimes C(K^r)\otimes C(L^r) & & \xrightarrow{U(\phi\otimes\phi)(1\otimes T\otimes 1)(\psi\otimes 1\otimes 1)} & & [C(K)\otimes C(L\ )]^r
\end{array}$$

The left-hand square commutes by the naturality of η and ξ. Since the diagonals are not involved inthe right-hand square, we can prove that it commutes up to $\Lambda\pi$-homotopy by an acyclic models argument, with K^r and L^r replaced by $K_1 \times \ldots \times K_r$ and $L_1 \times \ldots \times L_r$ so as to have domains given by representable functors for fixed $w \in W$. On zero simplices, the diagram commutes for any $w \in W$ and on $e_o \in W$, as the simplices vary, the diagram is Λ-homotopy commutative by a standard acyclic models argument. This starts the inductive construction of the desired homotopies, and the proof is completed precisely as was the proof of Lemma 7.1.

Corollary 7.7. If $(K, D) \in \mathcal{DA}$, then $\Gamma(K, D)$ is a Cartan object of $\mathcal{C}(\pi, \infty, \Lambda)$.

Proof. Since $D: K \to K \times K$ is commutative and associative, it is a morphism in $\mathcal{DA}$. Therefore the cup product (1) is a morphism in $\mathcal{C}(\pi, \infty, \Lambda)$, as required.

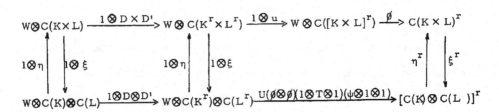

Lemma 7.8. If $(K,D) \in \mathcal{PA}$, $\Lambda = Z_p$, then $\Gamma(K,D)$ is an Adem object of $\mathcal{C}(p,\infty)$.

Proof. In the notations of Definition 4.1 (with $Y_o = Z_p\Sigma_{p^2}$), it suffices to prove that the following diagram is τ-homotopy commutative:

$$\begin{array}{ccccc}
W_1 \otimes W_2{}^p \otimes C^*(K)^{p^2} & \xrightarrow{w\otimes 1} & Y \otimes C^*(K)^{p^2} & \overset{\theta}{\searrow} & \\
\downarrow{\scriptstyle 1\otimes U} & & & & C^*(K) \\
W_1 \otimes (W_2 \otimes C^*(K)^p)^p & \xrightarrow{1\otimes \theta^p} & W_1 \otimes C^*(K)^p & \overset{\theta}{\nearrow} &
\end{array}$$

All maps θ are as defined in Definition 7.5; by dualization, it suffices to prove that the following diagram is τ-homotopy commutative

$$\begin{array}{ccccc}
W_1 \otimes W_2{}^p \otimes C(K) & \xrightarrow{w\otimes 1} & Y \otimes C(K) & \xrightarrow{\Delta} & C(K)^{p^2} \\
\downarrow{\scriptstyle T\otimes 1} & & & & \uparrow{\scriptstyle \Delta^p} \\
W_2{}^p \otimes W_1 \otimes C(K) & \xrightarrow{1\otimes\Delta} & W_2{}^p \otimes C(K)^p & \xrightarrow{U} & (W_2 \otimes C(K))^p
\end{array}$$

Let $\phi = (\varepsilon \otimes 1)\Phi$ and define $\alpha = \phi(w\otimes 1): W_1 \otimes W_2{}^p \otimes C(K^{p^2}) \longrightarrow C(K)^{p^2}$. Since $\Delta = \phi(1\otimes D)$, $\Delta(w\otimes 1) = \alpha(1\otimes 1\otimes D)$, $D: C(K) \longrightarrow C(K^{p^2})$. By the naturality of ϕ, the following diagram is commutative:

$$\begin{array}{ccccccc}
W_1 \otimes W_2{}^p \otimes C(K) & \xrightarrow{T\otimes D} & W_2{}^p \otimes W_1 \otimes C(K^p) & \xrightarrow{1\otimes\phi} & W_2{}^p \otimes C(K)^p & \xrightarrow{U} & (W_2 \otimes C(K))^p \\
\downarrow{\scriptstyle 1\otimes 1\otimes D} & & \downarrow{\scriptstyle 1\otimes 1\otimes D} & & \downarrow{\scriptstyle 1\otimes D^p} & & \downarrow{\scriptstyle (1\otimes D)^p} \\
W_1 \otimes W_2{}^p \otimes C(K^{p^2}) & \xrightarrow{T\otimes 1} & W_2{}^p \otimes W_1 \otimes C(K^{p^2}) & \xrightarrow{1\otimes\phi} & W_2{}^p \otimes C(K^p)^p & \xrightarrow{U} & (W_2 \otimes C(K^p))^p
\end{array}$$

Let $\beta = \phi^p U(1\times\phi)(T\times 1): W_1 \times W_2{}^p \times C(K^{p^2}) \longrightarrow C(K)^{p^2}$. By the diagram above, $\Delta^p U(1\times\Delta)(T\times 1) = \beta(1\times 1\times D)$. Thus it suffices to prove that α and β are τ-homotopic. Since α and β do not involve the diagonal, this can easily be shown by acyclic models precisely as in our previous proofs.

The following theorem summarizes properties of the Steenrod operations that are valid for arbitrary objects of the category $\mathcal{DA}$, $\Lambda = Z_p$. Of course, we use the notations of section 5 since we are dealing with cohomology.

Theorem 7.9. Let $(K,D) \in \mathcal{DA}$, $\Lambda = Z_p$. Then there exist natural homomorphisms P^s and, if $p > 2$, βP^s defined on each $H^q(K)$; degree $(P^s) = s$ if $p = 2$ and $\deg(\beta^\varepsilon P^s) = 2s(p-1)+\varepsilon$, $\varepsilon = 0$ or 1, if $p > 2$. These cohomology operations on $\mathcal{DA}$ satisfy the properties

(i) $\beta^\varepsilon P^s = 0$ if $s < 0$ or if $p = 2$ $(\varepsilon = 0)$ and $s > q$ or if $p > 2$ and $2s+\varepsilon > q$.

(ii) $P^s(x) = x^2$ if $p = 2$ and $s = q$; $P^s(x) = x^p$ if $p > 2$ and $2s = q$

(iii) If $(K,D) = (\tilde{K} \otimes Z_p, \tilde{D} \otimes Z_p)$, where $\tilde{K}$ is Z-free, then $\beta P^{s-1} = sP^s$ if $p = 2$ and βP^s is the composition of β and P^s if $p > 2$.

(iv) $P^s = \sum P^i \otimes P^{s-i}$ and $\beta P^s = \sum(\beta P^i \otimes P^{s-i} + P^i \otimes \beta P^{s-i})$ on $H^*(K \times L)$; the internal Cartan formula is satisfied in $H^*(K)$

(v) If $f: K' \to K$ and $g: K \to K''$ are morphisms in $\mathcal{DA}$ such that $gf = 0$, then $\sigma\beta^\varepsilon P^s = (-1)^\varepsilon \beta^\varepsilon P^s \sigma$, where $\sigma: H^q(K'') \to H^{q-1}(K')$ is the suspension associated with $C^*(K'') \to C^*(K) \to C^*(K')$.

(vi) If $L \subset K$ and $D(L) \subset L \times L$, then $\delta\beta^\varepsilon P^s = (-1)^\varepsilon \beta^\varepsilon P^s \delta$ where $\delta: H^q(L) \to H^{q+1}(K, L)$ is the connecting homomorphism.

(vii) The $\beta^\varepsilon P^s$ satisfy the Adem relations as stated in Corollary 5.1.

Proof. For (i), we must prove that $\beta^\varepsilon P^s = 0$ for $s < 0$ (the rest is the convention $e_i = 0$ for $i < 0$); by formulas (5.1) and (5.2), it suffices to show that $D_i(x) = 0$ for $i > (p-1)q$, $\deg(x) = q$. By (3) of Definition 7.5, it suffices to show that $\Delta(e_i \otimes k) = 0$ for $k \in C_{pq-i}(K)$. Now $\Delta = (\varepsilon \otimes 1)\Phi(1 \otimes D)$ and, by (iv) of Lemma 7.1, if $i > (p-1)q$, then

$$\Phi(e_i \otimes D(k)) \in \sum_{j<pq} W_{pq-j} \otimes [C(K)]^p_j \subset \operatorname{Ker}(\varepsilon \otimes 1).$$

(ii) and (iii) follow from Proposition 2.3; (iv) follows from Corollary 2.7, Lemma 7.6, and Corollary 7.7; (v) and (vi) follow from Theorem 3.3, noting for

(vi) that the suspension associated with $C^*(K/L) \to C^*(K) \to C^*(L)$ is the inverse additive relation to the connecting homomorphism δ; (vii) follows from Theorem 4.7 and Lemma 7.8.

By Theorem 3.4, the Kudo transgression theorem applies to appropriate spectral sequences involving objects of $\mathcal{DA}$ and, under the hypotheses of (iii) of the theorem, Proposition 6.8 applies to compute the higher Bocksteins on p-th powers of elements of $H^*(K)$, $(K, D) \in \mathcal{DA}$. In the next section, we shall show how to compoute P^0 for arbitrary objects $(K, D) \in \mathcal{DA}$ and shall give non-trivial examples to show that $P^0 \neq 1$ in general.

8. Simplicial sets and simplicial restricted Lie algebras

We shall here obtain the Steenrod operations on the cohomology of topological spaces, simplicial sets, and simplicial restricted Lie algebras, and shall consider the evaluation of P^0 on $H^*(K)$ for any $(K, D) \in \mathcal{DA}$, $\Lambda = Z_p$.

Let $\mathcal{S}$ denote the category of simplicial sets. For $K \in \mathcal{S}$, let $\tilde{K}$ denote the free simplicial Abelian group generated by K. Let Λ be a commutative ring and define a functor $A: \mathcal{S} \to \mathcal{DA}$ by letting $A(K) = \tilde{K} \otimes \Lambda$ with its natural diagonal D; here $\mathcal{DA}$ is as defined in Definition 7.3 and D is induced from the diagonal $k \to (k, k)$ on K. Composing A with Γ of Definition 7.5, we obtain a functor $\Gamma A: \mathcal{S} \to \mathcal{P}(\pi, \infty, \Lambda)$ for any $\pi \subset \Sigma_r$. Let $\mathcal{T}$ denote the category of topological spaces and let $S: \mathcal{T} \to \mathcal{S}$ be the total singular complex functor. Then $\Gamma AS: \mathcal{T} \to \mathcal{P}(\pi, \infty, \Lambda)$ is defined. If (K, L) is a simplicial pair, define $A(K, L) = \tilde{K}/\tilde{L} \otimes \Lambda$. Then ΓA is defined on the category $\mathcal{S}_2$ of simplicial pairs and ΓAS is defined on the catetogy $\mathcal{T}_2$ of topological pairs. Since the normalized cochains with coefficients in Λ of a simplicial pair (K, L) and of a topological pair (X, Y) may be defined as $C^*(K, L) = C^*(\tilde{K}/\tilde{L} \otimes \Lambda)$ and $C^*(X, Y) = C^*(SX, SY)$, the results of the previous section apply to the cohomology of simplicial and topological pairs.

For the remainder of this section, we take $\Lambda = Z_p$ and we let π be cyclic of order p. Via ΓA: $\mathcal{S}_2 \to \mathcal{P}(p, \infty)$, we have Steenrod operations P^s, $s \geq 0$, on $H^*(K, L)$ for all $(K, L) \in \mathcal{S}_2$, hence on $H^*(X, Y)$ for all $(X, Y) \in \mathcal{T}_2$. Of course, if $p = 2$, P^s is usually denoted by Sq^s. Theorem 7.9 gives all of the standard properties of the P^s except $P^0 = 1$. We now show that $P^0 = 1$ follows from the previously obtained properties of the P^s.

Proposition 8.1. P^0 is the identity operation and, if $p = 2$, P^1 is the Bockstein operation on the cohomology of simplicial (or topological) pairs.

Proof. Since $\beta P^0 = P^1$ if $p = 2$, it suffices to prove that $P^0 = 1$. If $(K, L) \in \mathcal{S}_2$, L non-empty, then $H^*(K, L) = H^*(K/L, P)$, where P is a point complex. Thus it suffices to prove that $P^0(x) = x$ for $x \in \tilde{H}^n(K) = H^n(K, P)$, since the result for L empty will follow trivially. If $K^{(n)}$ is the n-skeleton of K, then $\tilde{H}^n(K) \to \tilde{H}^n(K^{(n)})$ is a monomorphism, and we may thus assume that $K = K^{(n)}$. Then, by the Hopf Theorem [24, p. 431], there exists $f: K \to S^n$ such that $f^*(i_n^*) = x$, where $i_n^* \in \tilde{H}^n(S^n)$ is the fundamental class of the simplicial n-sphere. It therefore suffices to prove that $P^0(i_n^*) = i_n^*$. Now for any K, the suspension isomorphism $S^*: H^{q+1}(SK) \longrightarrow \tilde{H}^q(K)$ may be defined as the composite $H^{q+1}(SK) \to H^{q+1}(CK, K) \longrightarrow \tilde{H}^q(K)$, where CK is the simplicial cone of K, hence S^* commutes with the P^s. Since $S^*(i_n^*) = i_{n-1}^*$ for $n \geq 1$ and $P^0(i_0^*) = (i_0^*)^p = i_0^*$ (where i_0^* generates $\tilde{H}(S^0) = Z_p$), this proves the result.

We now use the fact that $P^0 = 1$ on the cohomology of simplicial sets to show how to compute P^0 on $H^*(K)$ for any object $(K, D) \in \mathcal{DA}$. In fact, we have the following addendum to Lemma 7.1 when W is the canonical $Z_p\pi$-free resolution of Z_p.

Lemma 8.2. Let $K_i \in \mathcal{A}$, $1 \leq i \leq p$, and let k_i be a q-simplex of K_i. Then, for any $\Phi: W \otimes C(K_1 \times \ldots \times K_p) \to W \otimes C(K_1) \otimes \ldots \otimes C(K_p)$ which satisfies

satisfies the conclusions of Lemma 7.1,

$$(\varepsilon\otimes 1)\Phi(e_{q(p-1)}\otimes k_1\otimes\ldots\otimes k_p) = (-1)^{mq}\nu(-q)^{-1}k_1\otimes\ldots\otimes k_p\,,$$

where $\nu(-q) = 1$ if $p = 2$ and $\nu(-2j+\varepsilon) = (-1)^j(m!)^{\varepsilon}$, $\varepsilon = 0$ or 1, if $p > 2$.

Proof. Let Δ_q be the fundamental q-simplex of $\Lambda\Delta[q]$. Since Δ_q is a Z_p-basis for $C_q(\Lambda\Delta[q])$, we clearly have that

(i) $\Phi(e_{q(p-1)}\otimes\Delta_q\otimes\ldots\otimes\Delta_q) \equiv \gamma e_o\otimes\Delta_q\otimes\ldots\otimes\Delta_q \mod \mathrm{Ker}(\varepsilon\otimes 1)$, $\gamma\in Z_p\pi$.

By the naturality of Φ (or by the proof of Lemma 7.1), (i) implies

(ii) $(\varepsilon\times 1)\Phi(e_{q(p-1)}\otimes k_1\otimes\ldots\otimes k_p) = \varepsilon(\gamma)k_1\otimes\ldots\otimes k_p$ for any $k_i\in(K_i)_q$.

To evaluate $\varepsilon(\gamma)$, let $i_q\in C_q(S^q)$ represent the fundamental class of $H_q(S^q)$; we may take $S^q = \Delta[q]/\dot{\Delta}[q]$ so that i_q is a basis for $C_q(S^q)$ and $i_q^*\in C^q(S^q)$ is well-defined. By (ii) and $D(i_q) = i_q\otimes\ldots\otimes i_q$, Definition 7.5 gives

(iii) $\theta(e_{q(p-1)}\otimes i_q^{*p})(i_q) = \alpha(i_q^{*p})[(\varepsilon\otimes 1)\Phi(e_{q(p-1)}\otimes i_q^p)] = (-1)^{mq}\varepsilon(\gamma)$.

Since $P^o\{i_q^*\} = \nu(-q)D_{q(p-1)}\{i_q^*\} = i_q^*$, $\nu(-q)\theta(e_{q(p-1)}\otimes(i_q^{*p}) = i_q^*$. Thus $(-1)^{mq}\nu(-q)\varepsilon(\gamma) = 1$ and the result is proven.

Corollary 8.3. Let $(K,D)\in\mathcal{DA}$. Write $D(k) = \sum k^{(1)}\otimes\ldots\otimes k^{(p)}\in C_q(K^p)$ for $k\in C_q(K)$, and regard each $k^{(i)}$ as an element of $C_q(K)$. Let $x\in C^q(K)$ be a cocycle. Then $P^o\{x\}$ is represented by that cocycle $y\in C^q(K)$ such that $y(k) = \sum x(k^{(1)})\cdots x(k^{(p)})\in Z_p$ for each $k\in C_q(K)$. In particular, if $D(k) = N\ell\in K^p$ for each $k\in K$, where $N = \sum\alpha^i\in Z_p\pi$, then $P^o = 0$ on $H^*(K)$.

Proof. By formulas (5.1) and (5.2), $y = \nu(-q)\theta(e_{q(p-1)}\otimes x^p)$ represents $P^o\{x\}$, and the result follows by an easy computation from Definition 7.5 and the lemma.

We now give a useful application of the theory for which the Steenrod operations satisfy the results of Theorem 7.9 and $P^o = 0$. Let $\mathcal{L}$ and $\mathcal{H}$ denote the categories of restricted Lie algebras and of primitively generated Hopf algebras over Z_p. Let $F: \mathcal{A} \rightarrow \mathcal{L}$ denote the free restricted Lie algebra functor, let $V: \mathcal{L} \longrightarrow \mathcal{H}$ denote the universal enveloping algebra functor, and let $P: \mathcal{H} \rightarrow \mathcal{L}$ denote the functor which assigns to $H \in \mathcal{H}$ its restricted Lie algebra of primitive elements. By a result of Milnor and Moore [19, Theorem 6.11], $PV(L) = L$ for $L \in \mathcal{L}$ and $VPH = H$ for $H \in \mathcal{H}$. By Theorems of Witt and Friedrich [9, Theorems 7 and 9, p.168-170], extended to restricted Lie algebras, if $K \in \mathcal{A}$ and $T(K)$ is the tensor algebra of K, then $V(FK) = T(K)$ in $\mathcal{H}$ and $FK = PT(K)$ in $\mathcal{L}$. These statements clearly remain valid for the categories $s\mathcal{A}$, $s\mathcal{L}$ and $s\mathcal{H}$ of simplicial objects in $\mathcal{A}$, $\mathcal{L}$, and $\mathcal{H}$ [see 15, Definition 2.1]. We shall need the following algebraic lemma.

Lemma 8.4. Let L be a restricted Lie algebra and let $IV(L) = \operatorname{Ker} \varepsilon$, $\varepsilon : V(L) \rightarrow Z_p$, be its augmentation ideal. Let $\psi: V(L) \longrightarrow V(L)^p = V(L) \otimes \dots \otimes V(L)$ denote the iterated coproduct. Then, for each $x \in IV(L)$, there exists $y \in V(L)^p$ such that $\psi(x) = Ny$.

Proof. Let $\pi: FK \rightarrow L$ represent L as a quotient of a free restricted Lie algebra. Then $V(\pi) = V(FK) \rightarrow V(L)$ is an epimorphism of Hopf algebras, and we may assume that $L = FK$. Clearly we may also assume that K is a finite dimensional Z_p-module. Since $T(K)$ admits a grading under which it is connected, [19, Proposition 4.20] implies that the p-th power operation ξ is zero on the augmentation ideal of the dual Hopf algebra $T(K)^*$. The cocommutativity of $T(K)$ implies that, for $x \in IT(K)$, $\psi(x)$ can be written in the form
$\psi(x) = Ny + \sum z_i \otimes \dots \otimes z_i$ in $T(K)^p$. By the triviality of ξ on $IT(K)^*$, each $z_i = 0$ and the result follows.

We now sketch a definitional framework for the study of homotopy invariants of simplicial restricted Lie algebras. Define a category $s\mathcal{L}_2$ as follows. The objects of $s\mathcal{L}_2$ are pairs (L, M) such that $L \in s\mathcal{L}$ and M is a restricted Lie ideal of L and the morphisms $f: (L, M) \rightarrow (L', M')$ in $s\mathcal{L}_2$ are morphisms

$f: L \to L'$ in $\mathcal{SL}$ such that $f(M) \subset M'$. Two such morphisms, f and g, are said to be Lie homotopic if there exist morphisms of restricted Lie algebras $h_i: L_q \to L'_{q+1}$, $0 \le i \le q$, such that $h_i(M_q) \subset M'_{q+1}$ and the identities (i) - (iii) of [15, Definition 5.1] are satisfied. Define the homotopy, homology, and cohomology groups of $(L, M) \in \mathcal{SL}_2$ by

(1) $\pi_*(L, M) = H_*(L/M)$ and

(2) $H_*(L, M) = H_*(IV(L/M))$ and $H^*(L, M) = H^*(IV(L/M))$

The homology and cohomology groups on the right sides of these equations are as defined at the start of section 7, with L/M and $IV(L/M)$ regarded as simplicial Z_p-modules. The argument of [15, Proposition 5.3] shows that Lie homotopic morphisms in $\mathcal{SL}_2$ induce the same morphisms on homotopy, homology, and cohomology. By [15, Theorem 22.1], $\pi_*(L, M)$ and $H_*(L, M)$ are, respectively, the homotopy groups of L/M and of $IV(L/M)$ regarded as simplicial sets. The Hurewicz homomorphism $h: \pi_*(L, M) \to H_*(L, M)$ may thus be defined as the map induced on homotopy from the inclusion $L/M \to IV(L/M)$. Since $IV(L/M) = IV(L)/IV(M)$, we have natural long exact homotopy, homology, and cohomology sequences on pairs $(L, M) \in \mathcal{SL}_2$, and h defines a natural transformation of long exact sequences. Note that $H_*(L, M)$ is the augmentation ideal of the Hopf algebra $H_*(V(L/M))$ if $H_*(V(L/M))$ is of finite type. Consider $FAS^n = F(\tilde{S}^n \otimes Z_p)$, where S^n is the simplicial n-sphere. It can be proven that $\pi_n(L)$ is the Z_p-module of Lie homotopy equivalence classes of morphisms $FAS^n \to L$ for $L \in \mathcal{SL}$ and that $H_*(FAS^n) \cong \tilde{H}_*(\Omega S^{n+1})$ is the augmentation ideal of the free commutative algebra on one primitive generator of degree n.

Our theory immediately yields Steenrod operations on $H^*(L, M)$.

<u>Theorem 8.5.</u> There exist natural homomorphisms P^s and , if $p > 2$, βP^s defined on $H^*(L, M)$ for $(L, M) \in \mathcal{SL}_2$. These operations satisfy the con-

clusions of Theorem 7.9 (except that the hypothesis of (iii) is not satisfied in general) and, in addition, the operation P^0 is identically zero.

Proof. We may regard $C_*(L, M) = C(IV(L/M))$ as an object of $\mathcal{DA}$, with diagonal $\overline{D} = \overline{\psi}$, the reduced coproduct as defined in Remarks 7.4(ii). Thus Theorem 7.9 applies directly, and $P^0 = 0$ follows from the previous lemma and corollary.

In [22], Priddy has given a different definition of $H_*(L)$ and $H^*(L)$. Let $\overline{W}$ be the functor from simplicial Z_p-algebras to $\mathcal{SA}$ defined by Moore [20]. If A is a simplicial Z_p-algebra, then $\overline{W}_0(A) = Z_p$ and $\overline{W}_q(A) = A_{q-1} \otimes \ldots \otimes A_0$, $q > 0$, as Z_p-modules. The face and degeneracy operators are as defined in [15, p. 87]. For $L \in \mathcal{SL}$, $\overline{W}V(L)$ is a simplicial cocommutative coalgebra with coproduct ψ given by

$$\psi(a_{q-1} \otimes \ldots \otimes a_0) = \sum (a'_{q-1} \otimes \ldots \otimes a'_0) \otimes (a''_{a-1} \otimes \ldots \otimes a'_0),$$

where $a_i \in V_i(L)$ satisfies $\psi(a_i) = \sum a'_i \otimes a''_i$. Priddy defines $H_*(L) = H_*(I\overline{W}V(L))$ and $H^*(L) = H^*(I\overline{W}V(L))$, where $I\overline{W}V(L)$ is regarded as a simplicial Z_p-module. For spectra, Priddy's definition and ours clearly differ only by a shift of degree; with his definition, $H^*(FAS^n) \cong \tilde{H}^*(S^{n+1}) = Z_p$. By Definition 7.3 and Remarks 7.4(ii), $(C(I\overline{W}V(L), \overline{\psi}) \in \mathcal{DA}$ and therefore Priddy's $H^*(L)$ alos admits Steenrod operations which satisfy the conclusions of Theorem 7.9 (except, in general, for (iii)) and $P^0 = 0$.

9. The dual homology operations; Nishida's theorem

For applications to loop spaces and to obtain a result used in the proof of the Adem relations, we shall discuss the homology operations P^s_* whose duals are the Steenrod operations on the mod p cohomology of a space X. Of course, $H^*(X) = H_*(X)^* = \mathrm{Hom}_{Z_p}(H_*(X), Z_p)$ and, if $H_*(X)$ is of finite type, $H_*(X) = H^*(X)^*$. Define P^s_* on $H_*(X)$ by $P^s = (P^s_*)^*$; P^s_* is clearly well-defined if $H_*(X)$ is of finite type, and either a direct limit argument or the next proposition imply that P_*^s is well-defined in general. P_*^s lowers degrees by s if $p = 2$ and by $2s(p-1)$ if $p > 2$. Our results on the P^s immediately yield the dual results for the P_*^s. We shall write the operations P_*^s on the left; the order of composition in the dual Adem relations must thus be reversed (that is, $H_*(X)$ is a left module over the opposite algebra of the Steenrod algebra). The following proposition was used in the proof of Lemma 4.6. Formula (2) of the proof is essentially Steenrod's definition [30] of the D_i.

Proposition 9.1. Let X be a space and let $d = \Phi(1 \otimes D): W \otimes_\pi C_*(X) \longrightarrow W \otimes_\pi C_*(X)^p$. Consider $d_*: H_*(\pi; H_*(X)) \longrightarrow H_*(\pi; H_*(X)^p)$. Let $x \in H_s(X)$. Then

(i) If $p = 2$, $d_*(e_r \otimes x) = \sum_k e_{r+2k-s} \otimes P_*^k(x) \otimes P_*^k(x)$.

(ii) If $p > 2$, $d_*(e_r \otimes x) = \nu(s) \sum_k (-1)^k e_{r+(2pk-s)(p-1)} \otimes P_*^k(x)^p$

$$- \delta(r)\nu(s-1) \sum_k (-1)^k e_{r+p+(2pk-s)(p-1)} \otimes P_*^k \beta(x)^p.$$

where $\nu(2j+\varepsilon) = (-1)^j (m!)^\varepsilon$ and $\delta(2j+\varepsilon) = \varepsilon$, $\varepsilon = 0$ or 1.

Proof. We may assume that $H_*(X)$ is of finite type. We shall compute $d^*: H^*(\pi; H^*(X)^p) \longrightarrow H^*(\pi) \otimes H^*(X)$ and then dualize. In the notations of Lemma 1.3, $H_*(X)^p \cong A \oplus Z_p\pi \otimes B$ as a π-module, and $H_*(\pi; H_*(X)^p) \cong H_*(\pi) \otimes A \oplus B$. It follows that $H^*(\pi; H^*(X)^p) \cong H^*(\pi) \otimes A^* \oplus B^*$. We claim first that $d^*(B^*) = 0$. To see this, we make explicit the isomorphism from B^* to the homology of $(W \otimes_\pi Z_p\pi \otimes B)^*$. For $y \in B^*$, define $\tilde{y} \in (W \otimes_\pi Z_p\pi \otimes B)^*$ by

$$\tilde{y}(w \otimes \alpha^i \otimes b) = \varepsilon(w) y(b) \quad \text{for } w \in W,\ 0 \leq i < p,\ b \in B.$$

Then y is a cocylce and $y \longrightarrow \tilde{y}$ induces the desired isomorphism. Define $\nu: Z_p\pi \longrightarrow Z_p$ by $\nu(1) = 1$ and $\nu(\alpha^i) = 0$, $1 \leq i < p$, and define $\bar{y} \in (W \otimes Z_p\pi \otimes B)^*$ for $y \in B^*$ by

$$\bar{y}(w \otimes \alpha^i \otimes b) = \varepsilon(w)\, \nu(\alpha^i) y(b) \quad \text{for } w \in W,\ 0 \leq i < p,\ b \in B.$$

Clearly $\tilde{y}(w \otimes \alpha^i \otimes b) = \overline{y}(N[w \otimes \alpha^i \otimes b])$. Therefore

$$d^*(\tilde{y})(w \otimes x) = \tilde{y}d_*(w \otimes x) = \overline{y}\, d_*(Nw \otimes x) = 0$$

for $w \in W$ and $x \in H_*(X)$ since $Nw \otimes x$ is a boundary in $W \otimes H_*(X)$ (because $d(e_{2i}) = Ne_{2i-1}$ and $d(T^{p-2}e_{2i+1}) = Ne_{2i}$ in W). This proves that $d^*(B^*) = 0$.

We next compute d^* on $H^*(\pi) \otimes A^*$. Let $y \in H^q(X)$ and $x \in H_{pq-i}(X)$. By Definition 7.5, we have the formula

(1) $\quad D_i(y)(x) = \theta_*(e_i \otimes y^p)(x) = (-1)^{iq} d^*(\varepsilon \otimes 1)^*(y^p)(e_i \otimes x)$

(where the isomorphism α from the tensor product of duals to the dual of tensor products has been omitted from the notation).

Let w_i be dual to e_i. Then $(\varepsilon \otimes 1)^*(y^p) = w_o \otimes y^p$ and therefore

$$d^*(w_o \otimes y^p)(e_i \otimes x) = (-1)^{iq} D_i(y)(x).$$

For any $z \in H^{pq-i}(X)$, the sign in the definition of α gives

$$(w_i \otimes z)(e_i \otimes x) = (-1)^{i(pq-i)} z(x).$$

Comparing these formulas, we see that

(2) $$d^*(w_o \otimes y^p) = \sum (-1)^i w_i \otimes D_i(y) .$$

To compute $d^*(w_j \otimes y^p)$ for $j > 0$, observe that if $\rho : W \longrightarrow \overline{W} = Z_p \otimes_\pi W$ is the natural epimorphism, then we have the commutative diagram:

$$\begin{array}{ccc}
W \otimes_\pi H_*(X) & \xrightarrow{\quad d_* \quad} & W \otimes_\pi H_*(X)^p \\
\downarrow{\scriptstyle \psi \otimes 1} & & \downarrow{\scriptstyle \psi \otimes 1} \\
(W \otimes W) \otimes_\pi H_*(X) & \xrightarrow{\quad 1 \otimes d_* \quad} & (W \otimes W) \otimes_\pi H_*(X)^p \\
\downarrow{\scriptstyle \rho \otimes 1 \otimes 1} & & \downarrow{\scriptstyle \rho \otimes 1 \otimes 1} \\
\overline{W} \otimes (W \otimes_\pi H_*(X)) & \xrightarrow{\quad 1 \otimes d_* \quad} & \overline{W} \otimes (W \otimes_\pi H_*(X)^p)
\end{array}$$

(The upper rectangle requires an easy acyclic models argument.) Dually, d^* is a morphism of $H^*(\pi)$-modules. Now $w_j(w_o \otimes y^p) = w_j \otimes y^p$ and, in $H^*(\pi)$, $w_j w_i = w_{i+j}$ if $p = 2$ or either i or j is even and $w_j w_i = 0$ if $p > 2$ and i and j are odd by formula (1.2). Therefore (2) implies the formulas

(3) if $p = 2$, $d^*(w_j \otimes y^p) = \sum_i w_{i+j} \otimes D_i(y)$ and

(4) if $p > 2$, $d^*(w_j \otimes y^p) = \sum_i w_{2i+j} \otimes D_{2i}(y) - \delta(j+1) \sum w_{2i+1+j} \otimes D_{2i+1}(y)$.

By formulas (5.1) and (5.2) and a reindexing, (3) and (4) become

(5) if $p = 2$, $d^*(w_j \otimes y^p) = \sum_k w_{j+q-k} \otimes P^k(y)$ and

(6) if $p > 2$, $d^*(w_j \otimes y^p) = \nu(-q)^{-1} \sum_k (-1)^k w_{j+(q-2k)(p-1)} \otimes P^k(y)$

$$- \delta(j+1)\, \nu(-q)^{-1} \sum_k (-1)^k w_{j+(q-2k)(p-1)-1} \otimes \beta P^k(y).$$

We now dualize. $d_*(H_*(\pi) \otimes H_*(X)) \subset H_*(\pi) \otimes A$ since $d^*(B^*) = 0$. For $x \in H_s(X)$, we may therefore write

$$d_*(e_r \otimes x) = \sum e_{r+s-pq} \otimes E_{qr}(x)^p, \quad E_{qr}(x) \in H_q(X).$$

Let $y \in H^q(X)$. Using the Kronecker pairing $<\ ,\ >$, we have

(7) $<w_{r+s-pq} \otimes y^p, d_*(e_r \otimes x)> = (-1)^{(r+s-q+m)q} < y, E_{qr}(x)>$.

Since $<P^k(y), x> = <y, P^k_*(x)>$, (5) implies that if $p = 2$, then

(8) $<d^*(w_{r+s-2q} \otimes y^2), e_r \otimes x> = <w_r \otimes P^{s-q}(y), e_r \otimes x> = <y, P_*^{s-q}(x)>$.

Thus $E_{qr}(x) = P_*^{s-q}(x)$ if $p = 2$ and, with $k = s-q$, this implies (i). Now assume that $p > 2$. By (6), $d^*(w_{r+s-pq} \otimes y^p)$ has a summand involving w_r only if $q = s-2k(p-1)-\varepsilon$, $k \geq 0$ and $\varepsilon = 0$ or 1, hence $E_{qr}(x) = 0$ for other values of q. For $q = s-2k(p-1)$, (6) gives

(9) $<d^*(w_{r+(2pk-s)(p-1)} \otimes y^p), e_r \otimes x> = \nu(-q)^{-1}(-1)^{k+rq} < y, P_*^k(x)>$.

By (7) and (9), $E_{qr}(x) = (-1)^{k+mq} \nu(-q)^{-1} P_*^k(x)$ if $q = s-2k(p-1)$; since $(-1)^{mq} \nu(-q)^{-1} = \nu(q) = \nu(s)$, this yields the first sum of (ii). Observe next that $<\beta y, x> = (-1)^{q+1} <y, \beta x>$ by the chain and cochain definitions of the Bockstein and the sign convention $\delta(f) = (-1)^{\deg f+1} fd$ used in defining $C^*(X)$. Now for $q = s-2k(p-1)-1$, (6) gives

(10) $\quad < d^*(w_{r+p+(2pk-s)(p-1)} \otimes y^p), e_r \otimes x > \; = \; \delta(r)\nu(-q)^{-1}(-1)^{k+r(q+1)+q} < y, P_*^k\beta(x) >$

By (7) and (10), $E_{qr}(x) = (-1)^{k+r+mq}\nu(-q)^{-1}\delta(r)P_*^{\,k}\beta(x)$ if $q = s-2k(p-1)-1$; since $\delta(r) = 0$ if r is even and $(-1)^{mq}\nu(-q)^{-1} = \nu(q) = \nu(s-1)$, this yields the second sum of (ii) and so completes the proof.

Remark 9.2. The proof above used no properties specific to topological spaces and so applies to compute

$$d_* = \Phi_*(1 \otimes D)_* : H_*(\pi; H_*(K)) \longrightarrow H_*(\pi; H_*(K)^p)$$

in terms of P_*^s and βP_*^s (where βP_*^s is defined by $(\beta P_*^s)^* = -\beta P^s$ if no Bockstein is present) for arbitrary objects $(K, D) \in \mathcal{B}\mathcal{A}$.

We now give a new proof of a result due to Nishida [21], which is essential to the computation of Steenrod operations in iterated loop spaces. Let $K(Z_p, 1) = E/\pi$ where π operates properly on the acyclic space E; by [14, IV 11], $C_*(E) = Z_p\pi \otimes C_*(E/\pi)$. Let $\sigma: E \longrightarrow E/\pi$ be the projection and let $f: W \longrightarrow C_*(E)$ be a π-morphism over Z_p. If $\overline{W} = Z_p \otimes_\pi W$, then f induces $\overline{f}: \overline{W} \longrightarrow C_*(E/\pi)$, and $\overline{f}\rho$ is homotopic to σf, $\rho: W \longrightarrow \overline{W}$. By Remarks 7.2, if η is the shuffle map, then we have the following homotopy commutative diagram for any space X:

$$\begin{array}{ccccc}
W \otimes_\pi C_*(X) & \xrightarrow{f \otimes 1} & C_*(E) \otimes_\pi C_*(X) & \xrightarrow{\eta} & C_*(E \times_\pi X) \\
\downarrow d & & \downarrow 1 \otimes D & & \downarrow 1 \times D \\
W \otimes_\pi C_*(X)^p & \xrightarrow{f \otimes \eta} & C_*(E) \otimes_\pi C_*(X^p) & \xrightarrow{\eta} & C_*(E \times_\pi X^p) \\
\downarrow \psi \otimes 1 & & \downarrow D \otimes 1 & & \downarrow D \times 1 \\
W \otimes W \otimes_\pi C_*(X)^p & \xrightarrow{\eta(f \otimes f) \otimes \eta} & C_*(E \times E) \otimes_\pi C_*(X^p) & \xrightarrow{\eta} & C_*(E \times E \times_\pi X^p) \\
\downarrow \rho \otimes 1 \otimes 1 & & \downarrow \sigma \times 1 \otimes 1 & & \downarrow \sigma \times 1 \times 1 \\
\overline{W} \otimes W \otimes_\pi C_*(X)^p & \xrightarrow{\eta(\overline{f} \otimes f) \otimes \eta} & C_*(E/\pi \times E) \otimes_\pi C_*(X^p) & \xrightarrow{\eta} & C_*(E/\pi \times E \times_\pi X^p)
\end{array}$$

Let $\mu_* = (\rho \otimes 1 \otimes 1)_*(\psi \otimes 1)_* : H_*(\pi; H_*(X)^p) \longrightarrow H_*(\pi) \otimes H_*(\pi; H_*(X)^p)$. The horizontal arrows are homology isomorphisms and we therefore have Steenrod operations P_*^s on $H_*(\pi; H_*(X))$, $H_*(\pi; H_*(X)^p)$, and $H_*(\pi) \otimes H_*(\pi; H_*(X)^p)$ such that d_*

and μ_* commute with the P_*^s. The following theorem uses d_* and μ_* to evaluate the P_*^s on $H_*(\pi; H_*(X)^p)$. Our result differs from Nishida's by a sign; the reason for this is that our formulas (2) and (6) in the proof above differ from the corresponding formulas in [30, p. 103 and p. 119]. We were pedantic about signs in the preceding proof because of this disagreement. We shall need the following identity on binomial coefficients in the proof of the theorem.

Lemma 9.3. $\sum_i (i, a-i)(n-i, i+b-n) = (n, a+b-n)$ for $a \geq 0$, $b \geq 0$, and $n \geq 0$.

Proof. The result is obvious if $b = 0$, when $i = n$ gives the only non-zero summand on the left. Using $(c-1, d) + (c, d-1) = (c, d)$, we find that the result for the triples $(a, b-1, n)$ and $(a, b-1, n-1)$ implies the result for the triple (a, b, n).

Theorem 9.4. Let X be a space, $x \in H_q(X)$. Then, in $H_*(\pi; H_*(X)^p)$,

(i) If $p = 2$, $P_*^s(e_r \otimes x^2) = \sum_i (s-2i, r+q-2s+2i)e_{r-s+2i} \otimes P_*^i(x)^2$.

(ii) If $p > 2$, $P_*^s(e_r \otimes x^p) = \sum_i (s-pi, [\frac{r}{2}]+qm-ps+pi)e_{r+2(pi-s)(p-1)} \otimes P_*^i(x)^p$

$$+ \delta(r)\alpha(q) \sum_i (s-pi-1, [\tfrac{r+1}{2}]+qm-ps+pi)e_{r+p+2(pi-s)(p-1)} \otimes P_*^i \beta(x)^p$$

where $\alpha(q) = \nu(q)^{-1}\nu(q-1) = -(-1)^{mq}m!$ and $\delta(2j+\varepsilon) = \varepsilon$, $\varepsilon = 0$ or 1.

Proof. We assume that $s > 0$, since the result is trivial for $s = 0$. If $r = 0$, then $e_0 \otimes x^p$ is in the image of $H_*(E \times X^p) \longrightarrow H_*(E \times_\pi X^p)$. In $H_*(E \times X^p)$, $P_*^s(e_0 \otimes x^p) = \sum e_0 \otimes P_*^{i_1}(x) \otimes \ldots \otimes P_*^{i_p}(x)$ summed over all p-tuples $(i_1, \ldots, i_p)$ such that $\sum i_p = s$. The sum of all terms with any $i_j \neq i_k$ lies in $e_0 \otimes NH_*(X)^p$ and is thus zero in $H_*(E \times_\pi X^p)$. Therefore $P_*^s(e_0 \otimes x^p) = 0$ unless $s = pt$, when $P_*^s(e_0 \otimes x^p) = e_0 \otimes P_*^t(x)^p$, which is in agreement with (i) and (ii). Recall that, by Definition 1.2 and the proof of Lemma 4.6, we have the following relations in $H_*(\pi)$.

(a) If $p = 2$, $P_*^i(e_j) = (i, j-2i)e_{j-i}$ and $\psi(e_r) = \sum e_j \otimes e_{r-j}$

(b) If $p > 2$, $P^i_*(e_j) = (i, [j/2] - pi)e_{j-2i(p-1)}$ and $\psi(e_r) = \sum \delta(r,j)e_j \otimes e_{r-j}$

where $\delta(r,j) = 1$ unless r is even and j is odd, when $\delta(r,j) = 0$.

If $q = 0$, then $d_*(e_r \otimes x) = e_r \otimes x^p$ and therefore

$$P^s_*(e_r \otimes x^p) = d_*(P^s_*(e_r) \otimes x) = P^s_*(e_r) \otimes x^p ;$$

by (a) and (b), the result holds in this case. We now proceed by induction on q and for fixed q by induction on r. Thus assume the result for $q' < q$ and for our fixed q and $r' < r$. Let $z = P^s_*(e_r \otimes x)$ and let z' denote the right side of the equation to be proven. Write $z - z' = \sum e_i \otimes y_i \in H_*(\pi; H_*(X)^p)$. We shall first prove that $\mu_*(z-z') = e_o \otimes (z-z')$; this will imply that $y_i = 0$ for all $i > 0$ since if i is maximal such that $y_i \neq 0$, then $e_i \otimes e_o \otimes y_i$ clearly occurs as a non-zero summand of $\mu_*(z-z')$. We shall then prove that $y_o = 0$ by explicit computation and so complete the proof. We give the details separately in the cases $p = 2$ and $p > 2$.

(i) $p = 2$. Since $\mu_*(z) = P^s_* \mu_*(e_r \otimes x^2)$, we find by (a), the Cartan formula in $H_*(\pi) \otimes H_*(\pi; H_*(X)^2)$, and the induction hypothesis on r, that

(c) $\mu_*(z) = P^s_*(\sum_j e_j \otimes e_{r-j} \otimes x^2) = \sum_{i,j} (i, j-2i)e_{j-i} \otimes P^{s-i}_*(e_{r-j} \otimes x^2)$, where

$$P^{s-i}_*(e_{r-j} \otimes x^2) = \sum_k (s-i-2k, r-j+q-2s+2i+2k)e_{r-j-s+i+2k} \otimes P^k_*(x)^2 \quad \text{if } j > 0.$$

The terms with $j = i > 0$ are zero since $(i, -i) = 0$. Applying the lemma to those (i,j) such that $j-i = \ell > 0$, with $a = \ell$, $b = r - \ell + q - s$, and $n = s - 2k$, we see that (c) reduces to the formula

(d) $\mu_*(z) = e_o \otimes z + \sum_{k,\ell>0} (s-2k, r+q-2s+2k)e_\ell \otimes e_{r-s+2k-\ell} \otimes P^k_*(x)^2 .$

A glance at the right side of (i) shows that $\mu_*(z-z') = e_o \otimes (z-z')$, hence $y_i = 0$ for $i > 0$. To compute y_o, observe that $P^o = 1$ and Proposition 9.1 imply

(e) $e_r \otimes x^2 = d_*(e_{r+q} \otimes x) + \sum_{k>0} e_{r+2k} \otimes P^k_*(x)^2 .$

$P_*^s d_* = d_* P_*^s$, the Cartan formula on $H_*(\pi) \otimes H_*(X)$, and Proposition 9.1 evaluate $P_*^s d_*(e_{r+q} \otimes x)$, and the induction hypothesis on q evaluates $P_*^s(e_{r+2k} \otimes P_*^k(x)^2)$ for $k > 0$. Carrying out these computations, we find that (e) implies the formula

$$\text{(f)} \qquad z = \sum_{k,\ell} (s-\ell, r+q-2s+2\ell) e_{r-s+2k+2\ell} \otimes P_*^k P_*^\ell (x)^2$$
$$+ \sum_{k>0,\ell} (s-2\ell, r+q+k-2s+2\ell) e_{r-s+2k+2\ell} \otimes P_*^\ell P_*^k (x)^2 .$$

In principle, (f) must imply (i) directly, but our argument with μ_* shows that we need only consider those terms involving e_o, with $2(k+\ell) = s-r$. Let $t = k+\ell-s$ and $c = q-k-\ell$; then these terms become

$$\text{(g)} \qquad \sum_{k \geq 0} (k-t, t+c-2k) e_o \otimes P_*^k P_*^{s+t-k}(x)^2 + \sum_{\ell < s+t} (c+\ell-s, s-2\ell) e_o \otimes P_*^\ell P_*^{s+t-\ell}(x)^2 .$$

By formula (f) of the proof of Theorem 4.7, rephrased as in section 5 and dualized (with the order of composition reversed under dualization since we are writing the operations P_*^s on the left), and by Remarks 4.8, (g) would be zero if $\ell = s+t$ were allowed in the second sum; thus (g) reduces to

$$\text{(h)} \qquad (c+t, -s-2t) e_o \otimes P_*^{s+t}(x)^2 = (q-s, r) e_o \otimes P_*^{1/2(s-r)}(x)^2 .$$

Since (h) is equal to the summand of z' involving e_o, it follows that $y_o = 0$.

(ii) $p > 2$. For brevity of notation, write $d = 2(p-1)$. As in the case $p = 2$, we find by (b) and induction on r that

$$\text{(i)} \qquad \mu_*(z) = \sum_{ij} \delta(r,j)(i,[j/2]-pi) e_{j-di} \otimes P_*^{s-i}(e_{r-j} \otimes x^p) \text{, where, if } j > 0,$$

$$P_*^{s-i}(e_{r-j} \otimes x^p) = \sum_k (s-i-pk, [\tfrac{r-j}{2}] + qm-ps+pi+pk) e_{r-j+d(pk-s+i)} \otimes P_*^k(x)^p$$
$$+ \delta(r-j)\alpha(q) \sum_k (s-i-pk-1, [\tfrac{r-j+1}{2}] + qm-ps+pi+pk) e_{r-j+p+d(pk-s+i)} \otimes P_*^k \beta(x)^p$$

The terms with $j = di > 0$ are zero. By the lemma, applied to those (i,j) such that $j-di = \ell > 0$, with $a = [\ell/2]$, $b = [\frac{r-\ell}{2}] + qm-s(p-1)$, and $n = s-pk$ for the first sum and $a = [\ell/2]$, $b = [\frac{r-\ell-1}{2}] + qm-s(p-1)$, and $n = s-pk-1$ for the second sum, (i) reduces to

$$\text{(j)}\quad \mu_*(z) = e_o \otimes z + \sum_{k,\ell>0} \delta(r,\ell)(s-pk,[\tfrac{r}{2}]+qm-ps+pk)e_\ell \otimes e_{r+d(pk-s)-\ell} \otimes P_*^k(x)^p$$

$$+ \sum_{k,\ell>0} \delta(r,\ell)\delta(r-\ell)\alpha(q)(s-pk-1,[\tfrac{r+1}{2}]+qm-ps+pk)e_\ell \otimes e_{r+p+d(pk-s)-\ell} \otimes P_*^k\beta(x)^p .$$

Now $\delta(r,\ell)\delta(r-\ell) = \delta(r)\delta(r+1,\ell)$, and it follows from a glance at (ii) that $\mu_*(z-z') = e_o \otimes (z-z')$. Thus $y_i = 0$ for $i > 0$. To compute y_o, observe that Proposition 9.1 implies that

$$\text{(k)}\quad e_r \otimes x^p = \nu(q)^{-1} d_*(e_{r+q(p-1)} \otimes x) - \sum_{k>0} (-1)^k e_{r+dpk} \otimes P_*^k(x)^p$$
$$+ \delta(r)\alpha(q) \sum_k (-1)^k e_{r+p+dpk} \otimes P_*^k\beta(x)^p .$$

Precisely as in the case $p = 2$, we can compute P_*^s on the right side of (k); carrying out this computation, we find

$$(\ell)\quad z = \sum_{k,\ell} (-1)^k (s-\ell,[r/2]+qm-ps+p\ell)e_{r+d(pk+p\ell-s)} \otimes P_*^k P_*^\ell (x)^p$$

$$- \delta(r)\alpha(q) \sum_{k,\ell} (-1)^k (s-\ell,[r/2]+qm-ps+p\ell)e_{r+p+d(pk+p\ell-s)} \otimes P_*^k \beta P_*^\ell (x)^p$$

$$- \sum_{k>0,\ell} (-1)^k (s-p\ell,[r/2]+k(p-1)+qm-ps+p\ell)e_{r+d(pk+p\ell-s)} \otimes P_*^\ell P_*^k (x)^p$$

$$- \delta(r)\alpha(q) \sum_{k>0,\ell} (-1)^k (s-p\ell-1,[\tfrac{r+1}{2}]+k(p-1)+qm-ps+p\ell)e_{r+p+d(pk+p\ell-s)} \otimes P_*^\ell \beta P_*^k (x)^p$$

$$+ \delta(r)\alpha(q) \sum_{k,\ell} (-1)^k (s-p\ell,[\tfrac{r+1}{2}]+k(p-1)+qm-ps+p\ell)e_{r+p+d(pk+p\ell-s)} \otimes P_*^\ell P_*^k \beta(x)^p .$$

Consider the first and third sums, with $r+d(pk+p\ell-s) = 0$. Let $t = k+\ell-s$ and $c = q-d(k+\ell)$. Then these two sums become

$$\text{(m)}\quad \sum_k (-1)^k (k-t,t+mc-pk)e_o \otimes P_*^k P_*^{s+t-k}(x)^p$$
$$- \sum_{\ell<s+t} (-1)^{s+t+\ell} (\ell+mc-s,s-p\ell)e_o \otimes P_*^\ell P_*^{s+t-\ell}(x)^p .$$

Consider the remaining sums of (ℓ), with $r+p+d(pk+p\ell-s) = 0$; r is odd, hence $\delta(r) = 1$. Let t be as above and let $c' = c-1$. Then these three sums become

(n) $$-\alpha(q) \sum_k (-1)^k (k-t, t+mc'-pk-1)e_o \otimes P_*^k \beta P_*^\ell (x)^p$$

$$- \alpha(q) \sum_{\ell < s+t} (-1)^{s+t+\ell} (\ell + mc'-s, s-p\ell-1)e_o \otimes P_*^\ell \beta P_*^{s+t-\ell} (x)^p$$

$$+ \alpha(q) \sum_\ell (-1)^{s+t+\ell} (\ell + mc'-s, s-p\ell)e_o \otimes P_*^\ell P_*^{s+t-\ell} \beta(x)^p .$$

By formulas (h) and (j) of the proof of Theorem 4.7 (with $\varepsilon = 0$), rephrased as in section 5 and dualized, we see that (m) and (n) would be zero if $\ell = s+t$ were allowed in the second sums. Therefore, by an easy verification, (m) and (n) reduce to the following expressions, where $i = k+\ell = s+t$.

(o) $(s-pi, \frac{r}{2} + qm - ps+pi)e_o \otimes P_*^i(x)^p$ with $dpi = ds-r$, and

(p) $\alpha(q)(s-pi-1, \frac{r+1}{2} + qm - ps + pi)e_o \otimes P_*^i \beta(x)^p$ with $dpi = ds-r-p$.

Clearly (o) is equal to the summand of z' involving e_o in its first sum and (p) is equal to the summand of z' involving e_o in its second sum. Thus $y_o = 0$ and the proof is complete.

10. The cohomology of $K(\pi, n)$ and the axiomatization of the P^s

We recall the structure of $H^*(K(\pi,n); Z_p) = H^*(\pi, n, Z_p)$ and compute completely the mod p cohomology Bockstein spectral sequence of $K(\pi,n)$ in this section. We also show (as should be well-known) that Serre's proof [23] of the axiomatization of the Sq^i using $K(Z_2, n)$ can be simply modified so as to apply in the case of odd primes. We shall consider only the cyclic groups $\pi = Z_{p^t}$, $1 \leq t \leq \infty$, where, by convention, $Z_{p^\infty} = Z$. We first fix conventions on admissible monomials relative to t.

Notations 10.1. (a) $p = 2$. For $I = (s_1, \ldots, s_k)$, we say that I is admissible if $s_i \geq 2s_{i+1}$ and $s_k \geq 1$. The length, degree, and excess of I are defined by $\ell(I) = k$, $d(I) = \sum s_j$, and, if $I = (s, J)$, $e(I) = s - d(J)$. Define

$P_t^I = P^{s_1} \dots P^{s_{k-1}} P_t^{s_k}$, where $P_t^s = P^s$ if $s \geq 2$, $P_t^1 = \beta_t$ if $t < \infty$, and $P_\infty^1 = 0$; thus, if $t = \infty$, we agree that admissibility requires $s_k \geq 2$. The empty sequence I is admissible, with length, degree, and excess zero, and I determines the identity operation.

(b) $p > 2$. For $I = (\varepsilon_1, s_1, \dots, \varepsilon_k, s_k, \varepsilon_{k+1})$, $\varepsilon_i = 0$ or 1, we say that I is admissible if $s_i \geq ps_{i+1} + \varepsilon_{i+1}$ and $s_k \geq 1$ or if $k = 0$, when $I = (\varepsilon)$. Define $\ell(I) = k$, $d(I) = \sum \varepsilon_j + \sum 2s_j(p-1)$, and, if $I = (\varepsilon, s, J)$, $e(I) = 2s + \varepsilon - d(J)$. Define $P_t^I = \beta^{\varepsilon_1} P^{s_1} \dots \beta^{\varepsilon_k} P^{s_k} \beta_t^{\varepsilon_{k+1}}$, where $\beta_t^0 = 1$ for all t, $\beta_t^1 = \beta_t$ for $t < \infty$, and $\beta_\infty^1 = 0$; thus, if $t = \infty$, we agree that admissibility requires $\varepsilon_{k+1} = 0$.

We now give a quick calculation of $H^*(Z_{p^t}, n, Z_p)$.

Lemma 10.2. $H^*(Z, 1, Z_p) = E(i_1)$ and $H^*(Z, 2, Z_p) = P(i_2)$. If $t < \infty$, then $H^*(Z_{2^t}, 1, Z_2) = P(i_1)$, with $\beta_t(i_1) = i_1^2$, and $H^*(Z_{p^t}, 1, Z_p) = E(i_1) \otimes P(\beta_t(i_1))$ if $p > 2$.

Proof. $K(Z, 1) = S^1$ and $K(Z, 2) = CP^\infty$, so the first statement is obvious. For the second statement, $H^*(Z_{p^t}, 1, Z_p) = H^*(Z_{p^t}; Z_p)$, and we can define a ΛZ_{p^t}-free resolution of Λ, with coproduct, precisely as in Definition 1.2 (with p there replaced by p^t) for any commutative ring Λ. The result follows by an easy computation.

Theorem 10.3. If $n \geq 2$ (or if $n = 1$ and either $p > 2$ or $t < \infty$), then $H^*(Z_{p^t}, n, Z_p)$ is the free commutative algebra generated by the following set: $\{P_t^I i_n \mid I$ is admissible and $e(I) < n$ or, if $p > 2$, $e(I) = n$ and $_1 = 1\}$. Moreover, $H^*(Z_{p^t}, n, Z_p)$ is a primitively generated Hopf algebra.

Proof. The lemma gives the result for $t < \infty$ and $n = 1$ and for $t = \infty$ and $n = 2$. Assume the result for $n-1$. Of course, the Serre spectral sequence $\{E_r\}$ of $K(Z_{p^t}, n-1) \longrightarrow E \longrightarrow K(Z_{p^t}, n)$, E acyclic, satisfies

$$E_2 = H^*(Z_{p^t}, n, Z_p) \otimes H^*(Z_{p^t}, n-1, Z_p) \text{ and } E_\infty = Z_p .$$

First, let $p = 2$; then, regarding squares as Steenrod operations, we see that we may rewrite the polynomial algebra $H^*(Z_{2^t}, n-1, Z_2)$, additively, as the exterior algebra $E(S)$, where

$$S = \{ P_t^I i_{n-1} \mid I \text{ is admissible and } e(I) < n \}.$$

By Theorem 3.4, $P_t^I i_{n-1}$ transgresses to $P_t^I i_n$. Define an abstract spectral sequence of differential algebras, $\{E'_r\}$, by letting $E'_2 = P(\tau S) \otimes E(S)$, where $\tau(S)$ is a copy of S with degrees augmented by one, and by requiring $s \in S$ to transgress to $\tau s \in \tau(S)$. Clearly $E'_\infty = Z_2$. Define a morphism of spectral sequences $f_r : E'_r \longrightarrow E_r$ by $f_2 = g \otimes 1$, where $g.P(\tau S) \longrightarrow H^*(Z_{2^t}, n, Z_2)$ is the morphism of algebras defined on generators by $g(\tau P_t^I i_{n-1}) = P_t^I i_n$; clearly commutation with the differentials determines f_r for $r > 2$. Since f_2^{0*} and f_∞ are isomorphisms, $f_2^{*0} = g$ is an isomorphism by the comparison theorem [14, p. 355].

Now let $p > 2$. We may rewrite $H^*(Z_{p^t}, n, Z_p)$, additively, as $E(S) \otimes Q(T)$, where Q denotes a truncated polynomial algebra ($x^p = 0$ for $x \in T$) and where

$$S = \{P_t^I i_{n-1} \mid I \text{ is admissible, } e(I) < n-1, \ d(I) + n \text{ even}\},$$

$$T = \{P_t^I i_{n-1} \mid I \text{ is admissible, } e(I) < n, \ d(I)+n \text{ odd}\}.$$

(Note that $e(I) \equiv d(I) \bmod 2$, hence $d(I)+n$ even and $e(I) = n-1$ is impossible.)

By Theorem 3.4, $P_t^I i_{n-1}$ transgresses to $(-1)^{d(I)} P_t^I i_n$ and, if $d(I) + n = 2q+1$, $P_t^I i_n \otimes (P_t^I i_{n-1})^{p-1}$ transgresses to $(-1)^n \beta P^q P_t^I i_n$. Define an abstract spectral sequence of differential algebras, $\{E'_r\}$, as follows. Let

$$E'_2 = [P(\tau S) \otimes E(\tau) \otimes P(\mu T)] \otimes [E(S) \otimes Q(T)]$$

(the bracketed expressions are the base and fibre, respectively). Here τS and τT are copies of S and T with degrees augmented by one and μT is a copy of T with degrees multiplied by p and then augmented by two. The differentials in $\{E'_r\}$ are specified by requiring $s \in S$ to transgress to $\tau s \in \tau S$, $t \in T$ to transgress to $\tau t \in \tau T$, and $\tau t \otimes t^{p-1}$ to transgress to $\mu t \in \mu T$. An easy computation demonstrates that $E'_\infty = Z_p$. Define a morphism of spectral sequences

$f_r : E'_r \longrightarrow E_r$ by $f_2 = g \otimes 1$, where $g: P(\tau S) \otimes E(\tau T) \otimes P(\mu T) \longrightarrow H^*(Z_{p^t}, n, Z_p)$ is the morphism of algebras defined on generators by $g(\tau P_t^I i_{n-1}) = (-1)^{d(I)} P_t^I i_n$ and $g(\mu P_t^I i_{n-1}) = -\beta P^q P_t^I i_n$ if $d(I) + n = 2q+1$. As in the case $p = 2$, the f_r for $r > 2$ are determined by commutation with the differentials, and g is an isomorphism by the comparison theorem. The last statement follows since, by the external Cartan formula, if X is an H-space and $x \in H^*(X)$ satisfies $\psi(x) = \sum x' \otimes x''$, then

$$\psi P^s(x) = \sum_{i+j=s} \sum P^i(x') \otimes P^j(x'') \text{ and } \psi\beta(x) = \sum(\beta(x') \otimes x'' + (-1)^{\deg x'} x' \otimes \beta(x'')).$$

Thus, since i_n and $\beta_t(i_n)$ are primitive, so are all of the $P_t^I i_n$.

We can now compute the mod p cohomology Bockstein spectral sequence $\{E_r\}$ of $K(Z_{p^t}, n)$. Recall that $\{E_r\}$ is a spectral sequence of differential algebras such that $E_1 = H^*(p^t, n, Z_p)$ and E_{r+1} is the homology of E_r with respect to β_r for $r \geq 1$. Since $H^*(p^t, n, Z)$ is a direct sum of cyclic groups with one generator of order p^r for each basis element of $\mathrm{Im}(\beta_r) \subset E_r$ and one generator of infinite order for each basis element of E_∞, the integral cohomology of $K(Z_{p^t}, n)$ is completely determined, additively, by $\{E_r\}$. If $t < \infty$ and $n = 1$, Lemma 10, 2 implies that $E_1 = E_t$ and $E_{t+1} = E_\infty = Z_p$, hence we need only consider the case $n \geq 2$.

<u>Theorem 10.4.</u> Let $n \geq 2$. Define a subset S of the set of generators for $E_1 = H^*(Z_{p^t}, n, Z_p)$ given in Theorem 10.3 by

(a) If $p = 2$, $S = \{P_t^I i_n \mid s_1$ and $d(I) + n$ are even and $\ell(I) > 0\}$.

(b) If $p > 2$, $S = \{P_t^I i_n \mid \varepsilon_1 = 0$, $d(I) + n$ is even, and $\ell(I) > 0\}$.

For $y \in S$, define $z(y) = \beta(y)y + P^{2q}\beta(y)$ if $p = 2$ and degree $(y) = 2q$ and define $z(y) = \beta(y)y^{p-1}$ if $p > 2$. Define an algebra $A_r(n, t)$ by

(c) $A_r(2n, \infty) = P\{i_n\}$ and $A_r(2n+1, \infty) = E\{i_n\}$.

(d) $A_r(2n, t) = P\{i_{2n}\} \otimes E\{\beta_t(2n)\}$ if $r \leq t < \infty$ and

$$A_r(2n,t) = P\{i_{2n}^{p^{r-t}}\} \otimes E\{z(i_{2n})i_{2n}^{p^{r-t}-p}\} \text{ if } r > t,$$

where $z(i_{2n}) = \beta(i_{2n})i_{2n} + P^{2n}\beta(i_{2n})$ if $p = 2$ and $t = 1$

and $z(i_{2n}) = \beta_t(i_{2n})i_{2n}^{p-1}$ if either $p > 2$ or $t > 1$

(e) $A_r(2n+1,t) = E\{i_{2n+1}\} \otimes P\{\beta_t(i_{2n+1})\}$ if $r \leq t < \infty$ and

$A_r(2n+1,t) = Z_p$ if $r > t$.

Then, if $r \geq 1$, $E_{r+1} = P\{y^{p^r} \mid y \in S\} \otimes E\{z(y)y^{p^r-p} \mid y \in S\} \otimes A_{r+1}(n,t)$,

$\beta_{r+1}(y^{p^r}) = z(y)y^{p^r-p}$ for $y \in S$, and $\beta_{r+t}(i_{2n}^{p^r}) = z(i_{2n})i_{2n}^{p^r-p}$.

__Proof.__ We first compute E_2 separately in the cases $p = 2$ and $p > 2$. Let $p = 2$ and define subsets T and U of the set of generators of E_1 by

$$T = \{P_t^I i_n \mid s_1 \text{ is even, } d(I)+n \text{ is odd, } e(I) < n-1, \text{ and } \ell(I) > 0\}$$
$$U = \{P_t^I i_n \mid d(I)+n \text{ is odd, } e(I) = n-1, \text{ and } \ell(I) > 0\}.$$

Recall that $\beta P^{s-1} = sP^s$ and observe that if $P_t^I i_n \in U$, then $I = (2q, J)$, where $d(J) + n = 2q+1$, and $\beta P_t^I i_n = (P_t^J i_n)^2$. Let C be the (additive) subcomplex of E_1 which is the tensor product of the following collections of subcomplexes:

(i) $P\{\beta(y)\} \otimes E\{y\}$ for $y \in T$, and

(ii) $P\{z^2\} \otimes E\{y\}$ for $y = P^{2q}z \in U$, $\deg(z) = 2q+1$.

Let IC be the positive degree elements of C. Then $H(IC) = 0$ under β, and therefore E_2 is isomorphic to $H(E_1/IC)$. If $y \in T \cup U$, then $P^{2q}y \in U$, $\deg y = 2q+1$, and therefore C is actually a subalgebra of E_1 and E_1/IC is a quotient differential algebra of E_1. It is easy to see that

$$E_1/IC = P\{y \mid y \in S\} \otimes E\{\beta(y) \mid y \in S\} \otimes A_1'(n,t),$$

where $A_1'(n,t)$ is the quotient of the polynomial algebra generated by i_n and, if $t < \infty$, $\beta_t(i_n)$ by the ideal generated by i_n^2 if n is odd or by $\beta_t(i_n)^2$ if n is even. Therefore

$$H(E_1/IC) = P\{y^2 \mid y \in S\} \otimes E\{\beta(y)y \mid y \in S\} \otimes A_2'(n,t),$$

where $A_2'(n,t) = A_2(n,t)$ unless n is even and $t = 1$, when $A_2'(n,t) = P\{i_n^2\} \otimes E\{\beta(i_n)i_n\}$. For $y \in S$ or $y = i_n$ if n is even and $t = 1$, $z(y) = \beta(y)y + P^{2q}\beta(y)$, $\deg y = 2q$, is a cycle in E_1 which projects to the cycle $\beta(y)y$ in E_1/IC. Since $z(y)^2$ bounds in E_1, it follows that E_2 has the stated form if $p = 2$. Next, let $p > 2$ and define a subset T of the set of generators of E_1 by

$$T = \{P_t^I i_n \mid \varepsilon_1 = 0,\ d(I) + n \text{ is odd, and } \ell(I) > 0\}.$$

Then, as a differential algebra, E_1 breaks up into the tensor product of the following collection of subalgebras:

(iii) $P\{y\} \otimes E\{\beta(y)\}$ for $y \in S$

(iv) $E\{y\} \otimes P\{\beta(y)\}$ for $y \in T$

(v) The free commutative algebra generated by i_n and, if $t < \infty$, $\beta_t(i_n)$.

The algebras in (iii) have homology $P\{y^p\} \otimes E\{z(y)\}$, those of (iv) are acyclic, and that of (v) has homology $A_2(n,t)$, hence E_2 is as stated. Now assume that E_{r+1} is as stated, $r \geq 1$ and any p. Then Proposition 6.8 computes β_{r+1} and E_{r+1} breaks up into the tensor product of $A_{r+1}(n,t)$ with subalgebras of the form $P\{x\} \otimes E\{\beta_{r+1}(x)\}$, where $x = y^{p^r}$, $y \in S$. This proves the result.

Finally, we prove the axiomatization of the P^s on topological spaces. Recall first that the Cartan formula and $P^0 = 1$ imply that the P^s commute with suspension [28, 30] and that we have shown in Proposition 8.1 that $P^0 = 1$ is implied by the commutation of P^0 with S^* and the fact that P^0 is the p-th power on a zero dimensional class. Thus the axioms we choose (for convenience of proof) are in fact redundant.

Theorem 10.5. There exists a unique family $\{P^s \mid s \geq 0\}$ of natural homomorphisms $H^*(X; Z_p) \to H^*(X; Z_p)$ such that $\deg(P^s) = s$ if $p = 2$, $\deg(P^s) = 2s(p-1)$ if $p > 2$, and

(i) P^0 is the identity homomorphism

(ii) $P^s(x) = x^p$ if $p = 2$ and $s = \deg x$ or $p > 2$ and $2s = \deg x$

(iii) $P^s(x) = 0$ if $p = 2$ and $s > \deg x$ or $p > 2$ and $2s > \deg x$

(iv) $P^s(x \otimes y) = \sum_{i+j=s} P^i(x) \otimes P^j(y)$ for $x \otimes y \in H^*(X \times Y)$

(v) $\sigma^* P^s = P^s \sigma^*$, where σ^* is the **suspension** of a fibration.

Proof. Suppose given $\{R^s \mid s \geq 0\}$ which also satisfy the axioms. If $x \in H^n(X, Z_p)$, then $x = f^*(i_n)$ for some $f: X \to K(Z_p, n)$, hence it suffices to prove that $P^s(i_n) = R^s(i_n)$. The result is obvious from (i), (ii), and (iii) if $n = 1$ or if $p > 2$ and $n = 2$. Assume that $P^s(i_{n-1}) = R^s(i_{n-1})$ for all s and consider $y = P^s(i_n) - R^s(i_n)$, $0 < s < n$ if $p = 2$ and $0 < 2s < n$ if $p > 2$. By (v), $\sigma^*(y) = 0$, where

$$\sigma^*: H^*(Z_p, n, Z_p) \to H^*(Z_p, n-1, Z_p).$$

If $p = 2$, σ^* is an isomorphism in degrees less than $2n$ and therefore $y = 0$. Let $p > 2$. As shown in the proof of Theorem 10.3, (i) and (iv) imply that both $P^s(i_n)$ and $R^s(i_n)$ are primitive. By Theorem 10.3, we see that

$$\{P^I i_n \mid I \text{ admissible, } e(I) \leq n\}$$

is a basis for the primitive elements of $H^*(Z_p, n, Z_p)$. The only elements of this set which are in $\operatorname{Ker} \sigma^*$ are p-th powers and elements of the form $\beta P^q(y)$, $\deg y = 2q+1$, which have degree $2pq + 2$. If n is odd, then y has odd degree and is thus zero. If n is even, then all primitive elements in $\operatorname{Ker} \sigma^*$ have degree at least pn, which is greater than the degree of y, and again $y = 0$.

11. Cocommutative Hopf algebras

In this section, we consider the following category $\mathcal{C}$. The objects of $\mathcal{C}$ are triples $C = (E, A, F)$ where A is a (Z-graded) cocommutative Hopf algebra over a commutative ring Λ, E is a right and F is a left (Z-graded) cocommutative A-coalgebra. Thus E and F are A-modules and cocommutative coalgebras (not necessarily unital or augmented), and their coproducts ψ are morphisms of A-modules. We say that C is unital if E and F are unital and augmented and their units and augmentations are morphisms of A-modules. A morphism $\gamma: C \longrightarrow C'$ in $\mathcal{C}$ is a triple $\gamma = (\alpha, \lambda, \beta)$, where $\lambda: A \longrightarrow A'$ is a morphism of Hopf algebras and $\alpha: E \longrightarrow E'$ and $\beta: F \longrightarrow F'$ are λ-equivariant morphisms of coalgebras; thus $\alpha(ea) = \alpha(e)\lambda(a)$ and $\beta(af) = \lambda(a)\beta(f)$ for $e \in E$, $a \in A$, and $f \in F$. We say that γ is unital if α and β are morphisms of unital augmented coalgebras. For C and C' in $\mathcal{C}$, define $C \otimes C' = (E \otimes E', A \otimes A', F \otimes F') \in \mathcal{C}$ and observe that

$$\psi = (\psi, \psi, \psi): C = (E, A, F) \longrightarrow (E \otimes E,\ A \otimes A,\ F \otimes F) = C \otimes C$$

is a morphism in $\mathcal{C}$; clearly ψ is unital if C is unital. Define homology and cohomology functors on the category $\mathcal{C}$ by

(1) $\quad H_{st}(C) = \mathrm{Tor}_{st}^{(A,\Lambda)}(E, F) \quad$ and $\quad H^{st}(C) = \mathrm{Ext}_{(A,\Lambda)}^{st}(E, F^*)$.

We shall define and study Steenrod operations on $H^*(C)$ when $\Lambda = Z_p$. The results here generalize work of Liulevicius [13].

In the following definitions, we recall the description of $H^*(C)$, with its product, in terms of the bar construction.

Definitions 11.1. For $C = (E, A, F) \in \mathcal{C}$, let $\overline{C} = (A, A, F) \in \mathcal{C}$. Let JA be the cokernel of the unit $\Lambda \longrightarrow A$. Define the bar construction $B(C)$ as follows. $B(C) = E \otimes T(JA) \otimes F$ as a Λ-module, where $T(JA)$ is the tensor algebra on JA. Write elements of $B(C)$ in the form $e[a_1|\dots|a_s]f$; such an element has homological degree s, internal degree $t = \deg e + \sum \deg a_i + \deg f$, and total degree s+t. Define $\varepsilon: B(C) \to E \otimes_A F$ and $d: B_{s,*}(C) \to B_{s-1,*}(C)$ by

(2) $\quad \varepsilon(e[\ ]f) = e \otimes f, \quad \varepsilon(e[a_1|\dots|a_s]f) = 0,$ and

(3)
$$\begin{aligned} d(e[a_1|\dots|a_s]f) &= -\bar{e}a_1[a_2|\dots|a_s]f \\ &\quad - \sum_{i=1}^{s-1} \bar{e}[\bar{a}_1|\dots|\bar{a}_{i-1}|\bar{a}_i a_{i+1}|a_{i+2}|\dots|a_s]f \\ &\quad - \bar{e}[\bar{a}_1|\dots|\bar{a}_{s-1}]a_s f, \quad \text{where } \bar{x} = (-1)^{1+\deg x} x. \end{aligned}$$

If $E = A$, then d is a morphism of left A-modules and $dS + Sd = 1 - \sigma\varepsilon$, where $\sigma: F \to B(\overline{C})$ and $S: B_{s,*}(\overline{C}) \to B_{s+1,*}(\overline{C})$ are defined by the formulas

(4) $\quad \sigma(f) = [\ ]f$ and $S(a[a_1|\dots|a_s]f) = [a|a_1|\dots|a_s]f$.

Clearly $d = 1 \otimes_A d$ on $B(C) = E \otimes_A B(\overline{C})$. By adjoint associativity,

$$\mathrm{Hom}_A(B(\overline{C}), E^*) \cong B^*(C) \cong \mathrm{Hom}_A(B(E, A, A), F^*).$$

Therefore (1) admits the equivalent reformulation

(5) $\quad H_*(C) = H(B(C))$ and $H^*(C) = H(B^*(C)) = \mathrm{Ext}_{(A,\Lambda)}(F, E^*)$.

<u>Definitions 11.2.</u> Let C and C' be objects of $\mathcal{C}$. Define the Alexander-Whitney map $\xi: B(C \otimes C') \to B(C) \otimes B(C')$ and the shuffle map $\eta: B(C) \otimes B(C') \to B(C \otimes C')$ by the formulas

(6)
$$\begin{aligned} &\xi(e \otimes e'[a_1 \otimes a_1'|\dots|a_s \otimes a_s']f \otimes f') \\ &= \sum_{k=0}^{s} (-1)^{\mu(k)} e[a_1|\dots|a_k]a_{k+1}\dots a_s f \otimes e'a_1'\dots a_k'[a_{k+1}'|\dots|a_s']f', \end{aligned}$$

where $\mu(k) = \deg e'(k + \deg a_1 \dots a_s f) + \sum_{i=1}^{k} \deg a_i' (k - i + \deg a_{i+1} \dots a_s f)$

$$+ \sum_{j=k+1}^{s} (1 + \deg a_j') \deg a_{j+1} \dots a_s f, \quad \text{and}$$

(7)
$$\begin{aligned} &\eta(e[a_1|\dots|a_s]f \otimes e'[a_{s+1}|\dots|a_{s+t}]f') \\ &= \sum_{\pi} (-1)^{\nu(\pi)} e \otimes e'[a_{\pi(1)}|\dots|a_{\pi(s+t)}]f \otimes f', \end{aligned}$$

where $a_i \in A$ if $i \le s$, $a_i \in A'$ if $i > s$, the sum is taken over all (s,t)-shuffles π (see [15, p. 17]), and

$$\nu(\pi) = \sum_{\pi(i) > \pi(s+j)} (1 + \deg a_i)(1 + \deg a_{s+j}).$$

The unnormalized bar construction $E \otimes T(A) \otimes F$ admits a structure of simplicial graded Λ-module under which ξ and η are in fact the classical normalized Alexander-Whitney and shuffle maps. Define $D = \xi B(\psi): B(C) \longrightarrow B(C) \otimes B(C)$. Then D gives $B(C)$ a structure of coassociative coalgebra; if C is unital, then $B(C)$ is unital and augmented. If $E = A$, then D coincides with the morphism of left A-modules defined inductively by

(8) $D([\ ]f) = \sum [\ ]f' \otimes [\ ]f''$ if $\psi(f) = \sum f' \otimes f''$, and

(9) $DS = SD$, where $S = S \otimes 1 + \sigma\varepsilon \otimes S$ on $B(\overline{C}) \otimes B(\overline{C})$.

Clearly D on $B(C) = E \otimes_A B(\overline{C})$ is the composite

$$E \otimes_A B(\overline{C}) \xrightarrow{\psi \otimes D} E \otimes E \otimes_A B(\overline{C}) \otimes B(\overline{C}) \xrightarrow{1 \otimes T \otimes 1} E \otimes_A B(\overline{C}) \otimes E \otimes_A B(\overline{C}).$$

We define the cup product on $B^*(C)$ to be the composite

(10) $\cup : B^*(C) \otimes B^*(C) \xrightarrow{\alpha} [B(C) \otimes B(C)]^* \xrightarrow{D^*} B^*(C)$.

We have the following analog of Lemma 7.1; a more precise analog (in terms of ξ) could also be proven, and an alternative proof by semi-simplicial rather than homological techniques is available.

Lemma 11.3. Let π be a subgroup of Σ_r and let W be a $\Lambda\pi$-free resolution of Λ such that $W_0 = \Lambda\pi$ with $\Lambda\pi$ generator e_0. Let $C \in \mathcal{C}$. Bigrade $W \otimes B(C)$ by $[W \otimes B(C)]_{st} = \sum_{i+j=s} W_i \otimes B_{jt}(C)$. Then there exists a morphism of bigraded $\Lambda\pi$-complexes $\Delta: W \otimes B(C) \longrightarrow B(C)^r$ which is natural in C and satisfies the following properties:

(i) $\Delta(w \otimes b) = 0$ if $b \in B_{0,*}(C)$ and $w \in W_i$ for $i > 0$

(ii) $\Delta(e_0 \otimes b) = D(b)$ if $b \in B(C)$, where D is the iterated coproduct

(iii) If $E = A$, then Δ is a morphism of left A-modules, where A operates on $W \otimes B(C)$ by $a(w \otimes b) = (-1)^{\deg w \deg a} w \otimes ab$.

(iv) $\Delta(W_i \otimes B_{st}(C)) = 0$ if $i > (r-1)s$.

Moreover, any two such Δ are naturally $\Lambda\pi$-homotopic.

Proof. Observe first that the cocommutativity of A ensures that (iii) is compatible with the π-equivariance of Δ. Observe next that it suffices to prove the result when $E = A$, since we can then define Δ on $W \otimes B(C) = E \otimes_A W \otimes B(\overline{C})$ to be the composite

$$E \otimes_A W \otimes B(\overline{C}) \xrightarrow{\psi \otimes \Delta} E^r \otimes_A B(\overline{C})^r \xrightarrow{U} [E \otimes_A B(\overline{C})]^r$$

where U is the evident shuffle. We define Δ on $W_i \otimes B_{st}(\overline{C})$ by induction on i and for fixed i by induction on s. Formula (i) defines Δ for $s = 0$ and all $i > 0$ and formula (ii) and π-equivariance defines Δ for $i = 0$ and all s. Let $i \geq 1$ and $s \geq 1$ and assume that Δ has been defined for $i' < i$ and for our given i and $s' < s$. Let $\{w_k\}$ be a $\Lambda\pi$-basis for W_i. By (iii) and π-equivariance, it suffices to define $\Delta(w \otimes S(y))$ for $w \in \{w_k\}$ and $y \in B_{s-1,*}(\overline{C})$. Let $S = \sum_{i=1}^{r} (\sigma\varepsilon)^i \otimes S \otimes 1^{r-i-1}$ on $B(\overline{C})^r$. Then $dS + Sd = 1 - (\sigma\varepsilon)^r$. We define

(v) $\Delta(w \otimes S(y)) = (-1)^{\deg w} S\Delta(w \otimes y) + S\Delta(d(w) \otimes S(y))$.

Observe that (v) is equivalent to (ii) on $w = e_o$ and that (v) is well-defined by the induction hypothesis. To verify that $d\Delta = \Delta d$, write (v) in the form $\Delta(1 \otimes S) = S\Delta(1 \otimes 1 + d \otimes S)$. Then:

$$\begin{aligned} d\Delta(1 \otimes S) &= dS\Delta(1 \otimes 1 + d \otimes S) = (1 - Sd)\Delta(1 \otimes 1 + d \otimes S) \\ &= [\Delta - S\Delta(d \otimes 1 + 1 \otimes d)](1 \otimes 1 + d \otimes S) \\ &= \Delta + \Delta(d \otimes S) - S\Delta(d \otimes 1) - S\Delta(1 \otimes d) + S\Delta(d \otimes dS) \\ &= \Delta + \Delta(d \otimes S) - S\Delta(d \otimes 1) - S\Delta(1 \otimes d) + S\Delta(d \otimes 1) - S\Delta(d \otimes Sd) \\ &= \Delta + \Delta(d \otimes S) - \Delta(1 \otimes dS) = \Delta + \Delta(d \otimes S) - \Delta + \Delta(1 \otimes dS) \\ &= \Delta(d \otimes 1 + 1 \otimes d)(1 \otimes S) \end{aligned}$$

(where no terms involving $\sigma\varepsilon$ are relevant by (i) and an easy verification). Thus (i), (ii), (iii), and (v), together with π-equivariance, provide an explicit construction of a natural morphism Δ of $\Lambda\pi$-complexes. To see that (iv) holds, observe that if $w \in \{w_k\} \subset W_i$ and $y = a_1[a_2|\dots|a_s]f$, then $\Delta(w \otimes S(y))$ is a linear combination in $B(\overline{C})^r$ of terms involving precisely the factors $a_i^{(j)}$ and $f^{(j)}$ in the $B(\overline{C})$, where $\psi(a_i) = \sum a_i^{(1)} \otimes \dots \otimes a_i^{(r)}$ and $\psi(f) = \sum f^{(1)} \otimes \dots \otimes f^{(r)}$ give the iterated coproducts. Thus no summand of $\Delta(w \otimes S(y))$ can have homological degree greater than rs. Since $\Delta(w \otimes S(y))$ has homological degree $i+s > rs$ if $i > (r-1)s$, this proves (iv). The uniqueness of Δ up to $\Lambda\pi$-homotopy follows easily by use of the contracting homotopy S on $B(\overline{C})^r$.

We now pass to the category $\mathcal{P}(\pi, \infty, \Lambda)$ of Defitions 2.1.

Definition 11.4. Let $C \in \mathcal{C}$. Let $\alpha: B^*(C)^r \to [B(C)^r]^*$ be the natural map and define a $\Lambda\pi$-morphism $\theta: W \otimes B^*(C) \to B^*(C)^r$ by the formula

(11) $\theta(w \otimes x)(k) = (-1)^{\deg w \deg x} \alpha(x)\Delta(w \otimes k)$, $w \in W$, $x \in B^*(C)^r$, $k \in B(C)$.

Since θ may be defined for $\pi = \Sigma_r$ and then factored through $j: W \to V$ as in Definition 2.1, and the resulting composite is naturally $\Lambda\pi$-homotopic to the original θ defined in terms of W, θ satisfies condition (ii) of Definition 2.1. By the lemma, formula (11) specializes to give

(12) $\theta(e_0 \otimes x) = D^*\alpha(x)$ for any $x \in B^*(C)^r$ and

(13) $\theta(w \otimes x) = \varepsilon(w)D^*\alpha(x)$ for any $x \in B^{0,*}(C)^r$ and $w \in W$.

By (10) and (12), θ satisfies condition (i) of Definition 2.1. Since θ is natural on morphisms in $\mathcal{C}$, we thus obtain a functor $\Gamma: \mathcal{C} \to \mathcal{P}(\pi, \infty, \Lambda)$ by setting $\Gamma(C) = (B^*(C), \theta)$ on objects and $\Gamma(\gamma) = B^*(\gamma)$ on morphisms. By (13), if C is unital in $\mathcal{C}$, then $\Gamma(C)$ is unital in $\mathcal{P}(\pi, \infty, \Lambda)$. If $\Lambda = Z_p$, π is cyclic of order p, and $C = \tilde{C} \otimes Z_p$ where $\tilde{C}$ is Z-free (that is, $\tilde{E}, \tilde{A}$, and $\tilde{F}$ are Z-free), then we agree to choose θ for C to be the mod p reduction of θ for $\tilde{C}$; $\Gamma(C)$ will thus be reduced mod p. Note that if $x \in B^*(C)$ has bidegree (s,t), then $\theta(w \otimes x)$ has bidegree $(s - \deg w, t)$. $W \otimes B^*(C)^r$ and $B^*(C)$ should be thought of as regarded by total degree in defining the functor Γ.

Observe that, by Definition 6.1, we now have $\cup_i$-products in $B^*(C)$ for any $C \in \mathcal{C}$. When $\Lambda = Z_p$, the results of Proposition 2.3 will clearly apply to the Steenrod operations in $H^*(C)$, and the following lemmas will imply the applicability of the external Cartan formula and the Adem relations.

Lemma 11.5. For any objects C and C' in $\mathcal{C}$, the following diagram is $\Lambda\pi$-homotopy commutative.

$$\begin{array}{ccc}
W \otimes B^*(C \otimes C')^r & \xrightarrow{\ \theta\ } & B^*(C \otimes C') \\
{\scriptstyle 1\otimes(\xi^*)^r}\uparrow\ \downarrow{\scriptstyle 1\otimes(\eta^*)^r} & & {\scriptstyle \xi^*}\uparrow\ \downarrow{\scriptstyle \eta^*} \\
W \otimes [B^*(C) \otimes B^*(C')]^r & \xrightarrow{\ \tilde{\theta}\ } & B^*(C) \otimes B^*(C')
\end{array}$$

Proof. It suffices to prove the $\Lambda\pi$-homotopy commutativity of the diagram

$$\begin{array}{ccc}
W \otimes B(C) \otimes B(C') & \xrightarrow{\ U(\Delta\otimes\Delta)(1\otimes T\otimes 1)(\psi\otimes 1\otimes 1)\ } & [B(C) \otimes B(C')]^r \\
{\scriptstyle 1\otimes\xi}\uparrow\ \downarrow{\scriptstyle 1\otimes\eta} & & {\scriptstyle \xi^r}\uparrow\ \downarrow{\scriptstyle \eta^r} \\
W \otimes B(C \otimes C') & \xrightarrow{\ \Delta\ } & B(C \otimes C')^r
\end{array}$$

and this diagram need only be studied with C and C' replaced by $\overline{C}$ and $\overline{C}'$. Since $B(\overline{C} \otimes \overline{C}')^r$ and $[B(\overline{C}) \otimes B(\overline{C}')]^r$ have obvious contracting homotopies, the result follows by an easy double induction like that in the proof of Lemma 11.3.

Corollary 11.6. If $C \in \mathcal{C}$, then $\Gamma(C)$ is a Cartan object of $\mathcal{C}(\pi, \infty, \Lambda)$

Lemma 11.7. If $C \in \mathcal{C}$, $\Lambda = Z_p$, then $\Gamma(C)$ is an Adem object of $\mathcal{C}(p, \infty)$.

Proof. Precisely as in the proof of Lemma 7.8, it suffices to prove the τ-homotopy commutativity of the following diagram:

$$\begin{array}{ccccc}
W_1 \otimes W_2{}^p \otimes B(C) & \xrightarrow{w \otimes 1} & Y \otimes B(C) & \xrightarrow{\quad\Delta\quad} & B(C)^{p^2} \\
\Big\downarrow {\scriptstyle T \otimes 1} & & & & \Big\uparrow {\scriptstyle \Delta^p} \\
W_2{}^p \otimes W_1 \otimes B(C) & \xrightarrow{1 \otimes \Delta} & W_2{}^p \otimes B(C)^p & \xrightarrow{\quad U \quad} & (W_2 \otimes B(C))^p
\end{array}$$

We need only consider this diagram with C replaced by $\overline{C}$, and, since $B(\overline{C})^{p^2}$ has a contracting homotopy, the result then holds by another easy double induction.

The following theorem summarizes the properties of the P^s and βP^s on $H^*(C)$ for $C \in \mathcal{C}$, $\Lambda = Z_p$. We shall be very precise as to grading since there is considerable confusion on this point in the literature. We are thinking of $H^*(C)$ as regraded by total degree in applying our general theory. An alternative formulation that is sometimes convenient will be given after the theorem.

Theorem 11.8. Let $C \in \mathcal{C}$, $\Lambda = Z_p$. Then there exist natural homomorphisms P^i and, if $p > 2$, βP^i defined on $H^*(C)$, with

(a) $P^i : H^{s,t}(C) \longrightarrow H^{s+i-t,\,2t}(C)$ if $p = 2$;

(b) $P^i : H^{st}(C) \longrightarrow H^{s+(2i-t)(p-1),\,pt}(C)$ and

$\beta P^i : H^{st}(C) \longrightarrow H^{s+1+(2i-t)(p-1),\,pt}(C)$ if $p > 2$.

These operations satisfy the following properties:

(i) $P^i = 0$ if $p = 2$ and either $i < t$ or $i > s+t$

$P^i = 0$ if $p > 2$ and either $2i < t$ or $2i > s+t$

$\beta P^i = 0$ if $p > 2$ and either $2i < t$ or $2i \geq s+t$

(ii) $P^i(x) = x^p$ if $p = 2$ and $i = s+t$ or if $p > 2$ and $2i = s+t$

(iii) If $C = \tilde{C} \otimes Z_p$, where $\tilde{C}$ is Z-free, then $\beta P^{i-1} = iP^i$ if $p = 2$ and βP^i is the composition of β and P^i if $p > 2$.

(iv) $P^j = \sum P^i \otimes P^{j-i}$ and $\beta P^j = \sum (\beta P^i \otimes P^{j-i} + P^i \otimes \beta P^{j-i})$ or $H^*(C \otimes C')$; the internal Cartan formula is satisfied in $H^*(C)$.

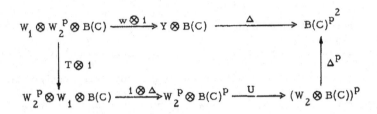

(v) If $\gamma: C' \longrightarrow C$ and $\phi: C \longrightarrow C''$ are unital morphisms in $\mathcal{C}$ such that $\phi\gamma = 0$ on the cokernels of the units, then $\sigma P^i = P^i\sigma$ and $\sigma\beta P^i = -\beta P^i\sigma$, where $\sigma: H^{st}(C'') \longrightarrow H^{s-1,t}(C')$ is the suspension.

(vi) The P^i and βP^i satisfy the Adem relations as stated in Corollary 5.1.

Proof. If $x \in H^{st}(C)$, then $D_i(x) = \theta_*(e_i \otimes x^p) \in H^{ps-i,pt}(C)$. The P^i and βP^i are defined by formulas (5.1) and (5.2), with x having its total degree $q = s+t$; thus (a) and (b) are valid. The vanishing of $P^i(x)$ for $i < t$ if $p = 2$ and of $\beta^\varepsilon P^i(x)$ for $2i < t$ if $p > 2$ follows from part (iv) of Lemma 11.3. The remainder of the theorem follows immediately from our general theory and the previous lemmas. For (v), note that the composite $B^*(C'') \xrightarrow{B^*(\phi)} B^*(C) \xrightarrow{B^*(\gamma)} B^*(C')$ is zero on the kernel of the augmentation $B^*(C'') \longrightarrow Z_p$. An alternative formulation of (v) in the non-unital case can easily be obtained.

In addition to (v), the Kudo transgression theorem, Theorem 3.4, applies to appropriate spectral sequences involving objects of $\mathcal{C}$. The hypothesis of (iii) is seldom satisfied in practice, and βP^i is generally an independent operation having nothing to do with any Bockstein. There is an alternative definition of the operations, which amounts to the following regrading of our operations. Define

(c) $\tilde{P}^i = Sq^i = P^{i+t}: H^{st}(C) \longrightarrow H^{s+i,2t}(C)$ if $p = 2$;

(d) $\tilde{P}^i = P^{i+t}: H^{s,2t}(C) \longrightarrow H^{s+2i(p-1),2pt}(C)$ and

$\beta\tilde{P}^i = \beta P^{i+t}: H^{s,2t}(C) \longrightarrow H^{s+2i(p-1)+1,2pt}(C)$ if $p > 2$.

This regrading is reasonable if $p = 2$, but has the effect of eliminating all operations on $H^{st}(C)$ for t odd if $p > 2$; of course, these operations are non-trivial since, if s and t are odd, the p-th power operation on $H^{st}(C)$ is non-trivial in general. The results of the theorem can easily be transcribed for the $\tilde{P}^i$ and $\beta\tilde{P}^i$; for example, the Adem relations are still correct precisely as stated but with all P^i and βP^i replaced by $\tilde{P}^i$ and $\beta\tilde{P}^i$. The motivation for the reindexing is just the desire to make $\tilde{P}^0$ the first non-trivial operation. This operation is of particular importance in the applications, and we now evaluate it.

Definition 11.9. Let $C = (E, A, F) \in \mathcal{C}$, where E, A, and F are positively graded and of finite type. Then $B^*(C)$ may be identified with $E^* \otimes T(IA^*) \otimes F^*$. Define $\lambda : B^{st}(C) \longrightarrow B^{s,pt}(C)$ by

$$\lambda(\varepsilon[\alpha_1|\dots|\alpha_s]\phi) = \varepsilon^p[\alpha_1{}^p|\dots|\alpha_s{}^p]\phi^p .$$

Then λ commutes with the differential and induces $\lambda_* : H^{st}(C) \longrightarrow H^{s,pt}(C)$. Of course, if $p > 2$, then $\lambda(\varepsilon[\alpha_1|\dots|\alpha_s]\phi) = 0$ if ε, α_i, or ϕ has odd degree and thus $\lambda_* = 0$ if t is odd.

Proposition 11.10. Let $C = (E, A, F) \in \mathcal{C}$, where E, A, and F are positively graded and of finite type. Let $x \in H^{s,t}(C)$ where t is even if $p > 2$. Then $\tilde{P}^o(x) = \lambda_*(x)$.

Proof. Let $y = e[a_1|\dots|a_s]f \in B_{s,pt}(C)$. A straightforward, but tedious, calculation demonstrates that

$$\Delta(e_{s(p-1)} \otimes y) = \sum (-1)^{ms}\, \nu(-s)^{-1}(e'[a_1'|\dots|a_s']f') + Nz,$$

where the sum is taken over the symmetric summands $e' \otimes \dots \otimes e'$, $a_i' \otimes \dots \otimes a_i'$, and $f' \otimes \dots \otimes f'$ of the iterated coproducts. (A moment's reflection on the case $C = (Z_p, Z_pG, Z_p)$, where G is a group, and a glance at Lemma 8.2 should convince the reader of the plausibility of this statement.) The result now follows easily from the definitions.

Remarks 11.11. If $p > 2$, then it can be shown by a tedious calculation that $-\beta\tilde{P}^o(x) = \langle x\rangle^p$ for $x \in H^{1,2t}(C)$, where $C = (Z_p, A, Z_p) \in \mathcal{C}$ and $\langle x\rangle^p$ is as defined in Remarks 6.9. It is possible that $-\beta\tilde{P}^s(x) = \langle x\rangle^p$ for $x \in H^{2s+1,2t}(C)$ and any $C \in \mathcal{C}$, but this appears to be difficult to prove.

Bibliography

1. J.F. Adams, On the structure and applications of the Steenrod algebra. Comment. Math. Helv., 1958.

2. J. Adem, The relations on Steenrod powers of cohomology classes. Algebraic geometry and topology. A symposium in honor of S. Lefschetz, 1957.

3. S. Araki and T. Kudo, Topology of H_n-spaces and H-squaring operations. Mem. Fac. Sci. Kyusyu Univ. Ser. A, 1956.

4. W. Browder, Homology operations and loop spaces. Illinois J. Math., 1960.

5. A Dold, Uber die Steenrodschen Kohomologieoperationen. Ann. of Math., 1961.

6. E. Dyer and R.K. Lashof, Homology of iterated loop spaces. Amer. J. Math., 1962.

7. D.B.A. Epstein, Steenrod operations in homological algebra. Invent. Math. I, 1966.

8. G. Hirsch, Quelques proprietes des produits de Steenrod. C.R. Acad. Sci. Paris, 1955.

9. N. Jacobson, Lie Algebras. Interscience Publishers. 1962.

10. S. Kochman, Ph.D. Thesis. University of Chicago, 1970.

11. D. Kraines, Massey higher products. Trans. Amer. Math. Soc. 1966.

12. T. Kudo, A transgression theorem. Mem. Fac. Sci. Kyusyu Univ. Ser. A , 1956.

13. A. Liulevicius, The factorization of cyclic reduced powers by secondary cohomology operations. Mem. Amer. Math. Soc., 1962.

14. S. MacLane, Homology. Academic Press, 1963.

15. J.P. May, Simplicial objects in algebraic topology. D. Van Nostrand
npany, 1967.

16. ________, The structure and applications of the Eilenberg-Moore
ctral sequences (to appear).

17. ________, Homology operations in iterated loop spaces (to appear).

18. ________, The cohomology of the Steenrod algebra (to appear).

19. J. Milnor and J.C. Moore, On the structure of Hopf algebras. Ann. of
h., 1965.

20. J.C. Moore, Constructions sur les complexes d'anneaux. Seminaire
ri Cartan, 1954/55.

21. G. Nishida, Cohomology operations in iterated loop spaces. Proc.
in Acad., 1968.

22. S. Priddy, Primary cohomology operations for simplicial Lie algebras.
ppear in Ill. J. Math.

23. J.P. Serre, Cohomologie modulo 2 des complexes d'Eilenberg-MacLane.
ment. Math. Helv., 1953.

24. E. Spanier, Algebraic Topology. McGraw Hill Book Company, 1966.

25. N.E. Steenrod, Products of cocycles and extensions of mappings.
of Math., 1947.

6. ________, Reduced powers of cohomology classes. Ann. of Math.,

7. ________, Homology groups of symmetric groups and reduced power
tions. Proc. Nat. Acad. Sci., U.S.A., 1953.

8. ________, Cyclic reduced powers of cohomology classes. Proc.
Acad. Sci., U.S.A., 1953.

9. ________, Cohomology operations derived from the symmetric group.
ment. Math. Helv., 1957.

30. N. E. Steenrod, Cohomology operations, Princeton University Press, 1962.

SOME PROPERTIES OF THE LOOP HOMOLOGY OF COMMUTATIVE COALGEBRAS*

by John C. Moore

§1. Recollection of fundamental definitions and properties.

Let k be a field. A (differential graded) coalgebra C over k is a differential graded vector space C over k together with morphisms of differential graded vector spaces $\Delta(C): C \to C \otimes C$ and $\varepsilon(C): C \to k$ such that $\Delta(C)$ (the diagonal or comultiplication of C) is associative and $\varepsilon(C)$ is a unit for $\Delta(C)$ ([2]). The coalgebra C is n-connected if $\varepsilon(C)_j$ is an isomorphism for $j \leq n$, where n is a nonnegative integer. A connected coalgebra is a 0-connected coalgebra, and a simply connected coalgebra is a 1-connected coalgebra.

In this paper only connected commutative coalgebras will be treated. Hence the word coalgebra will be taken to mean connected coalgebra with commutative comultiplication. It is perhaps worth recalling that the notion of commutativity for coalgebras (or algebras) is a diagrammatic notion involving the standard interchange of factors isomorphism for the tensor product of an ordered pair of differential graded vector spaces, and hence involves some signs.

To every coalgebra C there is assigned a (differential graded) Hopf algebra $\Omega(C)$ called the loop algebra of C. This may be described crudely as follows: For X a differential graded vector space, let $s^{-1}X$ denote the desuspension of X. Now if C is a coalgebra, there is a short exact sequence $0 \to k \to C \to J(C) \to 0$ of differential graded vector spaces. The underlying algebra of $\Omega(C)$ is the tensor algebra of $s^{-1}J(C)$, i.e. $T(s^{-1}J(C))$, and its diagonal is defined by demanding that for $n \in \mathbb{Z}$, and $x \in (s^{-1}J(C))_n$, the element x be primitive, i.e. $\Delta(x) = x \otimes 1 + 1 \otimes x$. The differential in $\Omega(C)$ comes in two parts d' and d'', the actual differential being $d' + d'' = d$, and the identity $d'd'' = -d''d'$ being satisfied. Both d' and d'' are compatible with the multiplicative and diagonal structure of $\Omega(C)$. The internal differential d' is the

*This research supported in part by NSF Grant GP-14590.

one obtained by identifying $\Omega(C)$ momentarily with $T(s^{-1}J(C))$ as a differential algebra, while d'' is the unique differential such that the diagram

$$\begin{array}{ccc} J(C) & \xrightarrow{J(\Delta(C))} & J(C) \otimes J(C) \\ \downarrow \tau & & \downarrow \tau \otimes \tau \\ \Omega(C) & \xrightarrow{d''} & \Omega(C) \otimes \Omega(C) \end{array}$$

is commutative where τ is the canonical morphism of degree -1 (note that $\tau \otimes \tau$ is of degree -2, whence the identity $d'd'' = -d''d'$). The special case where the differential of C is zero was described earlier ([5]), and as before the fact that d'' is compatible with the diagonal structure of $\Omega(C)$ depends on the fact that C is commutative. Observe that $\Omega(C)$ is connected if and only if C is simply connected.

If $f\colon C' \to C''$ is a morphism of coalgebras, there is induced a morphism of Hopf algebras, $\Omega(f)\colon \Omega(C') \to \Omega(C'')$, and $C \leadsto \Omega(C)$ is a functor from coalgebras to primitively generated (differential graded) Hopf algebras. The functor Ω is a coadjoint functor which is essentially equivalent to the fact that it is colimit preserving. These facts are essentially well known in case the characteristic of k is zero, and indeed for the case k = the field of rational numbers occurs as a minor part of the work of Quillen on rational homotopy theory [6]. The preceding adjointness relation will not be made explicit here since it is not important for this paper.

Recall that the category of coalgebras is pointed with point k. A sequence $C' \xrightarrow{f'} C \xrightarrow{f''} C''$ of coalgebras is a <u>fibration</u> if f' is the kernel of f'', and if forgetting the differential operators C is an injective C'' comodule. Another way of expressing the second part of the fibration condition is as follows: For p a positive integer, let F_pC'' be the subcoalgebra of C'' which as a differential graded vector space is the p skeleton of C'', and let $F_pC = C \mathbin{\square}_{C''} F_pC''$, and $F_p(C' \otimes C'') = C' \otimes F_pC''$. The condition that the sequence be a fibration now says that there exists $\theta\colon C \to C' \otimes C''$ a nondifferential isomorphism of filtered C'' comodules such that $E^0(\theta)\colon E^0(C) \to E^0(C' \otimes C'')$ is a differential isomorphism. Thus a fibration gives rise to a Serre spectral sequence with a good

E^1 term, namely $H(C') \otimes C''$ in the notation used above. Further if C'' is simply connected, then it can be shown that the filtration of C is perfect ([2], p. 215), thus the differential d^1 has the desired form, the Serre spectral sequence has the desired E^2 term, i.e. $E^2 = H(C') \otimes H(C'')$, $E^2_{p,q} = H(C')_q \otimes H(C'')_p$, and further the homology of the fibre $H(C')$ may be identified with the differential derived functor $\mathrm{Cotor}^{C''}(C,k)$ or, since C'' is commutative, with $\mathrm{Cotor}^{C''}(k,C)$.

Observe that the tensor product is the product in the category of coalgebras since coalgebras have been assumed commutative. Thus a (connected) Hopf algebra with commutative diagonal is just a group in the category of coalgebras, and a principal fibration $C' \xrightarrow{f'} C \xrightarrow{f''} C''$ is a fibration such that C' is a group and acts on C in a manner compatible with the projection f''. Indeed C' acts freely on C and the orbit coalgebra is C''. Principal fibrations may be characterized by these properties instead of those used in the definition. The filtration of the preceding paragraph is always perfect if the fibration is principal regardless of whether or not the base C'' is simply connected.

Now a coalgebra C is acyclic if $H(C) = k$, i.e. if $H(\eta)$: $k \to H(C)$ is an isomorphism, or equivalently $H(\varepsilon)$: $H(C) \to k$ is an isomorphism. If C is a simply connected coalgebra, then there is a canonical principal fibration $\Omega(C) \to E(C) \to C$ with $E(C)$ acyclic. A classical argument shows that if $\Omega'(C) \to E'(C) \to C$ is any other principal fibration with $E'(C)$ acyclic, then the Hopf algebras $H(\Omega(C))$ and $H(\Omega'(C))$ are isomorphic. Note $H(\Omega(C))$ is a primitively generated Hopf algebra since $\Omega(C)$ is primitively generated. Observe also that studying the loop homology of C is the same thing as studying $\mathrm{Cotor}^C(k,k)$, and that there is a canonical spectral sequence $E^2 = \mathrm{Cotor}^{H(C)}(k,k) \Longrightarrow H(\Omega(C))$ ([2], [4], and [5]).

Any principal fibration may be described using a twisting cochain [3]. The canonical acyclic fibration above for a simple connected coalgebra C has twisting cochain the composite $C \to J(C) \xrightarrow{\approx} s^{-1}(J(C)) \to \Omega(C)$.

One of the principal aims of this paper is to show that if $C' \to C \to C''$ is a fibration sequence of simply connected coalgebras, then the sequence of Hopf algebras $H(\Omega(C')) \to H(\Omega(C)) \to H(\Omega(C''))$ is exact ([4], [5]). In fact a somewhat more precise result will be obtained, and then used to obtain further information concerning the loop homology of simply connected coalgebras.

Intuitively one should think that the category of coalgebras is somewhat like the category of pointed topological spaces but somewhat simpler. Indeed as the work of Quillen ([6]) has shown this idea is realistic if the ground field is the field of rational numbers, and the results obtained here can be given a proper geometric interpretation in this case. The case of finite characteristic does not have such a simple geometric interpretation, but must be thought of as giving algebraic information regarding certain functors which occur as the E^2 term of certain spectral sequences relating fibrations and homological algebra.

Just as with spaces, if $f\colon C \to C''$ is a morphism of coalgebras, there is a commutative diagram

$$\begin{array}{ccc} C & \xrightarrow{\;f\;} & \\ {\scriptstyle\iota}\downarrow & & C'' \\ \tilde{C} & \xrightarrow{\;\tilde{f}\;} & \end{array}$$

such that $\tilde{f}$ is a fibration, and $H(\iota)$ is an isomorphism. This may be proved in a particularly simple fashion when C'' is a simply connected group. In that case $\tilde{C} = C \otimes E(C'')$, and let $\tilde{f}$ be the composite $C \otimes E(C'') \xrightarrow{f\otimes\pi} C'' \otimes C'' \xrightarrow{Q(C'')} C''$. Then if one lets ι be the morphism whose projection on the first factor is f, and whose projection on the second factor is the trivial morphism, one obtains a diagram having the desired properties.

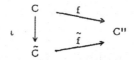

§2. The behavior of the loop homology of a simply connected fibration.

Theorem. If the characteristic of the ground field k is zero, and if $C' \xrightarrow{f'} C \xrightarrow{f''} C''$ is a fibration sequence of simply connected coalgebras, then

1) there is a bicomplete filtration on $H(\Omega(C'))$ such that $E^0 H(\Omega(C')) = \mathrm{Cotor}^{H(\Omega(C''))}(k, H(\Omega(C)))$, and such that $E^0_{0,*} H(\Omega(C'))$ is the image of the morphism $H(\Omega(f'))$ which is also the kernel of the morphism $H(\Omega(f''))$, and

2) the sequence of Hopf algebras

$$H(\Omega(C')) \xrightarrow{H(\Omega(f'))} H(\Omega(C)) \xrightarrow{H(\Omega(f''))} H(\Omega(C''))$$

is exact.

Proof. The spectral sequence such that $E^2 = \mathrm{Cotor}^{H(\Omega(C''))}(k, H(\Omega(C)))$ and $E^r \Longrightarrow H(\Omega(C'))$ ([2]) has the property that it is a spectral sequence of Hopf algebras with all primitive elements having either filtration 0 or -1 ([4], [5]). Hence $E^2 = E^\infty$ in this spectral sequence, and the theorem follows.

The preceding is just a restatement of Proposition 7.6 of [5] in a different context.

Theorem. If the characteristic of the field k is p odd, and if $C' \xrightarrow{f'} C \xrightarrow{f''} C''$ is a fibration sequence of simply connected coalgebras, then there is a short exact sequence

$$k \longrightarrow I \xrightarrow{\iota} N \xrightarrow{\pi} D \longrightarrow k$$

of Hopf algebras such that I is the image of $H(\Omega(f'))$, N is the kernel of $H(\Omega(f''))$, $P(D) = Q(D)$, and $P(D)_q = 0$ for $q \not\equiv -1 \bmod 2P$.

Proof. If B is the cokernel of $H(\Omega(f''))$, then there is a spectral sequence of Hopf algebras such that $E^2 = \mathrm{Cotor}^B(k,k) \otimes N$ and $E^r \Longrightarrow H\Omega(C)$ ([2], [4], and [5]). The form of the spectral sequence ([5]), then shows that I is a normal sub Hopf algebra of N and that the cokernel of the natural morphism $\iota : I \longrightarrow N$ is an exterior algebra D such that $P(D) = Q(D)$, and $P(D)_q = 0$ for $q \not\equiv \pm 1 \bmod 2P$.

There is a fibration sequence of Hopf algebras $F' \xrightarrow{g'} F \xrightarrow{g''} \Omega(C)$ such that F' is equivalent with $\Omega^2(C'')$ and F with $\Omega(C)$, and such that g'' is equivalent with $\Omega(f')$. Let L be the kernel of $H(g')$ and M the cokernel.

Note that the underlying algebra of L is free commutative. There is a spectral sequence of Hopf algebras such that $E^2 = M \otimes \mathrm{Tor}^L(k,k)$ and $E^r \Longrightarrow H\Omega(C)$ which maps naturally into another $''E^2 = \mathrm{Tor}^{H\Omega(F')}(k,k)$ $''E^r \Longrightarrow H\Omega(C'')$. Noting that M and I may be identified and that in both spectral sequences the only non-trivial differential is d^{p-1}, it is not difficult to see that there is a Hopf filtration on D such that $E^o(D)$ is the kernel of $k \otimes_I E^p \longrightarrow {''E^p}$. Now the form of the spectral sequence ([5], dualized) implies that $P(E^o(D))_q = 0$ unless $q \equiv -1 \bmod 2P$, and the theorem follows.

§3. Coalgebras, adjointness, and some elementary calculations of loop homology.

Let Coalg(k) denote the category of coalgebras described earlier, and let $\mathcal{C}^1(k)$ denote the full subcategory of the category of differential graded modules over k generated by those objects X such that $X_q = 0$ for $q < 1$. Now one may consider that $J(\ ): \mathrm{Coalg}(k) \to \mathcal{C}^1(k)$ is a functor, where for a coalgebra C, J(C) is the augmentation coideal of C considered as a differential graded vector space (§1). The functor J is a coadjoint functor, and we denote by $\Gamma(\ ): \mathcal{C}^1(k) \to \mathrm{Coalg}(k)$ its coadjoint $(\alpha,\beta): J \to \Gamma$, $\beta(C): C \to \Gamma J(C)$ for C a coalgebra.

If X is an object of $\mathcal{C}^1(k)$, then $\Gamma(X)$ may be described in the following way. Let $T_n(X)$ be the n-fold tensor product of X with itself for n a positive integer if $x_j \in X_{n_j}$ for $j = 1,\dots,n$, let $(x_1,\dots,x_n)$ denote the tensor product element in $T_m(X)$ where $m = \Sigma_{j=1}^n n_j$. Let $\tilde{T}(X) = \amalg_{n \geq 0} T_n(X)$ and let $\Delta: \tilde{T}(X) \to \tilde{T}(X) \otimes \tilde{T}(X)$ be the morphism such that $\Delta(x_1,\dots,x_n) = $ $= (x_1,\dots,x_n) \otimes 1 + 1 \otimes (x_1,\dots,x_n) + \Sigma_{j=1}^{n-1} (x_1,\dots,x_j) \otimes (x_{j+1},\dots,x_n)$. Now $\tilde{T}(X)$ together with Δ would be a coalgebra except that Δ is not commutative, and $\Gamma(X)$ is the maximal "subcoalgebra" of $\tilde{T}(X)$ together with Δ. Note that $\Gamma(X \pi Y) = \Gamma(X) \otimes \Gamma(Y)$ since due to the fact that Γ is an adjoint it preserves products. If $n \in \mathbb{Z}$, $n > 0$, and n is odd then if $X_q = 0$ for $q \neq n$, and X_n is 1-dimensional over k, then $\Gamma(X)_q = 0$ for $q \neq 0, n$, and $\Gamma(X)_0$, $\Gamma(X)_n$ are 1-dimensional over k unless the characteristic of k is 2 in which case $\Gamma(X)_q = 0$ for $q \not\equiv 0 \pmod n$ and $\Gamma(X)_q$ is 1-dimensional over k for $q \equiv 0 \pmod n$. If n is even and X is as above then $\Gamma(X)_q = 0$ for $q \not\equiv 0 \pmod n$ and $\Gamma(X)_q$ is 1-dimensional over k for $q \equiv 0 \pmod n$ and if γ_j denotes an appropriately chosen basis element of $\Gamma(X)_{jn}$ then $\Delta(\gamma_j) = \Sigma_{r+s=j} \gamma_r \otimes \gamma_s$, and one usually denotes γ_0 by 1.

Since the diagonal of $\Gamma(X)$ does not depend on the differential of X, the preceding paragraph describes $\Gamma(X)$ completely for X of finite type. Now one has the addition map $X \pi X \to X$ in $\mathcal{C}^1(k)$, and this defines a multiplication $\Gamma(X) \otimes \Gamma(X) \to \Gamma(X)$. If the characteristic of k is not 2, the underlying algebra of the Hopf algebra $\Gamma(X)$ is the tensor product of the divided polynomial algebra generated by the even part of X and the exterior algebra generated by the odd part of X. If the characteristic of k is 2, $\Gamma(X)$ is the divided polynomial algebra generated by X.

If the characteristic of k is zero, then $H\Gamma(X) = \Gamma(H(X))$, but this is not the case if the characteristic of k is $p \neq 0$. Suppose p is odd, and decompose X so that $X = H(X) \pi X' \pi X''$, X', X'' are acyclic, all nontrivial boundaries in X' are of odd degree, and all nontrivial boundaries in X'' are of even degree. Now $H(\Gamma(X)) = \Gamma(H(X)) \otimes H\Gamma(X') \otimes H\Gamma(X'')$, $H(\Gamma(X')) = k$, and if $X''_q = 0$ for $q \neq 2n, 2n+1$, $y \in X''_{2n}$, $z \in X''_{2n+1}$ form a basis for X'', and $dz = y$, then $H\Gamma(X'') = \Gamma[\gamma_p(y), \gamma_{p-1}(y)z]$. If $p = 2$, let $X = H(X) \pi X''$, X'' acyclic. Then $H\Gamma(X) = \Gamma(H(X)) \otimes H\Gamma(X'')$ and if $X''_q = 0$ for $q \neq n, n+1$, $y \in X''_n$, $z \in X''_{n+1}$, $dz = y$ are a basis of X'', then $H\Gamma(X'') = \Gamma[\gamma_2(y)]$.

The objects in $\mathrm{Coalg}(k)$ of the form $\Gamma(X)$ are the objects of an injective class. If $\mathrm{Coalg}^{o}(k)$ denotes the full subcategory of $\mathrm{Coalg}(k)$ generated by those objects with zero differential, then the objects in $\mathrm{Coalg}^{o}(k)$ of the form $\Gamma(X)$ (note this implies $d(X) = 0$), are the actual injectives in $\mathrm{Coalg}^{o}(k)$. Thus the preceding paragraph shows that if X is an object of $\mathcal{C}^1(k)$, then the object $H(\Gamma(X))$ of $\mathrm{Coalg}^{o}(k)$ is injective even though one needs to assume that the characteristic of k is zero in order to have that $H(\Gamma(X)) = \Gamma(H(X))$ in general.

Now if X is an object of $\mathcal{C}^1(k)$, let $S(X)$ denote the group over $\mathrm{Coalg}(k)$ such that as an algebra $S(X)$ is the free commutative algebra generated by X and the elements of X are primitive. Thus one obtains a functor $X \rightsquigarrow S(X)$ from $\mathcal{C}^1(k)$ to the groups over $\mathrm{Coalg}(k)$. Indeed $S(X)$ is primitively generated. If the characteristic of k is zero, then $HS(X) = SH(X)$. Suppose the characteristic of k is p odd. Decompose X so that $X = H(X) \pi X' \pi X''$, X', X'' are acyclic, all nontrivial boundaries in X' are of odd degree, and all nontrivial boundaries in X'' are of even degree. Now $HS(X) = S(H(X)) \otimes HS(X') \otimes HS(X'')$, $HS(X'') = k$, and if $X'_q = 0$ for $q \neq 2n-1, 2n$, $y \in X'_{2n-1}$, $z \in X'_{2n}$ form a base for X', and $dz = y$, then $HS(X') = S[yz^{p-1}, z^p]$. If the characteristic of k is two, let $X = H(X) \pi X''$, X'' acyclic. Then $HS(X) = S(H(X)) \otimes HS(X'')$, and $X''_q = 0$ for $q \neq n, n+1$, $y \in X''_n$, $z \in X''_{n+1}$, $dz = y$ are a base of X'', then $HS(X'') = S[z^2]$.

Now it follows that for any k, if X is an object of $\mathcal{C}^1(k)$, there exists an object W of $\mathcal{C}^1(k)$ with zero differential and containing $H(X)$ such that $HS(X) = S(W)$.

Proposition. If the characteristic of k is not two, and X is an object of $\mathcal{C}^1(k)$ such that $X_1 = 0$, then there is a principal fibration

$$S(s^{-1}X) \xrightarrow{\iota} E^{\#}(X) \xrightarrow{\pi} \Gamma(X)$$

in Coalg(k) with $E^{\#}(X)$ acyclic.

Proof. Let $\tau: \Gamma(X) \to S(s^{-1}X)$ be the composite $\Gamma(X) \longrightarrow J\Gamma(X) \xrightarrow{\beta(X)} X \xrightarrow{\approx} s^{-1}X \longrightarrow S(s^{-1}X)$. Observe that τ is a twisting function ([3]). The resulting principal fibration then satisfies the required conditions. This fact is essentially classical if $d(X) = 0$ (see [4] for example), and is easily checked in the general case using the preceding calculations.

The preceding fibration is in fact functorial on the simply connected part of $\mathcal{C}^1(k)$ and indeed may be assumed to take values in groups over Coalg(k).

Corollary. If the characteristic of k is not two, and X is an object of $\mathcal{C}^1(k)$ such that $X_1 = 0$, then $H(\Omega\Gamma(X)) \approx H(S(s^{-1}X)) \approx S(s^{-1}P(H(\Gamma(X))))$.

Here as usual $P(\;)$ denotes the primitive element functor.

Suppose now that the characteristic of k is two. For X an object of $\mathcal{C}^1(k)$, let $E(X)$ be the exterior algebra primitively generated by X, i.e. $E(X)$ is the quotient of $S(X)$ by the ideal generated by elements of the form x^2 where for some $n \in \mathbb{Z}$, $x \in X_n$. Calculations similar to those sketched earlier show that there is an object W of $\mathcal{C}^1(k)$ with zero differential containing $H(X)$ such that $H(E(X)) = E(W)$.

Proposition. If the characteristic of k is two, and X is an object of $\mathcal{C}^1(k)$ such that $X_1 = 0$, then there are principal fibrations

$$E(s^{-1}X) \to E^{\#}(X) \to \Gamma(X)$$
$$S(s^{-1}X) \to E^{b}(X) \to E(X)$$

with $E^{\#}(X)$, and $E^{b}(X)$ acyclic.

The fibrations are constructed using canonical twisting functions as in the preceding proposition. The acyclicity of $E^{\#}(X)$ and $E^{b}(X)$ is essentially classical.

Corollary. If the characteristic of k is two, and X is an object of $\mathcal{C}^1(k)$ such that $X_1 = 0$, then

$$H(\Omega\Gamma(X)) \approx H(E(s^{-1}(X))) \approx E(s^{-1}PH(\Gamma(X))) , \quad \text{and}$$
$$H(\Omega E(X)) \approx H(S(s^{-1}(X))) \approx S(s^{-1}PH(E(X))) .$$

§4. Some basic properties of loop homology.

Changing terminology slightly, if n is an integer, $n \geq 1$, an n-connected coalgebra C will mean a simply connected coalgebra C such that the coalgebra $H(C)$ is n-connected in the earlier sense. If C is an n-connected coalgebra, then there is defined the suspension morphism, $\sigma_*: QH\Omega(C) \to PH(C)$ which is a morphism of degree $+1$ from the undecomposables of $H\Omega(C)$ to the primitives of $H(C)$. It has the property that $\sigma_q: QH\Omega(C)_q \to PH(C)_{q+1}$ is an isomorphism for $q \leq 2n$, and a monomorphism for $q \leq 3n$.

Theorem. If C is an n-connected coalgebra, $n \geq 1$, then there is a diagram

$$\begin{array}{ccccc} & & C & & \\ & & \downarrow f & & \\ C(-1) & \xrightarrow{\iota(-1)} & \tilde{C} & \xrightarrow{\pi(-1)} & \Gamma(X) \end{array}$$

of simply connected coalgebras such that

1) f is a homology equivalence,

2) the row is a fibration sequence,

3) $C(-1)$ is $2n$-connected,

4) $k \to H\Omega(C(-1)) \to H\Omega(C) \to H\Omega(\Gamma(X)) \to k$ is a short exact sequence of Hopf algebras, and

5) the composite $QH\Omega(C(-1)) \xrightarrow{\sigma_*} PH(C(-1)) \xrightarrow{PH(\iota(-1))} PH(C)$ is zero.

Proof. Let $X \subset PH(C)$ be the image of the suspension morphism, and choose a retraction of graded vector spaces from $J(H(C))$ to X. This determines a morphism of coalgebras $g': H(C) \to \Gamma(X)$. There is now a diagram of coalgebras

$$\begin{array}{ccc} & & C \\ & & \downarrow \pi \\ H(C) & \xrightarrow{\iota} & Z'(C) \\ & \searrow^{g'} & \\ & & \Gamma(X) \end{array}$$

where $Z'(C)$ is the cocycle coalgebra of C (quotient of boundaries), and ι is the canonical monomorphism. Since $\Gamma(X)$ is injective in the category of coalgebras with zero differential, there is a morphism of coalgebras $g\colon Z'(C) \to \Gamma(X)$ such that $g\iota = g'$. Now making $g\pi$ into a fibration $\pi(-1)\colon \tilde{C} \to \Gamma(X)$ there results a commutative diagram

$$\begin{array}{ccccc} & & C & & \\ & & \downarrow f & \searrow^{g\pi} & \\ C(-1) & \xrightarrow{\iota(-1)} & \tilde{C} & \xrightarrow{\pi(-1)} & \Gamma(X) \end{array}$$

with row a fibration sequence of simply connected coalgebras and f a homology equivalence. There is a commutative diagram

$$\begin{array}{ccc} QH(\Omega(\tilde{C})) & \xrightarrow{QH\Omega(\pi(-1))} & QH\Omega(\Gamma(X)) \\ \downarrow \sigma_*(\tilde{C}) & & \downarrow \sigma_*(\Gamma(X)) \\ PH(\tilde{C}) & \xrightarrow{PH(\pi(-1))} & P\Gamma(X) = X \end{array} ,$$

and the suspension $\sigma_*(\Gamma(X))$ is an isomorphism of degree $+1$ by the calculation of the preceding paragraph. Since $PH(\pi(-1))$ is an epimorphism by construction it follows that $QH\Omega(\pi(-1))$ is an epimorphism, and thus $H\Omega(\pi(-1))$ is an epimorphism. Hence by the theorem of paragraph 2,

$$k \to H\Omega(C(-1)) \to H(\Omega(C)) \to H(\Omega\Gamma(X)) \to k$$

is a short exact sequence of Hopf algebras, where $H(\Omega(C))$ is identified with $H\Omega(\tilde{C})$ via $H\Omega(f)$. Thus the theorem is proved.

It is perhaps worth remarking that in the notation of the preceding theorem, there is a cartesian square

$$\begin{array}{ccc} C(-1) & \longrightarrow & E(\Gamma(X)) \\ \downarrow & & \downarrow \\ C & \longrightarrow & \Gamma(X) \end{array}$$

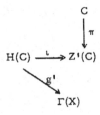

of coalgebras, and hence a fibration sequence $\Omega(\Gamma(X)) \to C(-1) \to C$, just as in geometry. If one uses the fibration sequence $S(s^{-1}X) \to E^{\#}(X) \to \Gamma(X)$ to obtain an acyclic fibration over $\Gamma(X)$, then one obtains a commutative diagram

$$\begin{array}{ccccc} \Omega(\Gamma(X)) & \longrightarrow & C(-1) & \longrightarrow & \Gamma(X) \\ \downarrow{\scriptstyle h'} & & \downarrow{\scriptstyle h} & & \downarrow{\scriptstyle 1_{\Gamma(X)}} \\ S(s^{-1}X) & \longrightarrow & C^{\#}(-1) & \longrightarrow & \Gamma(X) \end{array}$$

where the rows are principal fibrations, h', h are homology equivalences, and h' is multiplicative.

It was commented earlier in this section, that if C is an n-connected coalgebra, $n \geq 1$, then the suspension σ_q: $QH\Omega(C)_q \to PH(C)_{q+1}$ is a monomorphism for $q \leq 3n$, and an isomorphism for $q \leq 2n$. This is most easily proved using the fact that there is a spectral sequence $E^2 = \mathrm{Tor}^{H(\Omega(C))}(k,k) \Longrightarrow H_*(C)$ with classical convergence properties, an appropriate interpretation of the suspension morphism, $E^2_{1,*} = sQH(\Omega(C))$, and for $p > 1$, $E^2_{p,*}$ is $p(n+1)-1$ connected. Hence the properties of the suspension mentioned above do not use the fact that one is dealing with commutative coalgebras.

Using the notation of the preceding theorem, there is a commutative diagram

$$\begin{array}{ccccccc} QH\Omega(C(-1))_q & \longrightarrow & QH\Omega(C)_q & \longrightarrow & QH\Omega(\Gamma(X))_q & \longrightarrow & 0 \\ \downarrow{\scriptstyle \sigma'_q} & & \downarrow{\scriptstyle \sigma_q} & & \downarrow{\scriptstyle \sigma''_q} & & \\ PHC(-1)_{q+1} & \longrightarrow & PH(C)_{q+1} & \longrightarrow & P\Gamma(X)_{q+1} & \longrightarrow & 0 \end{array}$$

with exact rows, and such that σ'_q, σ''_q are isomorphisms for $q \leq 3n$. Thus the following proposition is proved.

<u>Proposition.</u> If C is an n-connected coalgebra, $n \geq 1$, then the suspension morphism

$$\sigma_q: QH\Omega(C)_q \to PH(C)_{q+1}$$

is an isomorphism for $q \leq 3n$.

<u>Corollary.</u> If

$$\begin{array}{ccccc} & & C & & \\ & & \downarrow{\scriptstyle f} & & \\ C(-1) & \xrightarrow{\iota(-1)} & \tilde{C} & \xrightarrow{\pi(-1)} & \Gamma(X) \end{array}$$

is a diagram of coalgebras as in the preceding theorem, then

1) $QH\Omega(\iota(-1))_q = 0$ for $q \leq 3n$, and

2) if the algebra $H\Omega(C)$ is commutative, then $C(-1)$ is 3n-connected if the characteristic of k is not two.

Proof. Part 1 follows immediately from the theorem, and the preceding proposition. Under the conditions of part 2, the sequence $0 \to QH\Omega\, C(-1) \to QH\Omega(C) \to QH\Omega(\Gamma(X)) \to 0$ is exact, and part 2 follows readily.

If the characteristic of k is two, a coalgebra C over k will be said to be of exterior type if for some differential graded vector space X, there exists a monomorphism $\iota\colon C \to E(X)$. Note in this situation it may be assumed without loss that $H(\iota)$ is also a monomorphism.

Theorem. If C is a n-connected coalgebra of exterior type over the field k of characteristic 2, $n \geq 1$, then there is a diagram

$$\begin{array}{ccccc} & & C & & \\ & & \downarrow{\scriptstyle f} & & \\ C(-1) & \xrightarrow{\iota(-1)} & \tilde{C} & \xrightarrow{\pi(-1)} & E(X) \end{array}$$

of simply connected coalgebras such that

1) f is a homology equivalence,

2) the row is a fibration sequence,

3) $C(-1)$ is a 2n-connected coalgebra of exterior type,

4) $k \to H\Omega(C(-1)) \to H\Omega(C) \to H\Omega(E(X)) \to k$ is a short exact sequence of Hopf algebras, and

5) the composite $QH\Omega(C(-1)) \xrightarrow{\sigma_*} PH(C(-1)) \xrightarrow{PH(\iota(-1))} PH(C)$ is zero.

Noting that the underlying coalgebra of any primitively generated Hopf algebra is of exterior type over a field of characteristic two, the proof proceeds in a manner similar to that of the preceding theorem.

Corollary. If under the conditions of the preceding theorem the algebra $H\Omega(C)$ is commutative, then $C(-1)$ is 3n-connected.

Since $H\Omega E(X)$ is free commutative, the sequence

$$0 \to QH\Omega C(-1) \to QH\Omega(C) \to QH(\Omega E(X)) \to 0$$

is exact when $H\Omega(C)$ is commutative.

REFERENCES

[1] Adams, J. F., On the cobar construction, Proc. Nat. Acad. Sci. U.S.A. 42, 409-412 (1956).

[2] Eilenberg, S., and Moore, J. C., Homology and fibrations I, Coalgebras, cotensor product and its derived functors, Commentarii Mathematici Helvetici 40, 199-236 (1966).

[3] Gugenheim, V. K. A. M., On the chain complex of a fibration, to appear.

[4] Moore, J. C., and Smith, L., Hopf algebras and multiplicative fibrations I, American Journal of Mathematics 90, 752-780 (1968).

[5] Moore, J. C., and Smith, L., Hopf algebras and multiplicative fibrations II, American Journal of Mathematics 90, 1113-1150 (1968).

[6] Quillen, D., Rational homotopy theory, Annals of Mathematics 90, 205-295 (1969).

COALGEBRAS OVER HOPF ALGEBRAS

by

Franklin P. Peterson[1]

§1. Introduction and Statement of Results.

Let A be a Hopf algebra[2] over a field k. Let M be a coalgebra over A, and let $\phi : A \to M$ be defined by $\phi(a) = a(1)$, where 1 is the counit of M. A classical theorem of Milnor-Moore [4] says that if $\operatorname{Ker} \phi = 0$, then M is a free A-module. In their work on Spin cobordism, Anderson-Brown-Peterson[2] generalized this theorem to the case where $\operatorname{Ker} \phi$ was a particular left ideal in A, when A was the mod 2 Steenrod algebra. Their theorem had very specific hypotheses, although it seemed clear that there should be a general algebraic theorem. It is the purpose of this paper to give such a general algebraic theorem, the proof of which is modeled after, and simpler than, the specific theorem in [2].

THEOREM. Let $B \subset A$ be a Hopf subalgebra, and assume B is finite dimensional as a vector space. Assume that there exist elements $Q_j \in B$, $j \in J$, such that $Q_j^2 = 0$ and such that a B-module P is free if and only if $H(P, Q_j) = 0$ all $j \in J$. Let M be a coalgebra over A as above, and assume $\operatorname{Ker} \phi = A\bar{B}$, where $\bar{B}$ = the positive dimensional elements of B. Let N be an A-module. Let $N^{(n)}$ denote the sub A-module of N generated, as an A-module, by all elements of dimension $\leq n$. Assume given an

[1]Partially supported by the N.S.F. and the U.S. Army Research Office (Durham).

[2]All modules are assumed to be graded, connected, and locally finite.

A-map $\theta : N \to M$ such that $\theta_* : H(N,Q_j) \to H(M,Q_j)$ is an isomorphism for all $j \in J$. Finally, assume that $N^{(0)} = A/A\bar{B}$ and that for every $x \in (N/N^{(n-1)})$ with $|x| = \dim x = n$, there, is a non-zero $b \in B$ such that $bx = 0$. Then θ is a monomorphism and $M \approx \theta(N) \oplus F$, where F is a free A-module, the isomorphism being one of A-modules.

Remarks. 1. The classical theorem follows by taking B to be the trivial Hopf subalgebra and $N = 0$. This has application to the case $A = \mathcal{A}$, the mod 2 Steenrod algebra, and $M = H^*(MO;Z_2)$.

2. The case $A = \mathcal{A}$, $B = \mathcal{A}_0 = \{1, Sq^1\}$, and $M = H^*(MSO;Z_2)$, gives a simple proof of part of Wall's determination of Ω_*[6].

3. The case $A = \mathcal{A}$, $B = \mathcal{A}_1 = \{1, Sq^1, Sq^2\}$, and $M = H^*(M\,Spin;Z_2)$, gives a generalization of the theorem of Anderson-Brown-Peterson mentioned above.

4. In the case $A = \mathcal{A}_p$, the mod p Steenrod algebra, $B = E(\beta, \beta\mathcal{P}^1 - \mathcal{P}^1\beta)$, and $M = H^*(MSPL;Z_p)$, the hypotheses of the theorem are known to be satisfied in a range of dimensions[5]. It is hoped that this theorem will apply here and help determine Ω_*^{PL}.

5. The hypotheses about the differentials Q_j have been proved for many interesting Hopf subalgebras of $\mathcal{A}$ by Adams and Margolis [1].

6. Results along a similar line have been recently obtained by Margolis [3]; his conclusion is that N and M are stably isomorphic under different assumptions.

§2. Proof of the Theorem.

Let $\bar{M} = M/\bar{A}M$, and let $p : M \to \bar{M}$ be the natural map. Let $Z \subset M$ be a sub vector space such that $p|Z$ is a monomorphism and $\bar{M} = p\,\theta(N) \oplus p(Z)$. That is, Z is to be an A-base for the free module F. Let $\tilde{N} = N \oplus (A \otimes Z)$, and extend θ to $\tilde{\theta} : \tilde{N} \to M$, an A-map, by sending $1 \otimes Z \to Z \subset M$. Let $\tilde{\theta}^{(n)} : \tilde{N}^{(n)} \to M^{(n)}$

We shall prove, by induction on n, that $\tilde{\theta}^{(n)}$ is an isomorphism. By construction, $\tilde{\theta}^{(n)}$ is an epimorphism. $\tilde{\theta}^{(0)}$ is an isomorphism because $\tilde{N}^{(0)} = A/A\bar{B}$ by hypothesis and $M^{(0)} = A/A\bar{B}$ because $\operatorname{Ker} \phi = A\bar{B}$. Assume $\tilde{\theta}^{(n-1)}$ is an isomorphism. Let $\lambda : \tilde{N}/\tilde{N}^{(n-1)} \to M/M^{(n-1)}$ be defined by $\tilde{\theta}$. Let P = the B-submodule of $\tilde{N}/\tilde{N}^{(n-1)}$ generated by n-dimensional elements. We now prove two lemmas.

LEMMA 2.1. $\lambda | P$ is a monomorphism.

LEMMA 2.2. Let $\{v_i\}$ be a vector space basis for P. Let $v \in \tilde{N}^{(n)}/\tilde{N}^{(n-1)}$. Then $v = \Sigma\, a_i v_i$ with $a_i = 0$ or $a_i \notin A\bar{B}$.

Proof of 2.1. A is a free B-module, by Milnor-Moore, hence $H(A,Q_j) = 0$. Thus $H(N,Q_j) \approx H(\tilde{N},Q_j)$, and $\tilde{\theta}_* : H(\tilde{N},Q_j) \to H(M,Q_j)$ is an isomorphism. Since $\tilde{\theta}^{(n-1)}$ is an isomorphism, by the 5-lemma we have $\lambda_* : H(\tilde{N}/\tilde{N}^{(n-1)}, Q_j) \to H(M/M^{(n-1)}, Q_j)$ is an isomorphism. λ is also an epimorphism. Let $K = \operatorname{Ker} \lambda$. Then $H(K,Q_j) = 0$ and K is a free B-module. Let q = the top dimension of B. Case 1. $K^n \cap P^n = 0$, $K \cap P \neq 0$. Let $x \in K \cap P$, $x \neq 0$. Then $bx \neq 0$, $|bx| \geq n + q + 1$, because K is a free B-module with no elements in dimension $\leq n$. However, $bx \in P$ and the top dimension of P is $\leq n + q$. Contradiction. Case 2. $K^n \cap P^n \neq 0$. Let $x \in (N/N^{(n-1)})^n$, $z \in Z^n$, such that $\lambda(x+z) = 0$. Then $\tilde{\theta}(x+z) \in M^{(n-1)}$ and $z = 0$ by construction of Z. Hence $x \in K^n$, but $bx = 0$ for some non-zero $b \in B$ by hypothesis. Since K is a free B-module, this is a contradiction. Thus $K \cap P = 0$ and lemma 2.1 is proved.

Proof of 2.2. Let $v = \Sigma\, a_i v_i$. If $a_i \in A\bar{B}$, then $a_i = \Sigma\, a'b'$ with $a' \notin A\bar{B}$, and $b' v_i \in P$. Replace $a_i v_i$ in the sum by $\Sigma\, a'(b' v_i)$. This proves lemma 2.2.

We now return to the proof of the theorem. To finish the induction we must show $\lambda \mid \tilde{N}^{(n-1)}/\tilde{N}^{(n-1)}$ is a monomorphism. Let $v \in \tilde{N}^{(n)}/\tilde{N}^{(n-1)}$ be such that $v \neq 0$, $\lambda(v) = 0$. Let $v = \Sigma\, a_i v_i$

as in lemma 2.2. Let $k = \max. |a_i|$, and let $a_{i_1}, \ldots, a_{i_r}$ have dimension k. Consider the image of v under the composite $\tilde{N}^{(n)}/\tilde{N}^{(n-1)} \xrightarrow{\lambda} M/M^{(n-1)} \xrightarrow{\psi} M \otimes M/M^{(n-1)}$. Using the fact that ψ is a map of A-modules, we obtain $0 = \sum_{j=1}^{r} a_{i_j}(1) \otimes \lambda(v_{i_j}) + \Sigma\, a''(1) \otimes v''$, where $|a''| < k$. This follows because $\psi\lambda(v_i) = 1 \otimes \lambda(v_i)$, $v_i \in P$ as $B(1) = 0$. By lemma 2.1, $\lambda|P$ is a monomorphism, hence $\{\lambda(v_{i_j})\}$ are linearly independent, and $a_{i_j}(1) = 0$. Thus $a_{i_j} \in A\bar{B}$ which is a contradiction.

BIBLIOGRAPHY

1. Adams, J.F., and Margolis, H.R.,"Modules over the Steenrod algebra", Topology, to appear.

2. Anderson, D.W., Brown, Jr., E.H., and Peterson, F.P., "The structure of the Spin cobordism ring", Ann of Math., 86(1967), 271-298.

3. Margolis, H.R., "Coalgebras over the Steenrod algebra", to appear.

4. Milnor, J., and Moore, J.C., "On the structure of Hopf algebras". Ann. of Math., 81(1965), 211-264.

5. Peterson, F.P., "Some results on PL-cobordism", Jour. of Math. of Kyoto Univ., 9(1969), 189-194.

6. Wall, C.T.C., "Determination of the cobordism ring", Ann. of Math., 72(1960), 292-311.

REPRESENTATION THEORY, SURGERY AND FREE ACTIONS OF FINITE GROUPS ON VARIETIES AND HOMOTOPY SPHERES

Ted Petrie

0. INTRODUCTION

The aims of this paper are:

(1) To relate the surgery exact sequence for manifolds with finite fundamental group π to an exact sequence involving the representation ring of π.

(2) To construct and distinguish smooth normal invariants for Lens spaces.

(3) To construct and distinguish free group actions on algebraic varieties.

(4) To relate (1), (2) and (3) to the normal invariants arising from free actions of metacyclic groups on homotopy spheres.

The most pleasing consequences of the paper occur in the last section where we connect up the preceding sections by relating free metacyclic actions on homotopy spheres to free actions on algebraic varieties, the latter actions being defined by complex representations of the metacyclic group. We show that the invariants of Section 1 are simply and explicitly determined by the functions defining the variety and the characters of the representation. Specifically, we show that the homotopy type of the orbit space (of a free metacyclic action on a homotopy sphere) is determined by this data and that all possible oriented homotopy types occur.

1. SURGERY THEORY AND REPRESENTATION THEORY

Let $\mathcal{C}$ be either the category of oriented smooth manifolds and maps or P.L. manifolds and maps. To an 2n-1-dimensional manifold M in $\mathcal{C}$ we attach the set $hS^{\mathcal{C}}(M)$ consisting of equivalence classes of pairs $[Z,f]$ with Z a $2n-1$-dimensional manifold in $\mathcal{C}$ and $f:Z \to M$ is a homotopy equivalence preserving orientation. Two pairs $[Z_1,f_1]$ and $[Z_2,f_2]$ are equivalent if there is a $\mathcal{C}$-isomorphism $\varphi:Z_1 \to Z_2$ so that $f_2\varphi$ is homotopic to f_1.

We also attach to M the set $\mathcal{N}^{\mathcal{C}}(M)$ consisting of pairs $[Z,f]$ with Z again a $2n-1$-dimensional manifold in $\mathcal{C}$ and $f \in \mathcal{C}$ a normal map $f:Z \to M$ i.e., deg. $f = 1$ and for some stable vector bundle (P.L. bundle) ξ, $f^*\xi$ is the stable normal bundle of Z in Euclidean space. In this set the equivalence relation is normal co-bordism, i.e., $[Z_1,f_1]$ and $[Z_2,f_2]$ are equivalent if there is a pair (W,F) with $\partial W = Z_1 \cup -Z_2$ and $F:(W,Z_1,Z_2) \to (M \times I, M \times 0, M \times 1)$, $F|Z_i = f_i$, $i = 0,1$ and $f^*\xi \times I$ is the stable normal bundle of W in Eunclidean space. There is an obvious map

$$d:hS^{\mathcal{C}}(M) \to \mathcal{N}^{\mathcal{C}}(M) \quad .$$

Suppose that $\pi_1(M) = \pi$ is a finite group and let $i:M \to B_\pi$ be a map into the classifying space of π which classifies the universal covering space $\widetilde{M}$ of M. Then there is a map

$$t:\mathcal{N}^{\mathcal{C}}(M) \to \Omega^{\mathcal{C}}_{2n-1}(B_\pi)$$

into the $2n-1$ $\mathcal{C}$-bordism group of B_π [5]. The definition is

$$t[Z,f] = [Z,if] - [M,i] \quad .$$

The Wall group $L_{2n}(\pi)$, [9] of the pair $(\pi,1)$ where $1:\pi \to Z_2$ is the trivial homomorphism, acts on the set $hS^{\mathcal{C}}(M)$ thereby defining a function $\omega:L_{2n}(\pi) \to hS^{\mathcal{C}}(M)$. Moreover there is an exact sequence of sets [8] and [9]:

(1.1) $$L_{2n}(\pi) \to hS^{\mathcal{C}}(M) \to \mathcal{N}^{\mathcal{C}}(M)$$

<u>Remark</u>. The set $\mathcal{N}^{\mathcal{C}}(M)$ is in 1-1 correspondence with the group of homotopy classes of maps of M to G/O, $[M,G/O]$ respectively, to G/PL, $[M,G/PL]$ when $\mathcal{C}$ is the smooth category, respectively the PL category [8].

We now relate this geometrically defined sequence to an exact sequence in representation theory. The natural way to describe a representation of an arbitrary finite group π is to describe a restriction to each cyclic subgroup. Formally, the natural map of the complex representation ring of π, $R(\pi)$ to $\Pi_{\pi'\subset\pi}R(\pi')$ π' cyclic is a monomorphism. It is natural to apply this philosophy to the study of group actions on manifolds.

Let $I(\pi)$ denote the kernel of the augmentation of $R(\pi)$ which takes each representation to its dimension. If the order of π is N and if Σ_π is the regular representation of π, then the powers $\{\lambda_\pi^k,\ k = 0,1,\ldots\} = S_\pi$ with $\lambda_\pi = N - \Sigma_\pi$ are in $I(\pi)$. The ring of fractions $S_\pi^{-1}R(\pi)$ consists of fractions a/b, $a \in R(\pi)$, $b \in S_\pi$ with $a_1/b_1 = a_2/b_2$ iff there is a $t \in S_\pi$ with $t(a_1b_2-a_2b_1) = 0$.

Introduce the function $A:hS^{\mathcal{C}}(M) \to \prod_{\pi'\in P} S_\pi^{-1}R(\pi')$ where P denotes the set of cyclic subgroups of π different from the identity. A variant of A was studied in [6]. The definition of A is this: Let $\pi' \in P$. The bordism group $\Omega_{2n-1}^{\mathcal{C}}(B_{\pi'})$ is annihilated by some power k of the order $|\pi'|$ of π'. This means that there is a manifold W supporting a free action of π' such that $\partial W = |\pi'|^k\tilde{M}$ as a free π' manifold. Likewise, if $[Z,f] \in hS^{\mathcal{C}}(M)$, then for some manifold W_1 supporting a free action of π', $\partial W_1 = |\pi'|^k f^*\tilde{M}$. Here $f^*\tilde{M}$ denotes the π' bundle over Z induced by f. The π' signatures of W and W_1 are elements of $R(\pi')$ and are written as $\mathrm{Sgn}(\pi',W)$ and $\mathrm{Sgn}(\pi',W_1)$. See [6]. The value of $\mathrm{Sgn}(\pi',W)$ at an element $g \in \pi'$ is written as $\mathrm{Sgn}(g,W)$.

We now define A by $A[Z,f] = \Pi_{\pi'\in P}[\mathrm{Sgn}(\pi',W_1) - \mathrm{Sgn}(\pi',W)]\lambda_{\pi'}^{-k}$ and $A[Z,f]_{\pi'} = [\mathrm{Sgn}(\pi',W_1) - \mathrm{Sgn}(\pi',W)]\lambda_{\pi'}^{-k}$.

In [6] we also introduced a homomorphism $\chi : L_{2n}(\pi) \to R(\pi)$. Composing with the natural map to $\Pi_{\pi' \in P} R(\pi')$ we obtain a homomorphism χ'. We proceed to define a function

$$\psi : \mathcal{N}^{c}(M) \to \prod_{\pi' \in P} S_{\pi}^{-1} R(\pi')/R(\pi')$$

yielding a commutative diagram

$$(1.2) \qquad \begin{array}{ccccc} L_{2n}(\pi) & \xrightarrow{\omega} & hS^{c}(M) & \longrightarrow & \mathcal{N}^{c}(M) \\ \downarrow{\scriptstyle \chi'} & & \downarrow{\scriptstyle A} & & \downarrow{\scriptstyle \psi} \\ \Pi_{\pi' \in P} R(\pi') & \longrightarrow & \Pi_{\pi' \in P} S_{\pi}^{-1} R(\pi') & \longrightarrow & \Pi_{\pi' \in P} S_{\pi}^{-1} R(\pi')/R(\pi') \end{array}$$

The key ingredients for the definition of ψ originated from a lecture of G. Segal (unpublished). Here are the ideas:

The complex equivariant K theory $K_{\pi}^{*}(\tilde{M})$ of a π space $\tilde{M}$ is a module over $R(\pi)$; so the module of fractions $S_{\pi}^{-1} K_{\pi}^{*}(\tilde{M})$ is defined. Since localization preserves exact sequences, $\tilde{M} \longrightarrow S_{\pi}^{-1} K_{\pi}^{*}(\tilde{M})$ defines a comhomology theory on $\tilde{M}$. Segal has defined a cohomology theory denoted by $K_{\pi}^{*}(\tilde{M}, S)$ which gives an exact triangle

$$(1.3) \qquad \begin{array}{ccc} K_{\pi}^{*}(\tilde{M}) & \longrightarrow & S^{-1} K_{\pi}^{*}(\tilde{M}) \\ {\scriptstyle \delta} \nwarrow & & \swarrow \\ & K_{\pi}^{*}(\tilde{M}, S) & \end{array}$$

with $K_{\pi}^{0}(pt, S) = S_{\pi}^{-1} R(\pi)/R(\pi)$.

<u>Caution</u>. The details of the Segal construction should be checked where π is not cyclic. This is the only case we really need in this paper. To be safe we assume π has odd order.

If π acts freely on $\tilde{M}$, then the action is induced by a mapping of $M = \tilde{M}/\pi$ into B_{π}. This makes $K^{*}(M)$, the ordinary complex K theory of M, a module over $K^{*}(B_{\pi}) \supset R(\pi)$ and

$K^*(M) \cong K_\pi(\tilde{M})$ as an $R(\pi)$ module. The augmentation ideal of $K^*(M)$ is nilpotent if M is a finite C.W. complex. This means that $I(\pi)^n K^*(M) = 0$ for some n and since $S_\pi \subset I(\pi)$ this implies that $S_\pi^{-1} K_\pi^*(\tilde{M}) = S_\pi^{-1} K^*(M) = 0$. It follows that $\delta: K_\pi^{-1}(\tilde{M}, S) \to K_\pi^0(\tilde{M})$ is an isomorphism.

To be quite specific, suppose that $\tilde{M}$ is a closed, oriented, simply connected smooth manifold of dimension $2n - 1$ supporting a free action of π as a group of orientation preserving diffeomorphisms. There is an index homomorphism [1] $I: K_\pi^{-1}(\tilde{M}, S) \to K_\pi^0(pt, S)$. An invariant of the action of π on $\tilde{M}$ is defined by $I\delta^{-1}(1) = \Delta(\pi, \tilde{M})$. Here $1 \in K_\pi^0(\tilde{M})$ is the identity. This is an invariant of the free π bordism class $[\tilde{M}]$ of $\tilde{M}$ in $\Omega_{2n-1}^{SO}(B_\pi)$ and $[\tilde{M}] \to \Delta(\pi, \tilde{M})$ defines an element

$$\Delta \in \mathrm{Hom}_{\Omega_*^{SO}(pt)}(\Omega^{SO}(B_\pi), K_\pi^0(pt, S)) .$$

Segal ties up his invariant with that of Atiyah-Singer [2] which we denote by $\sigma(\pi, \tilde{M})$. The connection is this:

(H) Suppose that there is a manifold W supporting an action of π (not necessarily free) such that $\partial W = \tilde{M}$ and the action of π restricted to ∂W coincides with the original action on $\tilde{M}$.

Under these circumstances $\sigma(\pi, \tilde{M})$ is defined. It is a function on the set $\pi - \{1\}$ and for $g \in \pi - \{1\}$

$$\sigma(g, \tilde{M}) = L(g, W) - \mathrm{Sgn}(g, W)$$

where $L(g,W)$ is a complex number depending on the characteristic classes of the tangent bundle and normal bundle of the fixed point set W^g of g. The function $g \to \mathrm{Sgn}(g,W)$ is a character. The function $L(g,W)$ is not arbitrary but may be regarded as an element of $S_\pi^{-1} R(\pi)$ after identifying $R(\pi)$ with the character ring of π. This means

$$\sigma(\pi, \tilde{M}) = L(\pi, W) \text{ in } S_\pi^{-1} R(\pi)/R(\pi) .$$

Theorem (Segal). If hypothesis (H) holds

$$\Delta(\pi,\tilde{M}) = L(\pi,\tilde{M}) \quad \text{in} \quad S_\pi^{-1}R(\pi)/R(\pi) \ .$$

This theorem is extremely useful for some calculations as we shall see.

It is easy to extend the definition of Δ to an element of $\mathrm{Hom}_{\Omega_*^{PL}}(\Omega^{PL}(B_\pi), S_\pi^{-1}R(\pi)/R(\pi))$. The group $\Omega_{2n-1}^{PL}(B_\pi)$ is annihilated by a power of the order N of π [5]. So if $[\tilde{M}] \in \Omega_{2n-1}^{PL}(B_\pi)$ then for some k, there is a PL manifold W with a free action of π such that $\partial W = N^k\tilde{M}$ set

$$\Delta(\pi,\tilde{M}) = (N-\Sigma_\pi)^{-k}\mathrm{Sgn}(\pi,W) \bmod R(\pi) \quad .$$

We are now in a position to define ψ. Let M be a compact manifold of dimension $2n-1$ with $\pi_1(M) = \pi$ and $\tilde{M}$ its universal covering space. Let f be a map of M to B_π classifying $\tilde{M}$. Then if $\pi' \subset \pi$ there is a map $\tilde{f}$ of $\tilde{M}/\pi'$ to $B_{\pi'}$ classifying the universal cover of $\tilde{M}/\pi'$, namely $\tilde{M}$. This data defines a map $r_{\pi'}: \Omega_{2n-1}^c(B_\pi) \to \Omega_{2n-1}^c(B_{\pi'})$ by

$$r_{\pi'}[M,f] = [\tilde{M}/\pi',\tilde{f}] \quad .$$

Let $\psi_{\pi'}$ be the composition

$$\mathcal{N}^{\mathcal{C}}(M) \xrightarrow{t} \Omega_{2n-1}^{\mathcal{C}}(B_\pi) \xrightarrow{r_{\pi'}} \Omega_{2n-1}^{\mathcal{C}}(B_{\pi'}) \xrightarrow{\Delta} S_\pi^{-1}R(\pi')/R(\pi') \ .$$

Then ψ is the product map $\Pi_{\pi'\in P}\psi_{\pi'}$.

The fact that the left hand square in (1.2) is commutative was established in [6]. The fact that the right hand square is commutative follows from the theorem of Segal, together with an explicit computation of generators for $\Omega_*^{SO}(B_{Z_n})$ and the definition of Δ involving the Atiyah Singer function $L(Z_n,W)$.

2. APPLICATIONS

The first application of these ideas is a study of the smooth normal invariants arising from actions of cyclic groups on homotopy spheres.

The complex representation ring (character ring) of the cyclic group Z_p is $Z[\gamma]/(\gamma^p - 1)$. If $\rho : Z_p \to U(m)$ is an honest representation of Z_p, then $\rho = \Sigma_{\nu=0}^{p-1} a_\nu \gamma^\nu$, $a_\nu \geq 0$, $\Sigma a_\nu = m$. We call ρ a fixed point free representation if $a_\nu = 0$ unless ν is a unit of Z_p written $\nu \in U(Z_p)$.

Henceforth ρ is assumed to be a fixed point free representation. Note that via ρ we have a fixed point free action of Z_p on the unit sphere in C^m. The notation $S(\rho)$ for the unit sphere in C^m seems appropriate since it calls attention to the particular Z_p action defined by ρ on the unit sphere. We also use the notation $L(\rho)$ for the Lens space $S(\rho)/Z_p$.

To each fixed point free representation ρ we can associate a unit of Z_p. If

$$\rho = \sum_{\nu \in U(Z_p)} a_\nu \gamma^\nu$$

we set $\deg \rho = \Pi_\nu \nu^{-a_\nu} \in U(Z_p)$.

Our aim is to construct normal invariants for the Lens spaces $L((m-1)\gamma + \gamma^a)$, $a \in U(Z_p)$ and to distinguish them by means of the ψ invariant. The construction is a generalization of an idea of W. Browder (unpublished). The germ of the idea was planted in a paper by Atiyah-Bott [7]. Let us proceed!

Suppose that k is an integer such that $k \cdot \deg \rho = a^{-1} \in U(Z_p)$. Choose integers r_i, $i = 1 \ldots m$ and t such that

$$(1) \quad r_i \equiv \nu^{-1} \quad \text{if} \quad \sum_{\mu \leq \nu} a_\mu \leq i < \sum_{\mu \leq \nu+1} a_\mu$$

$$(2) \quad ka \prod_{i=1}^{m} r_i + t_p = 1$$

Set $\rho_a = (m-1)\gamma + \gamma^a$ and define a degree one map $R(\vec{r}, k, \ )$ from $X_{k,\rho} = kL(\rho) \cup tS(\rho_a)$ to $L(\rho_a)$ by

(a) $R(\vec{r},k,\rho)[z_1:z_2: \;:z_m] = [z_1^{ar_1}, z_2^{r_2} \ldots z_m^{r_m}]$ for $z_1:z_2: \;:z_m$ in any of the k copies of $L(\rho)$.

(b) $R(\vec{r},k,\rho)|S(\rho_a)$ is the natural projection of $S(\rho_a)$ on $L(\rho_a)$.

There is a cobordism between $X_{k,\rho}$ and the connected sum $kL(\rho)\#tS(\rho_a) = kL(\rho)$ and the map extends over the cobordism to give a degree one map of $kL(\rho)$ to $L(\rho_a)$, a fact used below.

As $\widetilde{KO}(L(\rho))$ is p-primary, it follows readily that there is a bundle ξ over $L(\rho_a)$ such that $R(\vec{r},k,\rho)^*\xi$ is the stable normal bundle of $X_{k,\rho}$ in Euclidean space so $R(\vec{r},k,\rho)$ is a normal map [4] and $[X_{k,\rho}, R(\vec{r},k,\rho)] \in \mathcal{N}^{SO}(L(\rho_a))$.

The ψ values of these normal invariants are easily determined. Suppose, for simplicity, that p is prime. Then if ξ is a primitive p-th root of unity, the ring $Z[\xi][\frac{1}{1-\xi}]$ consists of fractions a/b with $a \in Z[\xi]$, b a power of $(1-\xi)$ and $S_\pi^{-1}R(\pi)/R(\pi) \cong Z[\xi][\frac{1}{1-\xi}]/Z[\xi]$. The value of Δ on $[S(\rho)] \in \Omega_{2m-1}^{SO}(B_\pi)$ is

$$\prod_\nu \left(\frac{1+\xi^\nu}{1-\xi^\nu} \right)^{a_\nu} .$$

This follows from the fact that $S(\rho)$ bounds the unit disk $D(\rho)$ in C^m and the only fixed point in the disk is the origin. Then if g is a chosen generator of Z_p,

$$L(g,D(\rho)) = \prod \left(\frac{1+\xi^\nu}{1-\xi^\nu} \right)^{a_\nu} .$$

From Segal's theorem we have $L(\pi,D(\rho)) = \Delta(\pi,S(\rho))$ in $S_\pi^{-1}R(\pi)/R(\pi)$. From this and the fact that $t[X_{k,\rho},R(\vec{r},k,\rho)] = k[S(\rho)] - [S(\rho_a)]$ we have

Proposition 2.1.

$$\psi[X_{k,} \; , R(\vec{r},k,\rho)]$$

$$= k\prod \left(\frac{1+\xi^\nu}{1-\xi^\nu} \right)^{a_\nu} - \left(\frac{1+\xi}{1-\xi} \right)^{m-1} \left(\frac{1+\xi^a}{1-\xi^a} \right)$$

Let's abbreviate the module $Z[\xi][\frac{1}{1-\xi}]/Z[\xi]$ by Λ. There remains the task of deciding when

$$k\prod\left(\frac{1+\xi^{\nu}}{1-\xi^{\nu}}\right)^{a_\nu} = k'\prod\left(\frac{1+\xi^{\nu}}{1-\xi^{\nu}}\right)^{a'_\nu} \quad \text{in } \Lambda .$$

Remark. At this point we should emphasize the fact that the function ψ is not additive. Moreover, I see no connection between the additive structure on $\mathcal{N}^{SO}(L(\rho_a))$ coming from the identification $\mathcal{N}^{SO}(L(\rho_a)) = [L(\rho_a),G/O]$ and $\psi(x+y) - \psi(x) - \psi(y)$ for $x,y \in \mathcal{N}^{SO}(L(\rho_a))$.

Since the order of $[L(\rho_a),\ G/PL] = \mathcal{N}^{PL}(L(\rho_a))$ is $p^{[\frac{2m-1}{4}]}$ and since ψ factors through $\mathcal{N}^{PL}(L(\rho_a))$ the number $p^{[\frac{2m-1}{4}]}$ is an upper bound on the number of distinct values $\psi[X_{k,\rho},\ R(\vec{r},k,\rho)]$.

An interesting geometrical application of the construction is the following:

Proposition 2.2. *If* $p = 3$ *or* 5 *every piecewise linear homotopy Lens space with fundamental group* $= Z_p$ *is smoothable.*

Proof. The statement is equivalent to showing that the map $[L(\rho_a),G/O]$ to $[L(\rho_a),G/PL]$ is onto. It suffices to produce $p^{[\frac{2m-1}{4}]}$ elements $[X_{k,\rho},R(\vec{r},k,\rho)] \in [L(\rho_a),G/O]$ with distinct values of ψ. This is easy for $p = 3$ and somewhat harder for $p = 5$ due to the number theory of the module Λ. We omit details.

Remark 1. D. Sullivan has much more information on the smoothability of P.L. Lens spaces (unpublished).

Remark 2. The normal invariants $[X_{k,\rho},R(\vec{r},k,\rho)]$ actually arise from a free Z_p action on an algebraic variety as we show in the next section.

3. ACTIONS ON ALGEBRAIC VARIETIES

The disk, respectively sphere, of radius η in a complex representation space M of π is denoted by D_η, respectively, S_η. Given m functions $f = (f_1,\ldots,f_m)$ with each f_i an element of the ring of π invariant complex polynomials on M, $P(M)^\pi$, we consider these varieties each invariant under the action of π defined by the representation:

(3.1) $V_f = \{v \in M | f_i(v) = \varepsilon_i,\ i = 1,2,\ldots,m\},\ \varepsilon_i > 0.$

(3.2) $W_f = V_f \cap D_\eta.$

(3.3) $K_f = V_f \cap S_\eta.$

For each prime p, dividing the order of π, select a cyclic subgroup of order p, Z_p, in the one conjugacy class of Sylow p subgroups. Let g_p be a fixed generator of Z_p. The representation restricted to Z_p is denoted by M_p and

$$M_p = M_p^{g_p} \oplus N_p$$

where

$$M_p^{g_p} = \{v \in M | g_p v = v\}.$$

Then N_p defines a fixed point free representation Δ_p of Z_p.

<u>Hypothesis</u>. In a neighborhood of the origin, the variety V_f intersects $\bigcup_p M_p^{g_p}$ transversally.

Under this hypothesis, η can be chosen so that π acts freely on K_f. We wish to study the class $[K_f] \in \Omega^{SO}(B_\pi)$ determined by the action of π of K_f. From the manner in which the action of π was constructed, an efficient manner for effecting this is to relate the natural map $\Omega^{SO}(B_\pi) \xrightarrow{r} \Pi_{p||\pi|}\Omega^{SO}(B_{Z_p})$ to the fixed point sets $M_p^{g_p} \cap W_f$ and the representations of Z_p on the tangent spaces of W_f at the fixed points in $M_p^{g_p} \cap W_f$.

It follows from the hypothesis, that the representation of Z_p on the tangent space of W_f at each of the a_p points in

$M_p^{g_p} \cap W_f$ is given by Δ_p. This implies that

(3.4) $r_{Z_p}[K_f] = a_p[S(\Delta_p)] \in \Omega^{SO}(B_{Z_p})$. In fact if an invariant disk is removed about each fixed point in W_f, the resulting manifold W' provides a bordism between K_f and $a_p S(\Delta_p)$.

Here are two good illustrations of this idea:

Example 1. The first is the case $\pi = Z_n$. If $\rho = \Sigma_{\nu \in U(Z_n)} a_\nu \gamma^\nu$ is an fixed point free representation of Z_n of dimension m, then the polynomial $f = z_1^n + z_2^n + \ldots + z_m^n + z_{m+1}^q$, $(n,q) = 1$ is invariant under the action of π on $\mathbb{C}^{m+1}$ defined by $\rho \oplus \gamma^0$ and

$$(3.5) \qquad [K_f] = q[S(\rho)] \in \Omega_{2m-1}(B_{Z_n}) .$$

If, in addition, n is prime,

$$(3.6) \qquad \Delta[K_f] = q \prod \left(\frac{1+\xi^\nu}{1-\xi^\nu} \right)^{a_\nu}$$

The normal invariant $[X_{q,\rho}, R(\vec{r},q,\rho)]$ is determined by $[K_f]$ as follows: For the given function f and representation ρ, the manifold K_f/Z_p is normally cobordant to $qL(\rho)$, the cobordism given by W'/Z_p. The normal map from $qL(\rho)$ to $L(\rho_a)$ $(a^{-1} = q\Pi \nu^{-a_\nu})$ extends over the cobordism to yield a normal map h of K_f/Z_p to $L(\rho_a)$. Thus we have :

Proposition 3.7. Let n be an odd integer and $\rho = \Sigma_{\nu \in U(Z_n)} a_\nu \gamma^\nu$ a representation of Z_n with $m = \Sigma a_\nu$. The function $f = \Sigma_{i=1}^m z_i^n + z_{m+1}^q$, $(n,q) = 1$ is invariant under the action of Z_n on C^{m+1} defined by $\rho \oplus \gamma^0$ and Z_n acts freely on K_f. This determines a normal invariant $[K_f/Z_n, h] \in \eta^{SO}(L(\rho_a))$, $a^{-1} = q\,\Pi \nu^{-a_\nu}$. Moreover, if n is prime

$$\psi[K_f/Z_n, h] = q \prod \left(\frac{1+\xi^\nu}{1-\xi^\nu} \right)^{a_\nu} - \left(\frac{1+\xi^a}{1-\xi^a} \right) \left(\frac{1+\xi}{1-\xi} \right)^{m-1}$$

Example 2. This example concerns the group

$$Z_{p,q} = \{x,y \mid x^p = y^q = 1,\ yxy^{-1} = x^\sigma\} .$$

Here p and q are odd primes and σ is a primitive q-th root of unity mod p. The representation space M for π in this case is obtained as the direct sum of an induced one dimensional free representation ∇_p for the normal cyclic subgroup generated by x and a one dimensional free representation for the quotient group Z_q of $Z_{p,q}$ by the normal subgroup generated by x. That is $M = i_*\nabla_p \oplus j^*\nabla_q$ where $1 \to Z_p \xrightarrow{i} Z_{p,q} \xrightarrow{j} Z_q \to 1$ is exact. Then $\dim_{\mathbb{C}} M = q + 1$ and the function

$$f = z_1^p + z_2^p + \ldots + z_q^p + z_{q+1}^{q^n}$$

satisfies the hypothesis. If the Z_p and Z_q characters of ∇_p and ∇_q are $\gamma_p^{\nu_p}$ and $\gamma_q^{\nu_q}$ then the characters φ and δ for N_p and N_q are

$$\varphi = \sum_{i=0}^{q-1} \gamma_p^{\nu_p \sigma^i}$$

$$\delta = \sum_{i=2}^{q-1} \gamma_q^{i\nu_q} + 2\gamma_q^{\nu_q}$$

so that

(3.8) $$r_{Z_p}[K_f] = q^n[S(\varphi)] \in \Omega_{2q-1}(B_{Z_q}) \quad .$$

(3.9) $$r_{Z_q}[K_f] = p[S(\delta)] \in \Omega_{2q-1}(B_{Z_q}) \quad .$$

We remark that for the polynomial f under consideration we may take $\eta = 1$ and ε a real number sufficiently near zero. Under these circumstances a_p and a_q are respectively the number of points $(0,0,\ldots,0,z)$ and $(z,z,\ldots,z,0)$ satisfying $z^{q^n} = \varepsilon$ and $qz^p = \varepsilon$ respectively. Clearly $a_p = q^n$ and $a_q = p$.

4. REMARKS ON FREE METACYCLIC ACTIONS ON HOMOTOPY SPHERES

Hypothesis: Let ∇_p and ∇_q be complex one dimensional free representations of Z_p and Z_q where p and q are odd primes. The characters of ∇_p and ∇_q are of the form $\gamma_p^t \in R(Z_p)$ and $\gamma_q^s \in R(Z_q)$ with $(t,p) = (s,q) = 1$. The polynomial

$$(4.1) \qquad f(z_1,z_2,\ldots,z_{q+1}) = z_1^p + z_2^p + \ldots + z_q^p + z_{q+1}^{q^n}$$

is invariant under the action of $Z_{p,q}$ defined by the $Z_{p,q}$ representation

$$(4.2) \qquad i_*\nabla_p \oplus j^*\nabla_q = V^{t,s} \quad .$$

In [7] we have shown that for some values of n and independent of t and q, the free $Z_{p,q}$ action on K_f defined by $V^{t,s}$ is freely cobordant to a free $Z_{p,q}$ action on a homotopy sphere Σ_f^{2q-1}. The precise condition on n is that the $Z(Z_{p,q})$ module $M = Z_{q^n}(\xi_p) = Z(\xi_p)/(q^n)$ (where ξ_p is a primitive p-th root of 1) determine zero in the reduced projective class group of $Z(Z_{p,q})$ as follows: Let F be a free $Z(Z_{p,q})$ module and

$$(4.3) \qquad 0 \to P \to F \to M \to 0$$

an exact sequence over $Z(Z_{p,q})$. Then P should be stably free.

We would like to extract as much information as possible from the invariants Δ and ψ to classify free $Z_{p,q}$ actions on homotopy spheres. To that end we should relate the actions on the K_f for various choices of t and s above, to the exact sequence (1.1).

First there is the problem of determining the different oriented homotopy types of manifolds X such that the universal covering space $\tilde{X}$ of X is a homotopy sphere and $\pi_1(X) = Z_{p,q}$. There are at most α distinct homotopy types where α is the number of elements of order pq in $H_{2q-1}(B_{Z_{p,q}}) = Z_{pq}$. In fact, the distinguishing invariant for such an X is the image of the fundamental class of X under a map of X to $B_{Z_{p,q}}$ which classifies $\tilde{X}$. In other words, the invariant is $\mu[X]$. Where $\mu:\Omega(B_\pi) \to H(B_\pi)$ is the natural transformation of homology theories.

Here is a near explicit description of manifolds realizing all homotopy types. We note that the $Z(Z_{p,q})$ module structure for $H_{q-1}(K_f)$ discussed in the preceding section certainly depended on $f = z_1^p + \dots + z_q^p + z_{q+1}^{q^n}$ but was independent of the representation $V^{t,s}$. Set

$$X_{t,s} = K_f/Z_{p,q}$$

$$X_{t,.} = K_f/Z_p$$

$$X_{.,s} = K_f/Z_q$$

where $Z_{p,q}$, Z_p and Z_q act on K_f via the representations $V^{t,s}$ $V^{t,s}$ restricted to Z_p, respectively, to Z_q.

Proposition 4.1: The element $X_{t,s} \in \Omega_{2q-1}(B_{Z_{p,q}})$ satisfies

$$r_{Z_p}[X_{t,s}] = q^n \left[S\left(\sum_{i=0}^{q-1} \gamma^{t\sigma^i}\right) \Big/ Z_p \right]$$

$$r_{Z_q}[X_{t,s}] = p \left[S\left(2\gamma^s + \sum_{i=2}^{q-1} \gamma^{is}\right) \Big/ Z_p \right]$$

Proof: This follows from Example 2, (3.8) and (3.9)

Corollary 4.2: The elements $[X_{t,.}] \in \Omega_{2q-1}(B_{Z_p})$ and $[X_{.,s}] \in \Omega_{2q-1}(B_{Z_q})$ satisfy

$$\mu[X_{t,.}] = q^n t^{-1} \in Z_p = H_{2q-1}(B_{Z_p})$$

$$\mu[X_{.,s}] = -ps^{-1} \in Z_q = H_{2q-1}(B_{Z_q}) .$$

Proof. There is a degree one map of $q^n[S(\Sigma_{i=0}^{q-1}\gamma^{t\sigma^i})/Z_p]$ to $L_p(\rho_a)$ and of $p\, S(2\gamma^s + \Sigma_{i=2}^{q-1}\gamma^{is})/Z_q]$ to $L_q(\rho_b)$ where $a = q^{-n}t \in U(Z_p)$ and $b = -p^{-1}s \in U(Z_q)$. This follows from Section 2. There are maps of $L_p(\rho_a)$ and $L_q(\rho_b)$ to $L_p(\rho_1)$ and $L_q(\rho_1)$ of degrees respectively congruent to a^{-1} mod p and b^{-1} mod q. This implies that

$$\mu\left[q^n\left[S\left(\sum_{i=0}^{q-1}\gamma^{t\sigma^i}\right)\Big/ Z_p\right]\right] = a^{-1}$$

$$\mu\left[p\left[S\left(2\gamma^s + \sum_{i=2}^{q-1}\gamma^{is}\right)\Big/ Z_q\right]\right] = b^{-1} \quad .$$

The result is now a consequence of the preceding proposition.

Corollary 4.3: If n is divisible by the order of the reduced projective class group of $Z(Z_{p,q})$, then $Z_{q^n}(\xi)$ determines zero in $\tilde{K}_0(Z(Z_{p,q}))$ and for each of the representations $V^{t,s}$ of $Z_{p,q}$, the spaces $X_{t,s}$ are cobordant to spaces $L_{t,s}$ such that $\pi_1(L_{t,s}) = Z_{p,q}$ and $\tilde{L}_{t,s}$ is a homotopy sphere. Moreover, $L_{t,.}$ and $L_{.,s}$ are homotopy Lens spaces $L(\rho_a)$ and $L(\rho_b)$ where

$$a = q^{-n}t \in Z_p$$

$$b = -p^t s \in Z_q \quad .$$

Corollary 4.4: As t and s range over the elements of orders p and q respectively in Z_p and Z_q we obtain α distinct oriented homotopy types $L_{t,s}$ with

$$\mu[L_{t,s}] = \mu[L_{t,.}] \oplus \mu[L_{.,s}] \in Z_p \oplus Z_q =$$

$$= Z_{(pq)} = H_{2q-1}(B_{Z_{p,q}}) \quad .$$

Proposition 4.5: Under the above hypothesis $Z_{t,.}$ and $Z_{.,s}$ determine normal invariants $[Z_{t,.},h_p] \in \mathcal{N}^{SO}(L_p(\rho_a))$ and $[Z_{.,s},h_q] \in \mathcal{N}^{SO}(L_q(\ _b))$. Moreover,

$$\Psi_{Z_p}[Z_{t,.},h_p] = q^n \prod_{i=0}^{q-1} \left(\frac{1 + \xi_p^{\sigma i}}{1 - \xi_p^{\sigma i}} \right)$$

$$- \left(\frac{1 + \xi_p^{a}}{1 - \xi_p^{a}} \right) \left(\frac{1 + \xi_p}{1 - \xi_p} \right)^{q-1}$$

$$\Psi_{Z_q}[Z_{.,s},h_q] = p \prod_{i=2}^{q-1} \left(\frac{1 + \xi_q^{i}}{1 - \xi_q^{i}} \right) \left(\frac{1 + \xi_q}{1 - \xi_q} \right)^{2}$$

$$- \left(\frac{1 + \xi_q^{b}}{1 - \xi_q^{b}} \right) \left(\frac{1 + \xi_q}{1 - \xi_q} \right)^{q-1} .$$

Proof: $[Z_{t,.},h_p] = [X_{t,.},h_p']$ and $[Z_{.,s},h_q] = [X_{.,s},h_q']$ where h_p' and h_q' are normal maps to $L_p(\rho_a)$ and $L_q(\rho_b)$ respectively. Now the universal covers of $X_{t,.}$ and $X_{.,s}$ are K_f with action of Z_p and Z_q given by the restriction of the action defined by $V^{t,s}$ to Z_p and Z_q. The result is now a consequence of Proposition 3.7.

REFERENCES

[1] Atiyah, M. and Segal, G. R., The index of elliptic operators: II, Ann. of Math. 87 (1968).

[2] Atiyah, M. and Singer, I., The index of elliptic operators: III, Ann. of Math. 87 (1968).

[3] Atiyah, M. and Bott, R., The Lefschetz fixed-point theorem for elliptic complexes: II, Ann. of Math. 88 (1968).

[4] Browder, William, Surgery on simply-connected manifolds, Notes, Princeton University, 1968.

[5] Conner, P. and Floyd, E., Differentiable Periodic Maps, Springer, 1964.

[6] Petrie, T., The Atiyah-Singer invariant, the Wall groups $L_n(\pi, 1)$ and the function $(te^x-1)/(te^x-1)$,

[7] ————, The existence of free metacyclic actions on homotopy spheres, to appear.

[8] Sullivan, D., Mimeographed notes from the geometric topology seminar at Princeton, 1968.

[9] Wall, C. T. C., Surgery on compact manifolds, to appear.

JOINT COBORDISM OF IMMERSIONS

Paul A. Schweitzer, S. J.

Considerable progress has been made in the study of immersions of differentiable manifolds in spheres, although much remains to be done. Much less is known about immersions in arbitrary closed differentiable manifolds. In this paper we study such immersions $f: M^m \to N^{m+k}$, modulo the relation of joint cobordism, a variant of Thom's concept of cobordism. Such joint cobordism allows both M^m and N^{m+k} to change by cobordisms V^{m+1} and W^{m+k+1}, provided that there is an immersion $V^{m+1} \to W^{m+k+1}$ extending f. (See §1 for precise details.) The cobordism classes of immersions form a group, denoted $I(m,k)$ in the unoriented case. $I(*,*) = \Sigma I(m,k)$ is a bigraded algebra over the unoriented bordism ring $\mathfrak{N}_* = \mathfrak{N}_*(\text{point})$.

By a modification of the Pontrjagin-Thom construction, we show that $I(m,k)$ is isomorphic to $\mathfrak{N}_{m+k}(QMO_k)$, the unoriented bordism of the space QMO_k (Theorem 2.1), provided $k > 0$. The proof fails in case $k = 0$. A similar isomorphism holds for manifolds with various structures (oriented, spin, and weakly almost complex manifolds, for example). We compute the algebra $\mathfrak{N}_*(QMO_k)$ by means of the extended squaring homology operations Q_i of [Kudo and Araki] in Section 3. Definitions and preliminaries are given in Section 1.

The present theory, which has also been considered in a recent paper [Uchida], differs from the theory of cobordism of immersions of [Wells] in that Wells treats immersions $f: M^m \to S^n$ and joint cobordisms $V^{m+1} \to S^{n+1} \times I$. For the restricted class of immersions in spheres, Wells' theory gives more precise results. Relationships with the theory of [Wells], the cobordism of embeddings (or pairs) of [Wall], and the cobordism of maps of [Stong] are given in (1.4) and (2.8).

I wish to thank the Centro de Investigación del I. P. N. of Mexico City for hospitality and support during early stages of this work, and Terry Wall for suggesting the use of bordism groups in this theory.

§1. DEFINITIONS AND BASIC PROPERTIES

Preliminaries.

All manifolds are compact and differentiable of class at least $\mathscr{C}^1$. ∂V denotes the boundary of the n-manifold $V = V^n$. The letters M and N denote closed manifolds.

A differentiable map of manifolds, $f\colon M^m \to N^n$, denoted (f,M,N), is an immersion if at each point $x \in M$ the differential map $df_x\colon T_xM \to T_{fx}N$ has rank m on the tangent space T_xM of M at x. Clearly the codimension $n-m$ of an immersion must be non-negative.

Definition. A joint cobordism of two immersions (f,M^m,N^n) and (f',M'^m,N'^n) is an immersion (g,V^{m+1},W^{n+1}) such that $\partial V = M \cup M'$, $\partial W = N \cup N'$ (disjoint unions), and the following diagram commutes:

$$\begin{array}{ccccc} M & \subset & V & \supset & M' \\ \downarrow{\scriptstyle f} & & \downarrow{\scriptstyle g} & & \downarrow{\scriptstyle f'} \\ N & \subset & W & \supset & N' \end{array}$$

When such a joint cobordism exists, we say that (f,M,N) is cobordant to (f',M',N'), written $(f,M,N) \sim (f',M',N')$.

Definition. Given an immersion $f\colon U \to W$ and a continuous mapping $g\colon V^n \to W$, we say that g is transverse regular to fU (written $g \pitchfork fU$) if for every $x \in U$ and $y \in V$ such that $fx = gy$, we have f is differentiable at x and $T_{fx}W = (df)T_xU + (dg)T_yV$.

It is convenient to assume that every joint cobordism (g,V,W) satisfies $g \pitchfork \partial W$, and this can always be achieved by modifying g by a small homotopy. Under this assumption, the proof of transitivity of the relation $\sim$, given $(f,M,N) \sim (f',M',N')$ and $(f',M',N') \sim (f'',M'',N'')$, reduces to attaching the two joint cobordisms along the common immersion (f',M',N'), and then straightening the resulting angles. Consequently it is easy to see that $\sim$ is an equivalence relation.

The Cobordism Algebra of Immersions.

Definition. $I(m,k)$ is the set of equivalence classes, modulo joint cobordism $\sim$, of all immersions of closed manifolds (f,M^m,N^{m+k}), $k,m \geq 0$. The equivalence class of (f,M,N) is denoted $[f,M,N]$.

Proposition 1.1. $I(*,*) = \Sigma_{m,k\geq 0} I(m,k)$ is a bigraded algebra over the unoriented bordism ring $\mathfrak{N}_*$, which is a subalgebra under the injection $[N] \overset{i}{\longmapsto} [id,N,N]$. There is also a distinct internal multiplication on $I(*,k)$ which makes it an $\mathfrak{N}_*$-algebra.

Proof. Addition and multiplication in $I(*,*)$ are given by the operations of disjoint union and Cartesian product, as follows:

$$[f,M^m,N^{m+k}] + [f',M'^m,N'^{m+k}] = [f \cup f',M^m \cup M'^m,N^{m+k} \cup N'^{m+k}]$$

$$[f,M^m,N^{m+k}] \times [f',M'^n,N'^{n+\ell}] = [f \times f',M^m \times M'^n,N^{m+k} \times N'^{n+\ell}] .$$

Note that the operations $+$ and $\times$ respect both gradings. Verification that $I(*,*)$ is a ring under $+$ and $\times$ is routine. The zero and unit are $[id,\phi,\phi]$ and $[id,*,*]$, where $*$ is a point.

The given morphism $i: \mathfrak{N}_* \longrightarrow I(*,*)$ is injective, since it has the left inverse $[f,M,N] \mapsto [N]$. Since i is easily seen to be a ring morphism, $I(*,*)$ is an algebra over $\mathfrak{N}_*$.

The internal product on $I(*,k)$, which preserves the codimension k, is given by

$$[f,M^m,N^{m+k}]\cdot[f',M'^n,N'^{n+k}] = [f\times id \cup id \times f',M^m \times N'^{n+k} \cup N^{m+k} \times M'^n,N^{m+k} \times N'^{n+k}].$$

Verification that $I(*,k)$ is an $\mathfrak{N}_*$-algebra under $+$ and $\cdot$ is straightforward. $\square$

Regular Homotopy.

Definition. A family $f_t: M \to N$ of immersions, $t \in I$, is a regular homotopy, and f_0 is regularly homotopic to f_1 ($f_0 \simeq_r f_1$), if the differential maps $df_t: TM \to TN$ form a homotopy.

Lemma 1.2. Let $F: M \times I \to N$ be a regular homotopy, with $f_t(x) = F(x,t)$ for $(x,t) \in M \times I$. Then F can be approximated by a differentiable regular homotopy $F': M \times I \to N$, such that the initial and final immersions f_0 and f_1 are unchanged.

The proof of this Lemma is an exercise in standard approximation techniques of differential topology, and is omitted.

Proposition 1.3. If two immersions $f_0, f_1: M^m \to N^{m+k}$ are homotopic by a regular homotopy, then $[f_0,M,N] = [f_1,M,N] \in I(m,k)$.

Proof. By the preceding lemma, we may assume that the regular homotopy $F: M \times I \to N$ connecting f_0 and f_1 is differentiable. Then the mapping $g: M \times I \to N \times I$, defined $g(x,t) = (F(x,t),t)$, is an immersion. Therefore $(g, M \times I, N \times I)$ is a joint cobordism of (f_0,M,N) and (f_1,M,N), and $[f_0,M,N] = [f_1,M,N]$. □

Relationships with Other Cobordism Theories.

By varying the restrictions on the maps $f: M^m \to N^{m+k}$ and on the joint cobordisms $g: V^{m+1} \to W^{m+k+1}$, the following theories are obtained:

(1) cobordism of imbeddings (or pairs) $E(m,k)$, when f and g are embeddings [Wall];

(2) cobordism of immersions in spheres (or, equivalently, in Euclidean spaces), $S(m,k)$ ($\mathfrak{N}_{m+k}(k)$ in Wells' notation), when f and g are immersions, $N^{m+k} = S^{m+k}$, and $W^{m+k+1} = S^{m+k} \times I$ [Wells];

(3) cobordism of maps $M(m,m+k)$, ($\mathfrak{N}(m,m+k)$ in Stong's notation), when f and g are only required to be continuous [Stong].

The bigraded structures $E(*,*)$, $S(*,*)$, and $M(*,*)$, like $I(*,*)$, are $\mathfrak{N}_*$-algebras, and there are canonical morphisms of $\mathfrak{N}_*$-algebras:

$$(1.4) \qquad \begin{array}{ccccccc} & \Sigma S(m,k) & & & & & \\ & \searrow & & & & & \\ \Sigma E(m,k) & \longrightarrow & \Sigma I(m,k) & \longrightarrow & \Sigma M(m,m+k) & \longrightarrow & \Sigma \mathfrak{N}_m \oplus \mathfrak{N}_{m+k} \end{array} .$$

All the morphisms except the last are defined by inclusion of equivalence classes, and the last is given by $[f,M,N] \mapsto ([M],[N])$.

<u>Joint Cobordism of Manifolds with Additional Structure</u>.

It is also possible to define joint cobordism for immersions of manifolds with additional structure, such as oriented, spin, and weakly almost complex manifolds. We now adapt the definitions and results of this section to cover such cases.

Suppose given maps $\pi_k \colon BH_k \longrightarrow BO_k$ and $\pi_G \colon BG \longrightarrow BO$, where BH_k and BG are spaces associated with the symbols H_k and G, and π_k and π_G are maps. (For example, if $G = H = SO$, Spin, or U, then BH_k, π_k and BG, π_G are the corresponding classifying spaces and maps.) A <u>G-structure</u> on a stable bundle ν over N is a lifting $\bar{\nu}$ of the classifying map $\nu\colon N \longrightarrow BO$ to BG,

$$\begin{array}{ccc} & \overset{\bar{\nu}}{\nearrow} & BG \\ & & \downarrow{\scriptstyle \pi_G} \\ N & \xrightarrow[\nu]{} & BO \end{array} .$$

A <u>G-structure</u> on a manifold N is a G-structure on the stable normal bundle of N in Euclidean space. The G-bordism group $\Omega_n^G(X)$, where X is a space, is defined in the usual way as the set of G-cobordism classes of maps $N \longrightarrow X$, where N has a G-structure. An <u>H_k-structure</u> on a k-plane bundle ν^k over M is a lifting $\bar{\nu}^k$

$$\begin{array}{ccc} & \overset{\bar{\nu}^k}{\nearrow} & BH_k \\ & & \downarrow \pi_k \\ M & \xrightarrow[\nu^k]{} & BO_k \end{array}$$

of the classifying map $\nu^k \colon M \to BO_k$.

Consider an immersion (f,V^m,W^{m+k}). Let $\nu_f = \nu(f,V,W)$, the normal k-plane bundle of V in W under f. Then a <u>(G,H_k)-structure</u> on (f,V,W) consists of (1) a G-structure $W \to BG$ on W, and (2) an H_k-structure $V \to BH_k$ on ν_f. A joint cobordism (g,V^{m+1},W^{m+k+1}) is a <u>joint (G,H_k)-cobordism</u> if it has a (G,H_k)-structure which induces given (G,H_k)-structures on both ends of the cobordism, (f,M^m,N^{m+k}) and (f',M'^m,N'^{m+k}). The resulting joint cobordism relation $\sim$ is again an equivalence relation.

<u>Definition</u>. $I(G,H_k;m,k)$ is the set of equivalence classes $[f,M^m,N^{m+k}]$ of all immersions of closed manifolds (f,M^m,N^{m+k}) with (G,H_k)-structures, modulo joint (G,H_k)-cobordism.

<u>Remark</u>. The internal multiplication $\cdot$ on $I(G,H_k;*,k)$ makes it an Ω_*^G-algebra. In order to make $I(G,H_*;*,*)$ an Ω_*^G-algebra under $+$ and $\times$, there must be maps $\mu_{k,\ell} \colon BH_k \times BH_\ell \to BH_{k+\ell}$ which are mutually homotopy associative and make the diagram

$$\begin{array}{ccc} BH_k \times BH_\ell & \xrightarrow{\mu_{k,\ell}} & BH_{k+\ell} \\ {\scriptstyle \pi_k \times \pi_\ell}\downarrow & & \downarrow{\scriptstyle \pi_{k+\ell}} \\ BO_k \times BO_\ell & \longrightarrow & BO_{k+\ell} \end{array}$$

homotopy commutative. Then the multiplication

$$\times \colon I(G,H_k;m,k) \otimes I(G,H_\ell;n,\ell) \to I(G,H_{k+\ell};m+n,k+\ell)$$

is defined. Thus Proposition 1.1 has the following analog.

<u>Proposition 1.1'</u>. The internal multiplication $\cdot$ makes $I(G,H_k;*,k)$ an Ω_*^G-algebra. Given $\{(BH_k,\pi_k) \mid k \geq 0\}$ and maps $\mu_{k,\ell}$ as described above, $I(G,H_*;*,*)$ is a bigraded algebra over the bordism ring Ω_*^G of manifolds with G-structures, which is a subalgebra under the injection $[N] \overset{i}{\longmapsto} [id,N,N]$.

We also have

<u>Proposition 1.3'</u>. Let (f_0,M^m,N^{m+k}) and (f_1,M^m,N^{m+k}) be immersions with (G,H_k)-structures. If $f_0 \simeq_r f_1$ by a regular homotopy which takes the (G,H_k)-structure of (f_0,M,N) into that of (f_1,M,N), then $[f_0,M,N] = [f_1,M,N] \in I(G,H_k;m,k)$.

<u>Proof</u>. The (G,H_k)-structure of (f_0,M,N) induces a (G,H_k)-structure on the immersion $(g, M \times I, N \times I)$ (cf. the proof of Proposition 1.3), which in turn induces the given (G,H_k)-structure on (f_1,M,N), because of the assumption that the regular homotopy preserves the (G,H_k)-structures. Therefore $(g, M \times I, N \times I)$ is a joint (G,H_k)-cobordism of (f_0,M,N) and (f_1,M,N). □

§2. THE BORDISM REPRESENTATION THEOREM

In this section, a modified form of the Pontrjagin-Thom construction is used to represent $I(m,k)$ as a bordism group of a certain space. The isomorphism is given in our main result, Theorem 2.1.

Let Ω and Σ denote the loop and reduced suspension functors on the category $\mathscr{C}$ of topological spaces with base points. Ω and Σ are adjoint functors under the natural equivalence $K(X,Y)\colon \mathrm{Hom}(\Sigma X,Y) \longrightarrow \mathrm{Hom}(X,\Omega Y)$ (where $\mathrm{Hom}(X,Y)$ is the set of based maps $X \to Y$). $K(X,Y)$ also sets up a one-to-one correspondence of homotopy classes, $[\Sigma X,Y] \longleftrightarrow [X,\Omega Y]$. There are natural transformations $A\colon 1 \longrightarrow \Omega\Sigma$ and $B\colon \Sigma\Omega \longrightarrow 1$ adjoint to the identity natural equivalences of Σ and Ω, respectively. A is injective, so

$\Omega^r A(\Sigma^r X)$: $\Omega^r\Sigma^r X \subset \Omega^{r+1}\Sigma^{r+1}X$, and we may define $Q(X) = \varinjlim_r \Omega^r\Sigma^r X$. Note that $\Omega Q \Sigma = Q$.

Suppose that spaces BG and BH_k and maps $\pi_G\colon BG \longrightarrow BO$ and $\pi_k\colon BH_k \longrightarrow BO_k$ are given, as in §1. Let MO_k and MH_k be the Thom spaces of the universal k-plane bundle $\xi_k\colon E(\xi_k) \longrightarrow BO_k$ and the induced bundle $\gamma_k = \pi_k^{-1}\xi_k\colon E(\gamma_k) \longrightarrow BH_k$. $\Omega_*^G(X)$ denotes the G-bordism group of a space X (cf. §1), which is a module over $\Omega_*^G \cong \Omega_*^G(\text{point})$.

<u>Theorem 2.1</u>. For $k > 0$, there is an isomorphism of Ω_*^G-modules

$$\alpha_k\colon I(G,H_k;*,k) \xrightarrow{\cong} \Omega_*^G(QMH_k)$$

which takes $I(G,H_k;m,k)$ onto $\Omega_{m+k}^G(QMH_k)$.

In particular, for immersions of unoriented manifolds,

$$\alpha_k\colon I(m,k) \xrightarrow{\cong} \mathfrak{N}_{m+k}(QMO_k) .$$

<u>Remark</u>. The morphism $\alpha_0\colon I(G,H_0;*,0) \longrightarrow \Omega_*^G(QMH_0)$ is defined, but our proof that α_k^{-1} exists depends on Hirsch's Theorem, which requires that $k > 0$. I do not know whether or not α_0 is an isomorphism.

<u>Construction of α_k.</u>

The morphism α_k is defined as follows. Suppose there is given

C0) <u>an immersion</u> $f_0\colon M^m \longrightarrow N^{m+k}$ <u>with normal bundle</u> $\nu = \nu(f_0,M,N)$, <u>with a</u> (G,H_k)-<u>structure</u>.

Identify N with $N \times 0 \subset N \times R^r$, where R^r is Euclidean r-space. The normal bundle of M in $N \times R^r$ under f_0 is $\nu \oplus r$, where r denotes the trivial r-plane bundle over M. Assume $r > m - k + 1$. Then there exists

C1) <u>an embedding</u> $f_1\colon M^m \subset N^{m+k} \times R^r$ <u>such that</u> $f_1 \simeq_r f_0$.

Since regular homotopy preserves the normal bundle of an immersion, the normal bundle of M in $N \times R^r$ under f_1 is also $\nu \oplus r$.

The exponential map (for some Riemannian metric on N) extends f_1 to an embedding of the disk bundle

$$\bar{f}_1 : D(\nu \oplus r) \to N \times R^r \subset (N \times R^r)^+ = \Sigma^r(N^+) ,$$

where M is identified with the zero cross-section of $\nu \oplus r$, and X^+ denotes the one point compactification of a space X. Let $q: D(\nu \oplus r) \to T(\nu \oplus r) = D(\nu \oplus r)/S(\nu \oplus r)$ be the identification map onto the Thom space of $\nu \oplus r$, and define a map

(2.2) $$\Sigma^r(N^+) \to T(\nu \oplus r)$$

by

$$y \mapsto \begin{cases} q\bar{f}_1^{-1}(y), & y \in \bar{f}_1(D(\nu \oplus r)) \\ * \quad , & y \in \Sigma^r(N^+) - \bar{f}_1(D(\nu \oplus r)) \end{cases}$$

where $* = q(S(\nu \oplus r))$ is the base point.

The bundle map $D(\nu \oplus r) \to D(\gamma_k \oplus r)$ covering the classifying map $M \to BH_k$ of ν induces a map of Thom spaces

(2.3) $$T(\nu \oplus r) \to T(\gamma_k \oplus r) = \Sigma^r T(\gamma_k) = \Sigma^r MH_k .$$

Taking the composition of (2.2) and (2.3), we get

C2) <u>a map</u> $f_2 : \Sigma^r(N^+) \to \Sigma^r MH_k$, <u>obtained from</u> f_1 <u>by the Pontrjagin-Thom construction just described</u>.

Finally, we get

C3) <u>a map</u> $f_3 : N \to QMH_k$, <u>which is the composition</u>

$$N \subset N^+ \xrightarrow{\bar{f}_2} \Omega^r \Sigma^r MH_k \subset QMH_k ,$$

<u>where</u> $\bar{f}_2$ <u>is homotopic to the rth adjoint of</u> f_2.

Now define

(2.4) $$\alpha_k([f_0, M^m, N^{m+k}]) = [N^{m+k}, f_3] \in \Omega^G_{m+k}(QMH_k) .$$

<u>Lemma 2.5</u>. $\alpha_k : I(G, H_k; m, k) \to \Omega^G_{m+k}(QMH_k)$ is well defined.

Proof. Let us call a quadruple of maps $f = (f_0, f_1, f_2, f_3)$ described by Ci) for $i = 0,1,2,3$ an admissible quadruple. For manifolds with boundary, $g = (g_0, g_1, g_2, g_3)$ is an admissible quadruple beginning with $g_0: V \to W$ provided the conditions Ci) are satisfied and in addition the quadruple $\partial g = (g_0|\partial V, g_1|\partial V, g_2|\Sigma^r(\partial W^+), g_3|\partial W)$ obtained by restriction to ∂V and ∂W is also admissible.

Consider two admissible quadruples $f = (f_0, f_1, f_2, f_3)$ and $f' = (f_0', f_1', f_2', f_3')$ such that there is a joint cobordism $(g_0, V^{m+1}, W^{m+k+1})$ between (f_0, M^m, N^{m+k}) and (f_0', M'^m, N'^{m+k}). To show that $\alpha_k([f_0, M, N])$ is well defined, it suffices to prove that there is an admissible quadruple g beginning with $(g_0, V^{m+1}, W^{m+k+1})$ such that $\partial g = f \cup f'$ (where the union is understood in the appropriate sense), for then the existence of (W, g_3) shows that $[N, f_3] = [N', f_3']$. The construction of g from g_0 is parallel to the construction of f from f_0, and explicit detail is omitted. □

Lemma 2.6. Given a map $g_3: W \to QMH_k$, there exists an admissible quadruple $g = (g_0, g_1, g_2, g_3)$ ending with g_3. Furthermore if an admissible quadruple f ending with $f_3 = g_3|\partial W$ is also given, g may be chosen so that $\partial g = f$.

Proof. The process of constructing an admissible quadruple given above is reversed. The rth adjoint of g_3 is a map

$$g_2: \Sigma^r(W^+) \to \Sigma^r MH_k .$$

C3) permits g_3 to vary by a homotopy, so we may modify g_2 so that $g_2|\Sigma^r(\partial W^+) = f_2$ and g_2 is transverse regular to the zero cross-section BH_k of $\gamma_k \oplus r$. Then let $V = g_2^{-1}(BH_k)$, and let $g_1: V \to W \times R^r = \Sigma^r(W^+) - *$ be the inclusion. V is a differentiable manifold, and g_1 is an embedding. If necessary, g_2 may be modified by a further homotopy, keeping $g_2|V$ fixed and "pushing out" from V, so that g_2 is defined by an embedding $D(\nu_{g_1}) \to W \times R^r$.

Thus C2) is satisfied.

The normal bundle of g_1 has the form $\nu_{g_1} = \nu \oplus r$ since it is a pullback of $\gamma_k \oplus r$. Thus the immersion g_1 has a normal r-frame. The following lemma completes the construction of g.

<u>Lemma 2.7</u>. Let $g_1: V^m \to W^{m+k} \times R^r$ be an immersion with a transversal r-field, and assume $k > 0$. Then there exists a regular homotopy $g_t: V \to W \times R^r$, $0 \le t \le 1$, such that $g_0(V) \subset W \times 0$. If in addition a regular homotopy $f_t: \partial V \to \partial W \times R^r$ is given such that $f_1 = g_1|\partial V$ and $f_0(\partial V) \subset \partial W \times 0$, then g_t may be chosen to be an extension of f_t.

This lemma follows from Hirsch's Theorem [Hirsch, §5], by a straightforward extension of the proof of Theorem 6.4 of [Hirsch], which treats the case $W^{m+k} = R^{m+k}$.

This completes the construction of g and the proof of Lemma 2.6. □

<u>Proof of Theorem 2.1</u>. By Lemma 2.5, α_k is well defined. By Lemma 2.6, α_k^{-1} exists, so α_k is an isomorphism.

The Ω_*^G-module structures of $I(G,H_k;*,k)$ and $\Omega_*^G(QMH_k)$, respectively, are given by

$$[N'] \otimes [f_0,M,N] \longmapsto [id \times f_0, N' \times M, N' \times N]$$

and

$$[N'] \otimes [N,f_3] \longmapsto [N' \times N, f_3 \circ p_2]$$

where $p_2: N' \times N \to N$ is the projection. Now it is easy to check that the admissible quadruple construction used to define α_k commutes with multiplication by N'. □

<u>Relationships with Other Cobordism Theories</u>.

The morphisms of diagram (1.4) connecting $I(m,k)$ with the cobordism theories $E(m,k)$ of embeddings, $S(m,k)$ of immersions in spheres, and $M(m,m+k)$ of maps can now be translated into bordism and homotopy using α_k and analogous isomorphisms. Diagram (2.8), a commutative diagram of $\mathfrak{N}_*$-modules in which all the vertical maps are isomorphisms when $k > 0$,

(2.8)
$$\begin{array}{ccccccc}
 & & \Sigma S(m,k) & \searrow & & & \\
 & & \downarrow\cong & & & & \\
\Sigma E(m,k) & \longrightarrow & & & \Sigma I(m,k) & \longrightarrow \ \Sigma M(m,m+k) \ \longrightarrow & \Sigma(\mathfrak{N}_m \oplus \mathfrak{N}_{m+k}) \\
\cong\downarrow & & \Sigma\pi_{m+k}(QMO_k) & \searrow & \alpha\downarrow\cong & \qquad\qquad \downarrow\cong & \downarrow = \\
\Sigma\mathfrak{N}_{m+k}(MO_k) & \longrightarrow & & & \Sigma\mathfrak{N}_{m+k}(QMO_k) & \longrightarrow \ \Sigma\mathfrak{N}_{m+k}(\underline{MO}_k) \ \longrightarrow & \Sigma(\mathfrak{N}_m \oplus \mathfrak{N}_{m+k})
\end{array}$$

gives the isomorphism of diagram (1.4) with the appropriate homotopy and bordism groups. The bordism group of the spectrum $\underline{MO}_k$ is defined $\mathfrak{N}_{m+k}(\underline{MO}_k) = \lim\limits_{r\to\infty} \mathfrak{N}_{m+k}(\Omega^r MO_{k+r})$. The isomorphisms for $S(m,k)$ and $M(m,m+k)$ are given (up to easy equivalences) in Theorem 3 of [Wells] and the Theorem of §3 of [Stong]. The isomorphism for $E(m,k)$ is easy to prove by the Pontrjagin-Thom approach.

Products on $I(G,H_*;*,*)$.

Suppose given spaces BH_k and maps $BH_k \longrightarrow BO_k$ and $BH_k \times BH_j \longrightarrow BH_{k+j}$ such that the diagram

(2.9)
$$\begin{array}{ccc}
BH_k \times BH_j & \longrightarrow & BH_{k+j} \\
\downarrow & & \downarrow \\
BO_k \times BO_j & \longrightarrow & BO_{k+j}
\end{array}$$

homotopy commutes, $k,j \geq 0$. Then (cf. Proposition 1.1) the Cartesian product defines a product

(2.10) $\qquad I(G,H_k;m,k) \otimes I(G,H_j;n,j) \longrightarrow I(G,H_{k+j};m+n,k+j)$.

The homotopy commutativity of (2.9) implies the existence of a bundle morphism $\gamma_k \times \gamma_j \longrightarrow \gamma_{k+j}$ (where γ_k is the pullback over

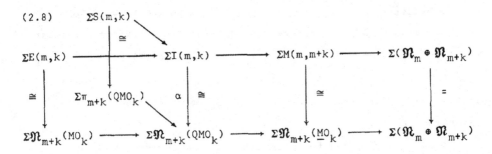

BH_k of the universal k-plane bundle ξ_k over BO_k). This bundle map induces a map of the corresponding Thom spaces $MH_k \wedge MH_j \rightarrow MH_{k+j}$. The functor Q preserves such pairings, so applying Q yields a map $QMH_k \wedge QMH_j \xrightarrow{\mu} QMH_{k+j}$.

Proposition 2.11. Under the morphism (isomorphism for $k > 0$)

$$\alpha = \sum_{k\geq 0} \alpha_k : I(G,H_*;*,*) \rightarrow \sum_{k\geq 0} \Omega_*^G(QMH_k)$$

the product (2.10) on $I(G,H_*;*,*)$ corresponds to the product induced by the pairing μ.

There is also an internal multiplication

$$I(G,H_k;m,k) \otimes I(G,H_k;n,k) \longrightarrow I(G,H_k;m+n,k)$$

given by

$$[f,M^m,N^{m+k}] \otimes [f',M'^n,N'^{n+k}] \mapsto [f \times id \cup id \times f', M \times N' \cup N \times M', N \times N'] .$$

Proposition 2.12. Under $\alpha_k: I(G,H_k;*,k) \rightarrow \Omega_*^G(QMH_k)$, this internal multiplication corresponds to the Pontrjagin product induced by the loop multiplication of QMH_k.

The proofs of (2.11) and (2.12) are straightforward.

§3. COMPUTATION OF I(m,k) FOR k > 0.

In this section the cobordism groups $I(m,k)$ of immersions of unoriented manifolds $M^m \rightarrow N^{m+k}$ are computed for $k > 0$. The structure of $I(*,k)$ as an $\mathfrak{N}_*$-algebra under the internal product of Proposition 2.12 is completely determined as a polynomial algebra over $\mathfrak{N}_*$. Throughout this section, H_*X and H^*X denote homology and cohomology with coefficient domain Z_2.

The computation of $I(*,k)$ involves the following steps:

1. $\tilde{H}_*MO_k$, which is well-known;
2. H_*QMO_k, which has been computed [Browder] in terms of $\tilde{H}_*MO_k$

and the homology extended squaring operations Q_i of [Kudo and Araki].

3. α_k: $I(*,k) \cong \mathfrak{N}_* QMO_k$, which is computed from $H_* QMO_k$ by the bordism spectral sequence, which collapses, $E^2 \cong E^\infty$.

The computation is summarized in Theorem 3.4.

Step 1. $\tilde{H}_* MO_k$.

$H^* BO_k \cong Z_2[W_1, \ldots, W_k]$, where $W_i \in H^i BO_k$ is the ith universal Stiefel-Whitney class. Using the Thom isomorphism $U \cup : H^n BO_k \cong \tilde{H}^{n+k} MO_k$ and the fact that $U^2 = U \cup W_n$, we obtain

$$\tilde{H}^* MO_k \cong W_k \cup Z_2[W_1, \ldots, W_k]$$

as algebras over Z_2. $\tilde{H}_* MO_k$ is the vector space dual of $\tilde{H}^* MO_k$. Monomials $q(W) = W_1^{q_1} \cdots W_k^{q_k}$ with $q_k > 0$ form a basis of $\tilde{H}^* MO_k$. If $[q(W)]$ denotes the dual of $q(W)$, then $\tilde{H}_* MO_k$ has the dual vector space basis

(3.1) $B_k = \{[q(W)] = [W_1^{q_1} \cdots W_k^{q_k}] \mid q_i \geq 0 \text{ for } 1 \leq i \leq k-1 \text{ and } q_k > 0\}$.

Step 2. $H_* QMO_k$.

$H_* \Omega^r \Sigma^r X$ is computed in terms of $H_* X$ in Theorem 3 of [Browder]. If $r \to \infty$, Browder's Theorem reduces to Proposition 3.2 below.

Over any r-fold loop space X (and more generally in any H_{r-1}-space) [Kudo and Araki] defined the extended squaring operations $Q_i: H_n X \longrightarrow H_{2n+i} X$ for $i < r$. Consequently, for any space X, all the operations $Q_i: H_n QX \longrightarrow H_{2n+i} QX$ are defined, $i \geq 0$. Call the iterated product $Q_I = Q_{i_1} \cdots Q_{i_s}$ monotone if the sequence $I = (i_1, \ldots, i_s)$ satisfies $0 < i_1 \leq i_2 \leq \cdots \leq i_s$, $s \geq 0$.

Proposition 3.2. (Browder). Let $\{x_j \mid j \in J\}$ be a Z_2-basis for the reduced homology $\tilde{H}_* X$. Then as Z_2-algebras $H_* QX$ is isomorphic to

$$Z_2[Q_I x_j \mid I \text{ monotone}, j \in J] ,$$

the polynomial algebra over Z_2 generated by the elements $Q_I X_j$.

For the particular case $X = MO_k$, we have

$$H_* QMO_k \cong Z_2[Q_I[q] \mid I \text{ monotone and } [q] \in B_k] .$$

Step 3. $\mathfrak{N}_* QMO_k$.

Proposition 3.3. Let X be an H-space with $H_*X \cong Z_2[a_j \mid j \in J]$ as algebras. Then $\mathfrak{N}_*X \cong \mathfrak{N}_* \otimes Z_2[a_j \mid j \in J] \cong \mathfrak{N}_*[a_j \mid j \in J]$ as $\mathfrak{N}_*$-algebras.

Proof. The Hurewicz map $h: \mathfrak{N}_*X \to H_*X$, defined by $h[N^n,f] = f_*[N] \in H_nX$ where $[N] \in H_nN$ is the fundamental class, is an epimorphism of algebras. Since H_*X is a polynomial algebra we may define an algebra morphism $i: H_*X \to \mathfrak{N}_*X$ such that $hi = id$ by choosing a lifting of the generators a_j.

In the unoriented bordism spectral sequence of X [Conner and Floyd, §§7 and 8], $E^2_{p,q} \cong H_pX \otimes \mathfrak{N}_q$ and iH_*X projects isomorphically onto $E^2_{*,0}$. This implies that all differentials d_r for $r \geq 2$ vanish on $E^r_{*,0}$, and therefore on $E^r_{*,*}$, since the spectral sequence is a spectral sequence of $\mathfrak{N}_*$-modules. Consequently the spectral sequence collapses, $E^2_{*,*} \cong E^\infty_{*,*}$, and the composition

$$\mathfrak{N}_* \otimes H_*X \xrightarrow{id \otimes i} \mathfrak{N}_* \otimes \mathfrak{N}_*X \to \mathfrak{N}_*X ,$$

where the second morphism gives the $\mathfrak{N}_*$-module structure, is an isomorphism of $\mathfrak{N}_*$-algebras. It follows that

$$\mathfrak{N}_*X = \mathfrak{N}_*[ia_j \mid j \in J] \cong \mathfrak{N}_*[a_j \mid j \in J]$$

as $\mathfrak{N}_*$-algebras. □

Combining the preceding steps, we can express the structure of the unoriented immersion cobordism algebra $I(m,k) \cong \mathfrak{N}_*QMO_k$ for $k > 0$.

Theorem 3.4. As a graded $\mathfrak{N}_*$-algebra, $k > 0$,

$$\mathfrak{N}_*QMO_k \cong \mathfrak{N}_*[Q_I[q] \mid I \text{ monotone and } [q] \in B_k]$$

Here B_k is a basis of $\tilde{H}_*MO_k$ (cf. 3.1), $I = (i_1,\ldots,i_s)$ is monotone if $0 < i_1 \le i_2 \le \cdots \le i_s$ where $s \ge 0$, and $Q_I = Q_{i_1}Q_{i_2}\cdots Q_{i_s}$ is an iterated extended squaring operation [Kudo and Araki]. If $\deg[q] = q$, then

$$\deg Q_I[q] = q \cdot 2^s + \sum_{j=1}^{s} i_j \cdot 2^{j-1} .$$

We recall that $\mathfrak{N}_*$ is isomorphic to $Z_2[V_2,V_4,V_5,V_6,V_8,\ldots]$, a polynomial algebra with one generator in each dimension not of the form $2^i - 1$.

Corollary 3.5. As a graded Z_2-algebra, $\mathfrak{N}_*QMO_k$ is isomorphic to the polynomial algebra

$$Z_2[V_n \mid 0 \le n \ne 2^i - 1] \otimes Z_2[Q_I[q] \mid I \text{ monotone and } [q] \in B_k] .$$

BIBLIOGRAPHY

W. Browder, Homology operations and loop spaces, Illinois J. Math. 4(1960) 347-357.

P. E. Conner and E. E. Floyd, Differentiable Periodic Maps, Springer, Berlin, 1964.

E. Dyer and R. K. Lashof, Homology of iterated loop spaces, Amer. J. Math. 84(1962) 35-88.

M. Hirsch, Immersions of manifolds, Trans. Amer. Math. Soc. 93(1959) 242-276.

T. Kudo and S. Araki, Topology of H_n-spaces and H_n-squaring operations, Mem. Fac. Sci. Kyūsyū Univ. Ser. A 10(1956) 85-120.

R. E. Stong, Cobordism of maps, Topology 5(1966) 245-258.

F. Uchida, Exact sequences involving cobordism groups of immersions, Osaka J. Math. 6(1969) 397-408.

C. T. C. Wall, Cobordism of pairs, Comment. Math. Helv. 35(1961) 136-145.

R. Wells, Cobordism groups of immersions, Topology 5(1966) 281-294.

THE THOM-MASSEY APPROACH TO EMBEDDINGS

by

EMERY THOMAS

Introduction.

Let M be an n-dimensional manifold (smooth, closed, connected) and k a positive integer. We consider in this paper the following questions:

(1) Does M embed in R^{n+k} ?

(2) If so, does the embedding have a (non-zero) normal vector field? That is, does the normal sphere-bundle of the embedding have a section?

Our approach to these questions is based on work of Thom [14] and Massey [7], [8], [9], the essential idea being that the cohomology ring of the sphere bundle of an embedding has special algebraic properties. In some instances one can use these properties to deduce directly that an embedding in R^{n+k} is not possible (note § 4). On the other hand, given the embedding one can use these properties to compute the obstructions to a section of the normal sphere-bundle, and thus obtain sufficient conditions for the embedding to have a normal vector field. Notice that by the theorem of M. Hirsch [3],

---

Research supported by a grant from the National Science Foundation.

> if M embeds in R^{n+k} with a normal vector field, then M immerses in R^{n+k-1} .

Thus we will sometimes be able to show that a manifold does <u>not</u> embed in R^{n+k} by knowing that it does not immerse in R^{n+k-1} .

The paper is divided into parts. In Part I we describe the Thom-Massey approach, and use this to compute "first level" invariants (traditionally called "secondary k-invariants"). This part is largely expository. Part II contains new material, as we show how to compute "higher level" invariants for the normal sphere-bundle of an embedding. As applications we obtain the following results.

<u>Theorem 1</u>. Let M be an n-dimensional spin manifold, $n \geqslant 11$, which embeds in R^{2n-s} , $3 \leqslant s \leqslant 5$. Assume that $H^4(M;Z)$ has no 2-torsion. If $n-s$ is odd and $P_1M \equiv 0 \bmod 16$, or if $n-s$ is even and $P_1M \equiv 0 \bmod 48$, then every such embedding has a normal vector field.

Recall that M is a spin manifold if $W_1M = W_2M = 0$, where W_iM denotes the ith Stiefel-Whitney class of M . Also, P_1M denotes the first Pontryagin class of M .

As a second application we will prove:

<u>Theorem 2</u>. Let M be an n-dimensional spin manifold, $n \geqslant 13$, which embeds in R^{2n-6} . If $H^3(M;Z_2) = H^7(M;Z_2) = 0$, and if $H^3(X;Z_3) = 0$ when n is even, then every such embedding has a normal vector field.

Using the result of Hirsch given above, we have:

<u>Corollary 1</u>. Let n be a positive integer such that $\alpha(n) = 3$. Then QP^n does not embed in R^{8n-6} .

Here $\alpha(n)$ denotes the number of ones in the dyadic expansion of n, and QP^n is the quaternionic projective space of real dimension $4n$. It is known that for such n, QP^n immerses in R^{8n-6} and embeds in R^{8n-4}.

PART I

§1. The Mayer-Vietoris Sequence. Let υ denote the normal bundle for an embedding of a compact n-manifold M in the sphere S^{n+q+1}, and let $\rho : SM \to M$ denote the q-sphere bundle associated to υ. By the embedding we may regard M and SM as subspaces of S^N, $N = n+q+1$. Let DM denote the disk bundle of υ and set

$$W = S^N - \text{interior } DM.$$

Thus,

$$SM = DM \cap W,$$
$$S^N = DM \cup W.$$

The triad $(S^N; DM, W)$ is excisive [2], and so we have a Mayer-Vietoris sequence

$$\ldots \to H^i(S^N) \to H^i(DM) \oplus H^i(W) \to^{\alpha} H^i(SM) \to H^{i+1}(S^N) \to \ldots,$$

where

$$\alpha(u, v) = i^*u - j^*v,$$

with

$$i: SM \subset DM, \quad j: SM \subset W,$$

denoting inclusions.

Let $\bar{\rho} : DM \to M$ be the projection; $\bar{\rho}$ is a homotopy equivalence and

$$\rho = \bar{\rho} o j : SM \to M .$$

Since $H^i(S^N) = 0$ for $0 < i \leqslant N-1$, and $H^N(W) = 0$ (W is an N-manifold with non-empty boundary), we have:

Theorem 1.1 (a). For $0 < i < n+q$, there is an isomorphism

$$H^i(M) \oplus H^i(W) \approx H^i(SM),$$

given by

$$(u,v) \to \rho^*u - j^*v.$$

(b). $$H^j(W) = 0 \qquad \text{for } j \geqslant n+q.$$

(For coefficients we take Z if M is orientable and Z_2 otherwise.)

This splitting of $H^*(SM)$ is one of the algebraic properties referred to in the Introduction. However, to obtain results we need to combine this splitting with information given by the Gysin sequence for ρ.

§2. The Gysin-Thom Sequence. Let $\rho: T \to B$ be a q-sphere bundle, $q \geqslant 1$. For convenience we identify B with the disk bundle associated to ρ, and hence we can consider the cohomology of the pair (B, T). (We take integer coefficients if ρ is orientable, mod 2 coefficients otherwise).

Proposition 2.1. Suppose that the Euler class of the bundle ρ is zero. Then, for $i \geqslant 0$, there is a short exact sequence,

$$0 \to H^i(B) \overset{\rho^*}{\to} H^i(T) \overset{\delta}{\to} H^{i+1}(B,T) \to 0 \quad .$$

This follows at once from the fact that the Thom class of ρ, regarded as a class in $H^{q+1}(B,T)$, restricts to the Euler class in $H^{q+1}(B)$; and $\tilde{H}^*(B,T)$ is a free module over $H^*(B)$ with the Thom class as generator.

We now use the sequence in 2.1 to compute the obstructions to a section of ρ. At this point we consider only first level invariants.

Let U denote the Thom class in $H^{q+1}(B,T)$. By exactness there is a class e in $H^q(T)$ such that $\delta e = U$; e is determined up to $\rho^* H^q(B)$. Using the Serre exact sequence [13], we have:

(2.2) Let $k: S^q \to T$ denote the fiber inclusion, and let $e \in H^q(T)$. Then

$$\delta e = U \iff k^* e \text{ generates } H^q(S^q) .$$

Let $B_i = BSO(i)$ if ρ is orientable, $B_i = BO(i)$ otherwise, $i \geqslant 2$. One then has the universal q-sphere bundle $\pi: B_q \to B_{q+1}$ $\eta: T \to B_q$, so that the following diagram commutes.

$$\begin{array}{ccc} T & \xrightarrow{\eta} & B_q \\ \rho \downarrow & & \downarrow \pi \\ B & \xrightarrow[\xi]{} & B_{q+1} \end{array} .$$

Now let $\chi \in H^{q+1}(B_{q+1})$ denote the universal Euler class. Regard χ as a map $B_{q+1} \to K_{q+1}$, where $K_{q+1} = K(Z, q+1)$ or $K(Z_2, q+1)$ depending on the orientability of ρ. Let $p: E \to B$ denote the principal fibration with χ as classifying map, so p has fibre $K_q = \Omega K_{q+1}$. Since $\pi^* \chi = 0$, there is a map $r: B_q \to E$ with $p \circ r = \pi$. Similarly, since $\chi(\xi) = 0$ by assumption, there is a map $f: B \to E$ with $p \circ f = \xi$.

By Thom-Wu,

$$Sq^i \chi = \chi \cup W_i , \quad i \geqslant 0 .$$

Corresponding to each such relation there is a mod 2 class $\alpha^i \in H^{q+i}(E)$. (See [6], [15] for details). We call the α^i 's <u>first</u>

level invariants.

We presume now that the lifting f is fixed, and we write

$$\alpha^i(\xi) = f^*\alpha^i \in H^{q+1}(B) .$$

One can show that $r^*\alpha^i = 0$, and hence if ρ has a section, $\alpha^i(\xi)=0$ for some choice of f.

We now use a fact that will be explained in detail in §5:

> For each choice of class $e \in H^q(T)$, with $\delta e = U$, there is associated a class of liftings
>
> $$\{f_e\}: B \to E .$$

We presume now that e is fixed, and that f is chosen to be one of the associated liftings f_e.

Theorem 2.3 (Liao-Massey). For $i \geqslant 1$,

$$\rho^*\alpha^i(\xi) = Sq^i(e) + e \cup W_i(\xi) .$$

We shall prove a generalization of this in §6.

§3. The morphisms λ, μ. We now combine the results of §§1, 2, taking the sphere bundle $T \to B$ to be the normal sphere bundle $SM \to M$ given in §1. Let

$$E^* = j^*\tilde{H}^*(W) \subset \tilde{H}^*(SM) .$$

By (1.1) and (2.1),

$$(3.1) \qquad \delta : E^* \approx H^*(M,SM) ;$$

note also that there is a unique choice of the class e so that e is in E^q. Using this choice of e, we define morphisms

$$\lambda: H^*(M) \to H^*(M) ,$$
$$\mu: H^*(M) \to E^* ,$$

by the equation: given $x \in H^*(M)$,

$$(3.2) \qquad \rho^*\lambda(x) + \mu(x) = e\cdot x .$$

These are morphisms of degree q, with the following properties.

Theorem 3.3. (1) μ is injective,

(2) $\lambda(\overline{w}_i M) = \alpha^i \upsilon \quad \mu(\overline{w}_i M) = Sq^i(e), \quad i \geq 1.$

(3) Suppose that $2q > n$, and let $x \in H^{n-q}(M)$.

Then,

$$\lambda(x) = x\cdot\overline{w}_q M .$$

(4) Let k be a positive integer such that $\overline{w}_i M = 0$, for $0 < i \leq k$. Then,

$$S_q^i\lambda = \lambda S_q^i , \qquad S_q^i\mu = \mu S_q^i , \qquad 0 \leq i \leq k .$$

Proof: Since $\delta\rho^* = 0$ by exactness, (3.2) gives

$$\delta\mu(x) = \delta(e\cdot x);$$

but δ is an $H^*(M)$-morphism, and hence

$$\delta\mu(x) = \delta e\cdot x = U\cdot x .$$

Therefore

$$\mu(x) = 0 \Rightarrow U\cdot x = 0 \Rightarrow x = 0 ,$$

which proves (1). Property (2) is simply a reformulation of (2.3). To prove (3), notice that

$$(3.4) \qquad e^2 = e\cdot\overline{w}_q M ,$$

by (2) and the fact that

$$Sq^q(e) = e^2, \lambda H^q(M) = 0 .$$

Thus, given $x \in H^{n-q}(M)$,

$$e \cdot \rho^* \lambda(x) = e^2 x - e \cdot \mu(x) = (e \cdot \overline{w}_q M) \cdot x - e \cdot \mu(x) .$$

Now $\deg e = q$, $\deg \mu(x) = n$, and hence by (1.1b), $e \cdot \mu(x) = 0$, since E is a subring of $H^*(SM)$. Thus applying δ we obtain:

$$U \cdot \lambda(x) = U \cdot (\overline{W}_q M \cdot x) ,$$

and hence (3) follows. (Note Massey [9].

Given the hypotheses of (4), notice that $Sq^i(e) = 0$, $0 < i \leqslant k$, by (1) and (2). Thus, by (3.2),

$$\rho^* Sq^i \lambda(x) + Sq^i \mu(x) = Sq^i(e \cdot x) = e \cdot Sq^i x ,$$

using the Cartan formula. But E is closed with respect to primary cohomology operations, and hence the above equation gives

$$\lambda(Sq^i x) = Sq^i \lambda(x), \mu(Sq^i x) = Sq^i \mu(x).$$

This completes the proof.

§4. Applications. We give several applications of the preceding material.

Theorem 4.1. (1) (Massey). Let M be an n-dimensional orientable manifold which embeds in S^{2n-1} . Then the embedding has a normal vector field $\Longleftrightarrow$ $\overline{w}_2 \, \overline{w}_{n-2} = 0$. (Thus if one such embedding has a normal vector field, all do.)

(2) (Adem-Gitler-Mahowald)

(a) Let $n = 2^r + 5$, $r \geqslant 4$. Then P^n does not embed in S^{2n-4} . (It is known to embed in S^{2n-3} .)

(b) Let n be a positive integer with $\alpha(n) = 2$. Then QP^n does not embed in S^{8n-5} . (It is known to embed in S^{8n-4} .)

Proof: (1) Let υ denote the (n-2)-sphere bundle associated to an embedding of M^n in S^{2n-1}. The Euler class of υ vanishes because 0 arises from an embedding. The remaining obstruction to a section of υ is the class $\alpha^2(\upsilon)$. (One shows that this is defined with zero indeterminacy.) By 3.3(2) and 3.3(3),

$$\alpha^2(\upsilon) = \lambda(\overline{W}_2 M) = \overline{W}_2 \cdot \overline{W}_{n-2} ,$$

from which the result follows.

(2) We prove only (a). Suppose that P^n embeds in S^{2n-4}. We show below that the normal sphere bundle of the embedding, υ, would then have a section. By Hirsch [3], this would imply that P^n immerses in S^{2n-5}. For $n = 2^r+5$, it is known that P^n does not immerse in S^{2n-5}, and hence it does not embed in S^{2n-4}.

By considering indeterminacy, one can show that the only obstruction to υ that can be non-zero is $\alpha^4(\upsilon)$. But $\overline{W}_4(P^n) = 0$, $n = 2^r+5$, and hence by 3.3(2), $\alpha^4(\upsilon) = 0$. Thus υ has a section, leading to a contradiction as above.

The proof of (b) is similar, using the fact that QP^n does not immerse in S^{8n-6} [12]. For details see [1].

We now give some results illustrating 3.3(4).

Proposition 4.2. Suppose that P^n embeds in S^{n+k+1}.

(a) If n and k are odd, or if $n \equiv 3 \mod 4$ and $k \equiv 2 \mod 4$, then $\alpha^i(P^n) = 0$, $i \geq 1$.

(b) If n odd and k even, then $\lambda(x^{2i}) = 0 \iff \lambda(x^{2i-1}) = 0$, $i \geq 1$. (Here x denotes the generator of $H^1(P^n)$.).

Proof. (a) We do only the case $n \equiv 3(4)$, $k \equiv 2(4)$. Then $\overline{W}_i P^n = 0$ for $i \not\equiv 0 \mod 4$, and hence we need only show that $\alpha^{4\ell} P^n = 0$, $\ell \geq 1$. We show this by proving that $\lambda H^{4i}(P^n) = 0$, $i \geq 1$.

By 3.2(4),

$$\lambda(x^{4i}) = \lambda Sq^i(x^{4i-i}) = Sq^i\lambda(x^{4i-2}) .$$

But degree $\lambda(x^{4i-2}) \equiv 0 \bmod 4$, and $Sq^2\ H^{4m}\ (P^n) = 0,\ m \geqslant 1$. Thus, $\lambda(x^{4i}) = 0$, and hence $\alpha^{4\ell}P^n = 0$.

The proof of (b) is similar and is left to the reader.

Using (4.2) we obtain

Corollary 4.3. (a) Let n be an integer $\geqslant 11$ with $n \equiv 3 \bmod 4$, and let $r = 1, 2$, or 4. Then every embedding of P^n in S^{2n-r} has a normal vector field.

(b) (Levine-Mahowald). Let $n = 2^r+1$, $r \geqslant 2$. Then, P^n does not embed in S^{2n-2}. (It is known that P^n does embed in S^{2n-1}).

Proof. (a) We do out only the case $r = 4$. By 4.2(a), $\alpha^2(\upsilon) = \alpha^4(\upsilon) = 0$. In the next section we consider the higher obstructions to a section of υ; it will be seen that these all vanish, either because υ is a spin bundle, or because of indeterminacy. Thus υ has a section. (For details see [6], [15]).

(b) Suppose that P^n embeds in S^{2n-2}. We show that this leads to a contradiction. Now

$$W(P^n) = (1 + x)^{n+1} = (1 + x)^{2r+2} ,$$

and so $\overline{W}(P^n) = (1 + x)^{2r-2}$. Thus

$$\overline{W}_{n-3}(P^n) \neq 0 .$$

By 4.2(b), we have

$$\lambda(x) = bx^{n-2} , \quad \lambda(x^2) = bx^{n-1} ,$$

for $b \in Z_2$. Notice that

$$\mu(x)\cdot\mu(x^2) = 0 , \qquad \text{by 1.1(b).}$$

Thus, by (3.2) and (3.4)

$$0 = (e\cdot x + bx^{n-2})(e\cdot x^2 + bx^{n-1}) = e^2x^3 = e\cdot\bar{w}_{x-3}\cdot x^3 = e.x^n .$$

But this is a contradiction, and hence P^n does not embed in S^{2n-2}.

Remark 1. Notice that 4.3(b) was obtained not using the Hirsch theorem. In fact P^n immerses in S^{2n-3}.

Remark 2. It is not always the case that a manifold embeds with a normal vector field as can be seen using 4.1(1). Moreover, consider complex projective space CP^n, where $n = 2^r + 1$, $r \geqslant 2$. Then, CP^n embeds in S^{4n-3}, with no normal vector field. Since $W_2(CP^n) = 0$, we cannot use 4.1(1). However, it is known that CP^n does not immerse in S^{4n-4}, and so using Hirsch's theorem, the embedding cannot have a normal vector field.

PART II

§5. Obstruction to a Section. Suppose we have an Hurewicz fibration

$$F \xrightarrow{i} T \xrightarrow{\pi} B \quad ,$$

where F, T, and B are complexes. Let u be a cohomology class in $H^n(F;J)$, with $J = Z$ or Z_p, p a prime, such that u is transgressive, and let $w \in H^{n+1}(B;J)$ be a class in the transgression of u. Let $C = K(J,n+1)$; as in [15; §§1-3] we then obtain the following commutative diagram, where p is the principal fibration with w as classifying map, and $pq = \pi$.

$$\begin{array}{ccc} F & \xrightarrow{i} & T \\ \downarrow u & & \downarrow q \\ \Omega C & \xrightarrow{j} & E \\ & & \downarrow p \\ & & B \xrightarrow{w} C . \end{array}$$

Now let X be a complex and $\xi: X \to B$ a map. We then have a commutative diagram

$$\begin{array}{ccc} F & = & F \\ k\downarrow & \lambda & \downarrow i \\ S & \to & T \\ \rho\downarrow & & \downarrow \pi \\ X & \underset{\xi}{\to} & B \end{array} ,$$

where ρ is the fibration induced from π by ξ.

Assume now that $\xi^* w = 0$. Then ξ lifts to a map $f: X \to E$ and our problem is to determine $\rho^* f^* \theta$, for a class $\theta \in H^*(E)$.

Suppose we are given two liftings, f_o, $f_1 : X \to E$. We say they are homotopic rel. ξ if there is a homotopy $f_t: X \to E$, $0 \leqslant t \leqslant 1$, such that $p\, f_t = \xi$. Let $\Lambda(\xi)$ denote the set of homotopy classes rel. ξ of liftings of ξ.

Consider the two maps $f\circ\rho$, $q\circ\lambda : S \to E$, where f is a lifting of ξ. Since

$$p\circ(f\circ\rho) = \xi\circ\rho = \pi\circ\lambda = p\circ(q\circ\lambda) ,$$

it follows that there is a map $e : S \to \Omega C$ such that

$$q\circ\lambda = e\cdot(f\circ\rho) ,$$

where the dot indicates the action of the principal fibration ρ. If f_o, t_1 are two liftings homotopic rel. ξ, and e_o, e_1 are the associated classes as above, then $e_o \simeq e_1$, as is easily seen. Thus we obtain a correspondence

$$\phi : \Lambda(\xi) \to H^n(S;J) .$$

Since u transgresses to w and $\xi^* w = 0$, it follows that $u \in k^* H^n(S;J)$. Let

$$\Gamma(u) = k^{*-1}(-u) \subset H^n(S;J).$$

<u>(5.1) Main Lemma.</u> $\phi(\Lambda(\xi)) \subset \Gamma(u)$. Moreover, if

$$\rho^* H^n(X;J) = \text{Kernel } k^* \text{ in dim. } n,$$

then

$$\phi(\Lambda(\xi)) = \Gamma(u) .$$

We give the proof at the end of the section.

Since $\rho^* \xi^* w = 0$, there is a map

$$\varepsilon : \Omega C \times S \to E$$

so that the following diagram commutes (r is the projection):

$$\begin{array}{ccc} \Omega C \times S & \xrightarrow{\varepsilon} & E \\ \downarrow r & & \downarrow p \\ S & \xrightarrow[\xi \circ \rho]{} & B \end{array} .$$

<u>Addendum (5.2).</u> Let $f : X \to E$ be a lifting of ξ and set $e = \phi(f)$; we regard e as a map $S \to \Omega C$. Then,

$$\varepsilon \circ (e,1) \simeq f \circ \rho : S \to E .$$

Before giving proofs we show how to use this information. Suppose we have constructed an n-stage Postnikov resolution of π; i.e., a diagram

$$(5.3) \quad T \xrightarrow{q_n} E_n \xrightarrow{p_n} E_{n-1} \to \cdots \xrightarrow{p_2} E_1 \xrightarrow{p_1} B ,$$

where $\pi = (p_1 \circ \cdots \circ p_n) \circ q_n$, and where each map p_{i+1} is the principal fibration induced by a map $\theta_i : E_i \to C_i$. (We think now of $B = E_o$, $w = \theta_o$ and hence $p_1 = p$, $E_1 = E$.)

Using the map $\pi : T \to B$, pull diagram (5.3) back to T , obtaining a sequence of spaces and maps

$$\to T_i \overset{\to}{\tau_i} T_{i-1} \to \cdots \to T_1 \overset{\to}{\tau_1} T$$

and maps $t_i : T_i \to E_i$ such that $p_i t_i = t_{i-1} \tau_i$. Since $\pi^* w = 0$, the space $T_1 = \Omega C x T$, up to homotopy type. Let $\mu_1 : \Omega C x E_1 \to E_1$ be the action map for the principal fibration E_1 , and set

$$\upsilon_1 = \mu_1 \circ (1 \times q_1) : \Omega C x T \to E_1 .$$

With the above identification it is easily seen that $\upsilon_1 = t_1$. (Note [15]).

Assume now that each space C_i is the product of (mod p) Eilenberg-MacLane spaces. Since $q^*_1 \theta_1 = 0$, we have that

$$\upsilon_1{}^* \theta_1 \in H^*(\Omega C x T, T) .$$

Up to homotopy, we can regard T as a subspace of each T_i , and

$$t_i{}^* \theta_i \in H^*(T_i, T) .$$

Let $\omega_i = t_i{}^* \theta_i$. Then, by McClendon [10], [16], each class ω_i determines a higher order twisted cohomology operation Ω_i , relative to the semi-tensor product

$$H^*(T;Z_p) \otimes A_p$$

where A_p denotes the mod p Steenrod algebra.

<u>Theorem (5.4)</u>. Let ξ lift to E_n , say to a map $f_n : X \to E_n$, and set $f_1 = p_2 \circ \ldots \circ p_n \circ f_n : X \to E_1$. Let $e = \phi(f_1) \in \Gamma(u)$. Then, $\rho^* f_n{}^* \theta_n = \Omega_n(e,\lambda)$.

As we will show in the proof, we regard $\Omega_n(e,\lambda)$ as a single class and not as a coset.

<u>Proof of Theorem 5.4.</u>

Consider the following commutative diagram:

$$\begin{array}{ccccc}
S_n & \xrightarrow{s_n} & T_n & \rightarrow & E_n \\
\downarrow{\scriptstyle \sigma_n} & & \downarrow{\scriptstyle \tau_n} & & \downarrow{\scriptstyle p_n} \\
\vdots & & \vdots & & \vdots \\
\downarrow & & \downarrow & & \downarrow \\
S_1 & \xrightarrow{s_1} & T_1 & \xrightarrow{t_1} & E_1 \\
\downarrow{\scriptstyle \sigma_1} & & \downarrow{\scriptstyle \tau_1} & & \downarrow{\scriptstyle p_1} \\
S & \xrightarrow[\lambda]{} & T & \xrightarrow[\pi]{} & B
\end{array}$$

Here the tower of spaces

$$\ldots \rightarrow S_i \xrightarrow{\sigma_i} S_{i-1} \rightarrow \ldots \rightarrow S_1 \xrightarrow{\sigma_1} S ,$$

is pulled back by λ from the tower over T. The crucial point is this: since $\pi \circ \lambda = \xi \circ \rho$, we have:

$$t_1 s_1 = \varepsilon_1 , \quad \text{and} \quad S_1 = \Omega C x S,$$

with $s_1 = 1 x \lambda$ (thinking of $T_1 = \Omega C x T$). Thus by (5.2), $t_1 s_1 (e,1) \simeq f_1 \circ \rho$. But S_n is induced over S_1 from E_n by $t_1 s_1$. Thus there is a map $g_n : S \rightarrow S_n$ such that

$$t_n s_n g_n \simeq f_n \rho ,$$

and

$$(e,1) \simeq \sigma_2 \circ \ldots \circ \sigma_n \circ g_n .$$

By definition,

$$g_n^* s_n^* \omega_n = \Omega_n (e,\lambda)$$

and hence

$$\rho^* f_n^* \theta_n = \Omega_n(e,\lambda) ,$$

as claimed.

Proof of the Main Lemma (5.1).

Consider the following commutative diagram:

$$\begin{array}{ccccccc} F & = & F & \overset{u}{\to} & \Omega C & = & \Omega C \\ k\downarrow & & i\downarrow & & \downarrow i & & \downarrow \\ S & \overset{\lambda}{\to} & T & \overset{q}{\to} & E & \overset{r}{\to} & PC \\ \rho\downarrow & & \downarrow\pi & & \downarrow p & & \downarrow \\ X & \underset{\xi}{\to} & B & = & B & \underset{w}{\to} & C , \end{array}$$

where r denotes projection. The class $-e$ is defined to be the loop:

$$-e = (rq\lambda) \vee (rf\rho)^{-1} .$$

Thus,

$$(-e)\circ k = (rq\lambda k) \vee (rf\rho k)^{-1} .$$

But $\rho k = *$ and $f(*) = *$. Thus,

$$(-e)\circ k = (rq\lambda k) \vee * \simeq rq\lambda k = u ,$$

by the above diagram. Hence,

$$k^*(e) = - u ,$$

as claimed.

To prove (5.2), notice that

$$\varepsilon\circ(*,1) \simeq g\lambda ,$$

and so

$$\varepsilon\circ(e,1) \simeq e\cdot(q\lambda) .$$

But the map e was chosen so that

$$e\cdot(q\lambda) \simeq f\rho,$$

which completes the proof of (5.2).

§6. Evaluation of Obstructions.

We illustrate Theorem 5.4 by evaluating some obstructions for sphere bundles. For $i \geq 2$, set $B_i = B\,SO(i)$, and for a fixed integer q , consider the universal q-sphere bundle

$$\pi : B_q \to B_{q+1} .$$

In the diagram below, we give a Postnikov resolution (mod 2) of π , through dimension $q + 7$ (as in 5.3).

$$\begin{array}{llrcl}
B_q & & & & \\
\downarrow r_3 & & & & \\
E_3 & \xrightarrow{\gamma^4} & K_{q+4} & = & C_3 \\
\downarrow p_3 & & & & \\
E_2 & \xrightarrow{(\beta^3,\beta^4,\beta^7)} & K_{q+3} \times K_{q+4} \times K_{q+7} & = & C_2 \\
\downarrow p_2 & & & & \\
E_1 & \xrightarrow{(\alpha^2,\alpha^4)} & K_{q+2} \times K_{q+4} & = & C_1 \\
\downarrow p_1 & & & & \\
B_{q+1} & \xrightarrow{\chi} & K(Z,q+1) & = & C_0
\end{array} \qquad (6.1)$$

Here $K_i = K(Z_2, i)$, $i \geq 1$; the degree of each class α, β, γ is given by adding q to the superscript on the class. Let $j_i : \Omega C_i \to E_{i+1}$, $0 \leq i \leq 3$, denote the fiber inclusion. The obstructions restrict to the fiber as follows. (We set ι_i = fundamental class of K_i .)

$$\begin{aligned}
j_0^* \alpha^m &= Sq^m \iota_q , \qquad m = 2,4. \\
j_1^* \beta^3 &= Sq^2 \iota_{q+1} \otimes 1 , \\
j_1^* \beta^4 &= Sq^2 Sq^1 \iota_{q+1} \otimes 1 + 1 \otimes Sq^1 \iota_{q+3} , \\
j_1^* \beta^7 &= Sq^6 \iota_{q+1} \otimes 1 + 1 \otimes Sq^4 \iota_{q+3} , \\
j_2^* \gamma^4 &= Sq^2 \iota_{q+2} \otimes 1 \otimes 1 + 1 \otimes Sq^1 \iota_{q+3} \otimes 1 .
\end{aligned} \qquad (6.2)$$

Now let $\xi : X \to B_{q+1}$ be a q-sphere bundle over a complex X. Given conditions on the characteristic classes of ξ, we compute the operations Ω_n occuring in (5.4), for the resolution (6.1).

Consider the following set of higher order cohomology operations, defined on integral cohomology classes:

$$\begin{aligned} &\Phi_3: Sq^2Sq^2 = 0 , \\ &\Phi_4: (Sq^2Sq^1)Sq^2 + Sq^1Sq^4 = 0 , \\ &\Phi_7: Sq^6Sq^2 + Sq^4Sq^4 = 0 , \\ &\Psi_4: Sq^2\Phi_3 + Sq^1\Phi_4 = 0 . \end{aligned} \tag{6.3}$$

Let $\rho: S \to X$ be a q-sphere bundle ξ over X. Suppose that $\chi(\xi) = 0$, so that there is a lifting f_1 of ξ to E_1. As in §5, let $e = \phi(f_1) \in H^q(S ; Z)$.

<u>Theorem 6.4.</u>

(a) $\rho^*\alpha^m(\xi) = Sq^m(e) + e\cdot w_m(\xi)$, $m=2,4$.

Suppose that $(\alpha^2\xi, \alpha^4\xi) = 0$, and $w_2\xi = 0$. Then,

(b) $\rho^*\beta^3(\xi) = \Phi_3(e)$.

Suppose also that $w_4\xi = 0$.

(c) If $H^7(X;Z_2) = 0$, then

$$\rho^*\beta^7(\xi) = \Phi_7(e).$$

Suppose, finally, that $H^4(X;z)$ has no 2-torsion. Then:

(d) $\rho^*\beta^4(\xi) = \Phi_4(e) + e\cdot(\frac{P_1\xi}{4})$.

(e) If $(\beta^3\xi, \beta^4\xi) = 0$ and if

$$P_1\xi \equiv 0 \bmod 8, \text{ then}$$

$$\rho^*\gamma^4\xi = \Psi_4(e) + e\cdot(\frac{P_1\xi}{8}).$$

Proof. Part (a) is simply a special case of Theorem 5.4, since the operation Ω in this case is the primary operation

$$w_m \otimes 1 + 1 \otimes Sq^m \in H^*(B_{q+1}) \otimes A_2 ,$$

$m = 2,4$. (See [15]).

For part (b) we take as our universal sphere fibration the bundle B Spin(q) → B Spin (q+1). Also, we need to introduce the Postnikov resolution for the sphere S^q :

$$S^q \overset{\sigma}{\to} Q_2 \overset{b_2}{\to} Q_1 \overset{b_1}{\to} K(Z,q) . \qquad (6.5)$$

Here b_1 is the principal fibration with (Sq^2, Sq^4) as classifying map, b_2 the principal fibration with (ϕ_3, ϕ_4) as classifying map. (ϕ_i is a representative for the operation Φ_i, $i = 3,4,7$).

Consider now diagram 5.5 taking T = B Spin (q), B = B Spin (q+1). Then, $T_1 \equiv K(Z,q) \times T$. Since T is 3-connected, through dimension (q+3)

$$(T_2, T) \equiv (Q, x T, T) ,$$

and

$$\omega_2 = \phi_3 \otimes 1 .$$

This proves part (b). (Note Mosher [11].)

To prove parts (c) and (d), we denote by

$$K(Z_2, 3) \overset{i_n}{\to} C_n \overset{\gamma_n}{\to} B\ Spin\ (n), \quad n \geq 5 ,$$

the principal fibration with w_4 as classifying map. In (5.5) we now take $T = C_q$, $B = C_{q+1}$. Thus $w_2 = w_4 = 0$ in T , and so $T_2 \equiv Q_1 \times T$. Using the fact that b_1^* (in 6.5) is injective in dimensions greater than q , together with (6.2), (6.3), one shows that

$$t_2^* \beta^7 = \phi_7 \otimes 1 + \iota_q \otimes u_7 ,$$

where $u_7 \in H^7(T)$. Since we assume that $H^7(X;Z_2) = 0$, part (c) now follows.

To prove (d), recall that there is a class $Q_1 \in H^4(B\ \mathrm{Spin}\ (n);Z)$ such that

$$Q_1 \bmod 2 = w_4 , \qquad 2Q_1 = P_1 ,$$

(see [17]). Thus by construction there is a generator $R_1 \in H^4(C_n; Z) = Z$,

such that

$$2R_1 = \gamma_n^* Q_1 , \quad (i_n^* R_1) \bmod 2 = Sq^1 \iota_3 .$$

Since C_q is 3-connected and $T_2 \equiv Q_1 \times T$, it follows from (6.2) and (6.3) that

$$t_2^* \beta^4 = \phi_4 \otimes 1 + \iota_q \otimes b\bar{R}_1 , \tag{6.6}$$

where $b \in z_2$ and $\bar{R}_1 = R_1 \bmod 2$. We prove (d) by showing that the coefficient $b = 1$. (Note that $4R_1 = P_1$.)

Let M_n , respectively N_n , denote the Thom complex for the canonical bundle over $B\ \mathrm{Spin}\ (n)$, respectively, C_n , $n \geqslant 5$; let U_n , respectively V_n , denote the respective Thom class. The calculation of the coefficient b follows at once from:

$$\Phi_4(V_n) = V_n \cdot \bar{R}_1 . \tag{6.7}$$

Assuming (6.7) we complete the proof of (d). Take $X = Cq$, and let $\rho: S \to X$ denote the bundle ξ over X induced from the canonical bundle over C_{q+1} . We regard the Thom complex of ξ as the pair (X,S); likewise, we think of N_{q+1} as (C_{q+1}, Cq) . The bundle ξ gives a map of Thom complexes

$$\xi' : (X,S) \to (C_{q+1}, C_q) ,$$

and $U = \xi'^* V_{q+1}$ is the Thom class for ξ . Let $e \in H^q(S)$ be the class given in 5.4. By (6.6) it follows that

$$\rho^*\beta^4(\xi) = \Phi_4(e) + b\, e\cdot\bar{R}_1(\xi)\ .$$

Since $\delta e = U$, by (2.2), we have:

$$0 = \Phi_4(U) + b\, U\cdot\bar{R}_1(\xi)$$

$$= \xi'^*(\Phi_4(V_{q+1}) + bV_{q+1}\cdot\bar{R}_1)\ .$$

Thus $b = 1$, using (6.7). This completes the proof of (d).

Proof of (6.7). We "expand" the construction of the universal example Q_1, as shown below (note [16]):

$$Q_1 \overset{r}{\to} Y_{q+1} \overset{s}{\to} K(Z,q+1)\ , \qquad b_1 = r\circ s.$$

Here s has Sq^2 as classifying map and r has $Sq^4\circ s$. There is a unique map $f\colon M_{q+1} \to Y_{q+1}$ such that $f^*s^*\iota_{q+1} = U_{q+1}$. Let $g : P_{q+1} \to M_{q+1}$ be the fibration over M_{q+1} induced by f from the fibration r. Since $g^*H^{q+5}(M_{q+1};Z_2) = 0$, it follows that

$$w = \bar{g}^*\Phi_4 \quad \text{generates} \quad H^{q+5}(P_{q+1};Z_2)\ ,$$

where $\bar{g} : P_{q+1} \to Q_1$ covers f. Let $c\colon N_{q+1} \to M_{q+1}$ be the natural map of the Thom complexes.

$$c^*(U_{q+1}\cdot w_4) = c^*(Sq^4U_{q+1}) = 0\ ,$$

and so c lifts to a map $h : N_{q+1} \to P_{q+1}$. We show that

$$h^*\colon H^{q+5}(P_{q+1}) \approx H^{q+5}(N_{q+1})\ ,$$

which will then complete the proof of (6.7).

Since $Sq^4U_{q+1} = (U_{q+1}\cdot Q_1) \bmod 2$, it follows that

$$g^*(U_{q+1}\cdot Q_1) = 2W\ ,$$

where W generates $H^{q+5}(P_{q+1};Z) = Z$, and $\bar{W} \bmod 2 = w$. Since $2(V_{q+1}\cdot R_1) = c^*(U_{q+1}\cdot Q_1)$, it follows that

$$2(V_{q+1}\cdot R_1) = 2h^*(W)\ .$$

But $H^{q+5}(N_{q+1};Z)$ is torsion-free, and so

$$h^*W = V_{q+1} \cdot R_1 = \text{generator of } H^{q+5}(N_{q+1};Z) .$$

This completes the proof of (6.7).

Proof of 6.4(e). The proof is similar to that of (d); we omit the details.

§7. Applications. We combine §§3 and 6 to prove the theorems given in the Introduction. These theorems are a consequence of the following result.

Theorem 7.1. Let M be an n-dimensional spin manifold which embeds in S^{n+q+1}, $q > n/2$. Let υ denote the normal sphere bundle of the embedding. Then

(a) $\alpha^2(\upsilon) = 0 , \quad \beta^3(\upsilon) = 0 .$

Suppose also that $W_4 M = 0$.

(b) If $H^7(M;Z_2) = 0$, then $\beta^7(\upsilon) = 0$.

Suppose finally that $H^4(M ; Z)$ has no 2-torsion. Then

(c) $\lambda(\frac{P_1 M}{4}) = \beta^4(\upsilon)$.

(d) If $P_1 M \equiv 0 \bmod 8$, then $\lambda(\frac{P_1 M}{8}) = \gamma^4(\upsilon)$.

Proof: We do out the details only for (b) and (c). Consider the following commutative diagram:

$$\begin{array}{ccccc}
 & & & & C_q \\
 & & & & \downarrow q \\
Q_1 \times SM & & \xrightarrow{\varepsilon_2} & & E_2 \\
\downarrow b_1 \times 1 & & & & \downarrow p_2 \\
K(Z,q) \times SM & & \xrightarrow{\varepsilon_1} & & E_1 \\
\downarrow r & & & & \downarrow p_1 \\
SM & \xrightarrow[\rho]{} & M & \xrightarrow[\upsilon]{} & C_{q+1}
\end{array}$$

This is obtained from (5.5), setting $\varepsilon_i = t_i s_i$, $i = 1,2,$ and using the fact that $\upsilon\rho = \pi\lambda$, and that $S_2 \equiv Q_1 \times SM$, since $W_2M = W_4M=0$.

By (5.2) there is a lifting $f_1 : M \to E_1$ such that $p_1 \circ f_1 = \Psi$, and $f_1 \circ \rho \simeq \varepsilon_1 \circ (e,1)$. Here $e \in H^q(SM;Z)$ is chosen so that $e \in E$.

<u>Claim (7.2)</u>. There exist maps $g_1 : W \to Q_1$, $f_2 : M \to E_2$ such that $p_2 \circ f_2 = f_1$, and $b_1 \circ g_1 \circ j \simeq e$, $f_2 \circ \rho \simeq \varepsilon_2 \circ (g_2 \circ j, Z)$. (Here $j:SM \to W$ is the inclusion.)

Assuming (7.2), the proof of 7.1(b) and (c) follows at once. For we now have:

$$\Phi_7(e) = j^* g_1^* \phi_7 , \Phi_4(e) = j^* g_1^* \phi_4 ,$$

and so by 6.4(2),(d),

$$\beta^7(\upsilon) = \lambda(0) = 0 , \quad \beta^4(\upsilon) = \lambda(\frac{P_1 M}{4}) .$$

<u>Proof of 7.2</u>. Since $e \in E$ there is a class $w \in H^q(W;Z)$ such that $e = j^* w$. Now

$$Sq^2 e = Sq^4 e = 0 ,$$

by 3.3(2), (since $W_2M = W_4M = 0$) , and so

$$Sq^2 w = Sq^4 w = 0 ,$$

since j^* is injective. Thus w, thought of as a map $W \to K(Z,q)$, lifts to a map $\bar{g} : W \to Q_1$. Moreover, since $\alpha^2(\upsilon) = \alpha^4(\upsilon) = 0$, also by 3.3(2), f_1 lifts to a map $\bar{f} : M \to E_2$. But

$$p_2 \circ \bar{f} \circ \rho \simeq p_2 \circ \varepsilon_2 \circ (g \circ j, 1) ,$$

and so there is a class

$$v \in H^{q+2}(SM) \oplus H^{q+4}(SM)$$

such that

$$v \cdot (\bar{f} \circ \rho) \simeq \varepsilon_2 \circ (g \circ j, 1) ,$$

where the dot indicates the action of the principal fibration p_2 . By 1.1 there are classes $x \in H^{q+2}(M) \oplus H^{q+4}(M)$, $y \in H^{q+2}(W) \oplus H^{q+4}(W)$ such that

$$v = \rho^* x - j^* y .$$

Set

$$f_2 = x \cdot \bar{f} , \quad g = y \cdot \bar{g} .$$

(The action used in defining g is from the principal fibration b_1.)

Using the properties of principal fibrations, one easily checks that these maps satisfy (7.2). Thus the proof is complete.

Proof of Theorem 1. The mod 2 obstructions to a section of the normal bundle of the embedding $M^n \subset S^{2n-s}$, $3 \leq s \leq 5$ are

$$\chi, \alpha^2, \alpha^4, \beta^3, \beta^4, \gamma^4 .$$

But by Theorem (7.1), given the hypotheses of Theorem 1, these obstructions all vanish. Moreover, the single mod 3 obstruction also vanishes. (Note [6]).

Proof of Theorem 2. The mod 2 obstructions for a section, given the embedding $M^n \subset S^{2n-6}$, are:

$$\chi, \alpha^2, \alpha^4, \beta^3, \beta^4, \beta^7, \gamma^4 .$$

Now χ vanishes because the bundle arises from an embedding, and α^2, β^3 vanish (note 7.1(a)), since M is spin. β^7 vanishes by 7.1(b), and α^4, β^4 , and γ^4 vanish because these classes lie in $H^{n-3}(M)$, which is zero by duality and by hypothesis. Finally, the mod 3 invariant vanishes, as in Theorem 1.

BIBLIOGRAPHY

1. J. Adem, S. Gitler and M. Mahowald. Embedding and immersion of projective spaces, Bol. Mat. Mex., 10 (1965), 84-88.

2. S. Eilenberg and N. Steenrod. Foundations of Algebraic Topology, Princeton Univ. Press, Princeton, 1952.

3. M. Hirsch. Immersions of Manifolds, Trans. Amer. Math. Soc., 93 (1959), 242-276.

4. J. Levine. Embedding and immersion of real projective space, P.A.M.S., 14 (1963), 801-803.

5. M. Mahowald. On the embeddability of the real projective space, P.A.M.S., 13 (1962), 763-764.

6. ________. On obstruction theory in orientable fiber bundles, Trans. A.M.S., 110 (1964), 315-349.

7. W. Massey. On the cohomology ring of a sphere bundle, J. Math. and Mech., 7 (1958), 265-289.

8. ________. On the embeddability of the real projective spaces, Pacific J. Math., 9 (1959), 783-789.

9. ________. Normal vector fields on manifolds, P.A.M.S., 12 (1961), 33-40.

10. J. McClendon, Higher order twisted cohomology operations, Invent. Math., 7 (1969), 183 -214.

11. R. Mosher. The product formula for the third obstruction, Pacific J. Math, 17 (1968), 573-578.

12. B. J. Sanderson and R. L. E. Schwarzenberger. Non-immersion theorems for differentiable manifolds, Proc. C.P.S., 59 (1963), 319-322.

13. J. P. Serre. Homologie singulière des espaces fibrés, Ann. Math., 54 (1951), 425-505.

14. R. Thom. Espaces Fibrés en sphères et carrés de Steenrod, Ann. Sci. Ecole Norm. Sup., 69 (1952), 109-182.

15. E. Thomas. Seminar on fiber spaces, Lecture Notes in Mathematics No. 13, Springer-Verlag, New York, 1966.

16. ________. Postnikov invarients and higher order cohomology operations, Ann. Math., 85 (1967), 184-217.

17. ________. On the cohomology of the classifying space for the stable spinor group, Bol. Mat. Mex., 7 (1962), 57 - 69.

COHOMOLOGY OPERATIONS AND HOMOTOPY COMMUTATIVE H-SPACES

Alexander Zabrodsky

0. Introduction. In this study, we investigate the action of the Steenrod algebra and some secondary operations on the mod 2 cohomology of homotopy commutative H-spaces.

In [7] such relations for finite (primitively generated) cohomology of (not necessarily homotopy commutative) H-spaces are studied. The main result is given by

[7] Theorem 1.1: Let X be an H-space with $H^*(X,Z_2)$ being finite and primitively generated. If $x \in PH^v(X,Z_2)$ where $v = 2^{t_1} + 2^{t_2} + \dots + 2^{t_k} - 1, 0 < t_1 < t_2 < \dots < t_k$, then $x \in Sq^{2^{t_1}} Sq^{2^{t_2}} \dots Sq^{2^{t_k}} PH^{2^{t_k}-1}(X,Z_2)$.

As it can be seen, these relations tie cohomology classes with classes of lower dimensions by the action of the Steenrod algebra. Here we shall see that in the case of homotopy commutative H-spaces (not necessarily with finite cohomology), the relations tie cohomology classes with classes of higher dimensions. More precisely, we prove:

Theorem 2.1: Let X,μ be a homotopy commutative H-space. If $H^*(X,Z_2)$ is the exterior algebra on odd dimensional generators, then $Sq^{4n}: PH^{4n+1}(X,Z_2) \longrightarrow PH^{8n+1}(X,Z_2)$ is a monomorphism ($PH^*(X,Z_2)$-- the module of primitives).

To demonstrate how these relations actually "tie up" the cohomology of X, we outline the proof of the following proposition:

Corollary 4.2: Let X,μ be an H-space satisfying the hypothesis of 2.1. If, in addition, $Sq^2 PH^3(X,Z_2) \neq 0$, then there exists a mapping $h: X \longrightarrow SU(\{2\})$ yielding a monomorphism $h^*: H^*(SU(\{2\}),Z_2) \longrightarrow H^*(X,Z_2)$ where $SU(\{2\})$ is the mod 2 SU, i.e., there exist mappings $\psi_1: SU \longrightarrow SU(\{2\})$ and $\psi_2: SU(\{2\}) \longrightarrow \prod_{n=1}^{\infty} K(Z,2n+1)$, ψ_1 yielding an isomorphism of mod 2 cohomology and ψ_2 an isomorphism of mod p (p-odd) cohomology.

With the hypothesis of 2.1 together with the assumption that X is homotopy associative, we prove:

Corollary 3.5: There exists a secondary operation $\varphi_{4n}: PH^{4n-1}(X,Z_2) \longrightarrow PH^{8n-1}(X,Z_2)/Sq^2$ associated with the relation excess $(Sq^2Sq^{4n-1} + Sq^{4n}Sq^1) > 4n$ satisfying $\ker \varphi_{4n} \subset \mathrm{im}\, Sq^2$.

1. The a and c Operations. Most of the ideas in this chapter are mainly due to Stasheff [5] and [6]. Some of them are related to the Kudo-Araki ([4]), Browder ([1]) and the Dyer-Lashof ([2]) operations.

Let X,μ and X',μ' be homotopy associative H-spaces, i.e., $\mu(1\times\mu) \sim \mu(\mu\times 1)$ and $\mu'(1\times\mu') \sim \mu'(\mu'\times 1)$. Let $f: X \longrightarrow X'$ be an H-mapping: $\mu'\circ(f\times f) \sim f\circ\mu$ (rel $X\vee X$). If $Y \subset X\times\mathcal{E}X'$ ($\mathcal{E}X'$--the space of paths in X' starting at the unit e') is the "fiber" of f, i.e., $Y = \{(x,\varphi) \mid f(x) = \varphi(1)\}$, then Y admits an H-structure $\mu_Y: Y\times Y \longrightarrow Y$ given by $\mu_Y[(x_1,\varphi_1),(x_2,\varphi_2)] = \mu(x_1,x_2), \mathcal{E}\mu'(\varphi_1,\varphi_2) + F(x_1,x_2)$ where $F(x_1,x_2)$ is the path connecting $\mu'(f(x_1),f(x_2))$ with $f\circ\mu(x_1,x_2)$ induced by the homotopy $\mu'\circ f\times f \sim f\circ\mu$. The mapping $\Pi: Y \longrightarrow X$ $(\Pi(x,\varphi) = x)$ is obviously multiplicative: $\Pi\circ\mu_Y = \mu(\Pi\times\Pi)$.

It can be seen that the obstruction for the existence of a homotopy $\mu_Y(1\times\mu_Y) \sim \mu_Y(\mu_Y\times 1)$ covering the given homotopy $\mu(1\times\mu) \sim \mu(\mu\times 1)$ is given by a class $a = a(f)\in[X\wedge X\wedge X, \Omega X']$ represented by a mapping $\bar{a}: X\times X\times X \longrightarrow \Omega X'$ where $\bar{a}(x,y,z)$ is obtained by a translation of the closed path in X' illustrated by:

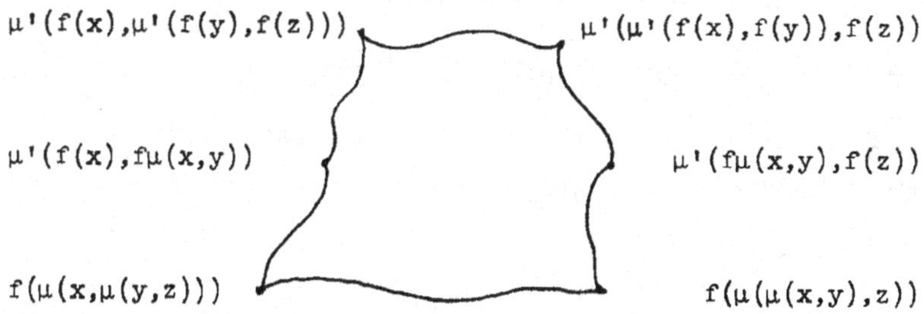

If $f':X',\mu' \longrightarrow X'',\mu''$ is another H-mapping, X'',μ''-homotopy associative, then it can be easily seen that:

1.1 Proposition:

(a) $a(f'\circ f) = (\Omega f')_{\#}a(f) + (f\wedge f\wedge f)^{\#}a(f')$

(b) If $f:Y \longrightarrow Z$ is any mapping, then $a(\Omega f) = 0$.

Similarly, if X,μ, X',μ' and X'',μ'' are homotopy commutative i.e., $\mu\sim\mu\circ T$, $\mu'\sim\mu'\circ T$, $\mu''\sim\mu''\circ T$ $(T(x,y) = y,x)$, then one gets in a similar way a class $c(f)\epsilon[X\wedge X, \Omega X']$ represented by the translation of the closed path in X' given by:

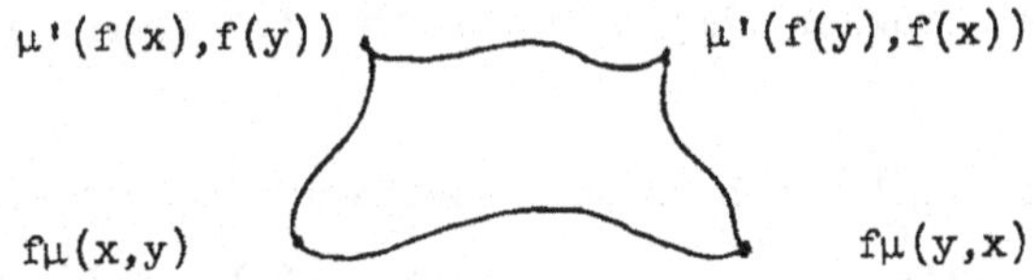

This class measures the obstruction for the existence of a homotopy $\mu_Y \sim \mu_Y\circ T$ covering the homotopy $\mu \sim \mu\circ T$. One has:

1.2 Proposition:

(a) $c(f'\circ f) = (\Omega f')_{\#}c(f) + (f\wedge f)^{\#}c(f')$

(b) If $h:Y \longrightarrow Z$ is an H-mapping, then $c(\Omega h) = 0$.

Finally, one can see that if $X'' = \prod_{j=1}^{k} K(Zp,n_j)$, $f'\circ f\sim *$ and hence, the fiber of $f'\circ f$ is given by $Y' \sim X\times\Omega X''$; then the multiplication $\mu_{Y'}$ in Y' induces a "twisted" multiplication in $X\times\Omega X''$ given by a "twisting function" $w:X\wedge X \longrightarrow \Omega X''$: $\mu_{Y'}[(x,\lambda),(x',\lambda')] = \mu(x,x'), \lambda\cdot\lambda'\cdot w(x,x')$. $[w]\epsilon\sum_{j=1}^{k} H^{n_j-1}(X\wedge X,Z_2)$ is essentially given by:

$[w] = \overline{p}^*\otimes\overline{p}^*\sum_{j=1}^{k}\overline{\mu}^*{}_{Y'}p_j^*(\iota_{n_j-1})$ where $\overline{p}:X\times\Omega X'' \longrightarrow X$ and

$p_j:\Omega X'' —— K(Z_p,n_j-1)$ are the projections and

$\iota_{n_j-1}\epsilon H^{n_j-1}(K(Z_p,n_j-1),Z_p)$ is the fundamental class. Moreover:

$a(f'\circ f) = (1\otimes\overline{\mu}^* - \overline{\mu}^*\otimes 1)[w]$

$c(f'\circ f) = (1-T^*)[w]$.

1.3 Remark: One should actually consider a homotopy associative (commutative) H-space as an H-space X,μ together with a fixed homotopy $\mu(1\times\mu) \sim \mu(\mu\times 1)$ $(\mu\sim\mu\circ T)$. An H-mapping should be considered as a mapping f together with a fixed homotopy $\mu'\circ f\times f\sim f\circ\mu$. The classes $a(f)$ $(c(f))$, if considered as being defined on mappings admitting a homotopy $\mu'\circ(f\times f) \sim f\circ\mu$ and letting this homotopy to vary, $a(f)$ $(c(f))$ become classes in

$a(f)\in[X\wedge X\wedge X,\ \Omega X']/\mathrm{im}[(1\times\mu)^{\#}-(1\times p_1)^{\#}-(1\times p_2)^{\#}-(\mu\times 1)^{\#}+(p_1\times 1)^{\#}+(p_2\times 1)^{\#}]$

$(p_i: X\times X \longrightarrow X$ being the projection) $(c(f)\in[X\wedge X,\ \Omega X']/\mathrm{im}(1+T^{\#}))$.

2. Applications: The Action of the Steenrod Algebra.

2.1 Theorem: Let X,μ be a homotopy commutative H-space. Suppose $H^*(X,Z_2)$ is an exterior algebra on odd dimensional generators. Then: $Sq^{4n}: PH^{4n+1}(X,Z_2) \longrightarrow PH^{8n+1}(X,Z_2)$ is a monomorphism. ($PH^*(X,Z_2)$ is the module of primitives in $H^*(X,Z_2)$.)

Proof: Suppose $x\in PH^{4n+1}(X,Z_2)$, $Sq^{4n}x = 0$. Consider the following diagram:

$$\begin{array}{ccc}
 & K(Z_2,8n)\times K(Z_2,4n+1) & \\
 \swarrow^{j_1} & & \searrow^{j} \\
X\times K(Z_2,8n)\times K(Z_2,4n+1) & \xrightarrow{\ \tilde f\ } & Y \\
\downarrow & & \downarrow \pi \\
X & \xrightarrow{\ f\ } & K(Z_2,4n+1) \\
 & & \downarrow h \\
 & & K(Z_2,8n+1)\times K(Z_2,4n+2)
\end{array}$$

where $h^*\iota_{8n+1} = Sq^{4n}\iota_{4n+1}$, $h^*\iota_{4n+2} = Sq^1\iota_{4n+1}$, $f^*\iota_{4n+1} = x$, and where Y is the "fiber" of h. As $h\circ f\sim *$, the fiber Y' of $h\circ f$ and $X\times K(Z_2,8n)\times K(Z_2,4n+1) = Y_1'$ are homotopy equivalent (as spaces), while the multiplication in Y' induces a "twisted" multiplication $\tilde\mu$ in Y_1' given by a twisting function $w: X\wedge X \longrightarrow K(Z_2,8n)\times K(Z_2,4n+1)$.

Now, as $Sq^2Sq^{4n} + Sq^{4n+1}Sq^1 = Sq^{4n+2}$, there exists a class $v\in H^{8n+2}(Y,Z_2)$ with $j^*v = Sq^2\iota_{8n}\otimes 1 + 1\otimes Sq^{4n+1}\iota_{4n+1}$ and

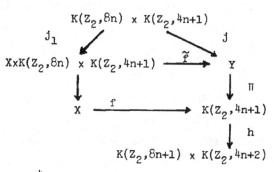

$\overline{\mu}_Y^* v = \Pi^*\otimes\Pi^*(\iota_{4n+1}\otimes\iota_{4n+1})$. Further, (by 1.1) $c(h\circ f) = (f\otimes f)^* c(h) + h_* c(f)$ and $c(h) = 0$ while $h_* c(f) = (Sq^{4n}, Sq^1)c(f)$ and as $Sq^{4n} c(f) = c(f)^2 = 0$ and $Sq^1 c(f) = 0$, $c(h\circ f) = 0$ and hence: $(1-T^*)[w] = 0$ $([w]\in H^{8n}(X\wedge X, Z_2) \oplus H^{4n+1}(X\wedge X, Z_2))$.

Now, considering $\tilde{f}$ as a mapping $Y_1' \longrightarrow Y$,
$\tilde{f}^* v = v'\otimes 1\otimes 1 + 1\otimes Sq^2\iota_{8n}\otimes 1 + 1\otimes 1\otimes Sq^{4n+1}\iota_{4n+1}$ and
$\overline{\tilde{\mu}}^*\tilde{f}^* v = (p_1^*\otimes p_1^*)\overline{\mu}^* v' + (Sq^2, Sq^{4n+1})[w]$ where $v'\in H^{8n+2}(X, Z_2)$. As $Sq^{4n+1}H^{4n+1}(X, Z_2) = 0$, $(Sq^2, Sq^{4n+1})[w] = (Sq^2, 0)[w]$. It can be easily seen (as $Sq^1 = 0$) that $Sq^2(\ker(1-T^*)) \subset \operatorname{im}(1+T^*)$; hence, $(Sq^2, 0)[w]\in \operatorname{im}(1+T^*)$. Now, $\overline{\tilde{\mu}}^*\tilde{f}^* v = \tilde{f}^*\otimes\tilde{f}^*\overline{\tilde{\mu}}^* v = \tilde{f}^*\Pi^*\iota_{4n+1}\otimes\tilde{f}\Pi^*\iota_{4n+1} =$
$$= (p_1^*\otimes p_1^*)(f^*\otimes f^*)(\iota_{4n+1}\otimes\iota_{4n+1}) = (p_1^*\otimes p_1^*)\, x\otimes x.$$

It follows that $\overline{\mu}^* v' \equiv x\otimes x \bmod \operatorname{im}(1+T^*)$ and thus, for every $y\in PH_{4n+1}(X, Z_2)$ $\langle x, y\rangle = \langle v', y^2\rangle = 0$ (as there are no even dimensional primitives in $H_*(X, Z_2)$ $y^2 = 0$). Hence, x annihilates primitives and as $PH^*(X, Z_2) \longrightarrow QH^*(X, Z_2)$ is a monomorphism, $x = 0$.

In chapter 4 we show how the fact that $Sq^{4n}: PH^{4n+1} \longrightarrow PH^{8n+1}$ is a monomorphism implies much further relations in $H^*(X, Z_2)$.

3. Applications: Secondary Operations.

3.1 Lemma: Let X, μ be an H-space and let G be a group. Given $f: X \longrightarrow G$, denote by $D_f: X\wedge X \longrightarrow G$ the deviation of f from being an H-mapping: $D_f(x, y) = f(\mu(x, y))\cdot f(y)^{-1} f(x)^{-1}$. Then $c(\Omega f)$ can be represented by a mapping $\Omega X\wedge\Omega X \longrightarrow \Omega^2 G$ corresponding to the composition $(\Sigma\Omega X)\wedge(\Sigma\Omega X) \xrightarrow{\mathcal{E}\wedge\mathcal{E}} X\wedge X \xrightarrow{D_f} G$ ($\mathcal{E}: \Sigma\Omega X \longrightarrow X$--the evaluation map).

Proof: The loop multiplication in X can be given by $(\lambda_1+\lambda_2)(t) = \Omega\mu(\lambda_1(2t), \lambda_2(2t-1))$ (with the understanding
$$\lambda_i(t) = \lambda_i(0) \quad \text{if } t \le 0$$
$$\lambda_i(t) = \lambda_i(1) \quad \text{if } t \ge 1).$$
The homotopy $C_X: \lambda_1+\lambda_2 \sim \lambda_2+\lambda_1$ is given by $C_X(\lambda_1, \lambda_2, s)[t] = \mu(\lambda_1(2t-s), \lambda_2(2t-1+s))$. Similar formulas hold for

the loop addition in ΩG. Thus, $c(\Omega f)$ can be represented by a mapping $\overline{C}:\Omega X\wedge\Omega X \longrightarrow \Omega^2 G$ given by: $\overline{C}(\lambda_1,\lambda_2)(s,t) =$
$f\circ\mu(\lambda_1(2t-s),\lambda_2(2t-1+s))\cdot f(\lambda_1(2t-s)^{-1}f(\lambda_2(2t-1+s))^{-1}$.
If $\varphi:R\times R \longrightarrow R\times R$ is given by $\varphi(s,t) = 2t-s, 2t-1+s$, then $\varphi\sim 1$ via a homotopy keeping $R\times R - I\times I$ mapped into itself. Hence, $\overline{C}\sim\overline{C}'$ where $\overline{C}'$ is given by: $\overline{C}'(\lambda_1,\lambda_2)(s,t) =$
$f\circ\mu(\lambda_1(s),\lambda_2(t))\cdot f(\lambda_1(s))^{-1}f(\lambda_2(t))^{-1} = D_f(\lambda_1(s),\lambda_2(t))$; $\overline{C}'$ is the desired representation.

Definition: Let (X,μ) be a homotopy commutative H-space. Let $x\in PH^n(X,Z_p)$. The class $c(x)\in H^{n-1}(X\wedge X,Z_p)/\mathrm{im}(1+T^*)$ is the image of the class $c(f)$ where $f:X \longrightarrow K(Z_p,n)$ satisfies $f^*\iota_n = x$.

As the change in the H-structure of f by a mapping $D:X\wedge X \longrightarrow \Omega K(Z_p,n) = K(Z_p,n-1)$ alters $c(f)$ by $(1+T^*)D^*\iota_{n-1}$, $c(x)$ depends only on x. One can easily see that for every $\alpha\in G(2)$, $c(\alpha x) = \alpha c(x)$ and if $g:X \longrightarrow Y$ is an H-mapping into a homotopy commutative H-space, then $c(g^*y) = [c(g)]^*\sigma^*y + g^*\otimes g^*c(y)$ for every $y\in PH^*(Y,Z_p)$. ($\sigma^*:H^*(Y,Z_p) \longrightarrow PH^*(\Omega Y,Z_p)$ is the cohomology suspension.)

As a direct consequence of 3.1, one has:

3.2 Proposition: Let X,μ be an H-space, $x\in H^*(X,Z_p)$. Then $c(\sigma^*x)$ can be represented by $(\sigma^*\otimes\sigma^*)\overline{\mu}^*x$.

3.3 Proposition: Let $a_i,b_i\in G(2)$. Suppose $\Sigma a_i b_i = Sq^n$ and excess $(b_i) < n-1$. Let $h:K(Z_2,n-2) \longrightarrow \prod_i K(Z_2,n-2+|b_i|)$ given by $h^*\iota_{n-2+|b_i|} = b_i\iota_{n-2}$ and let $\Omega[\prod_i K(Z_2,n-2+|b_i|)] \xrightarrow{j} Y \xrightarrow{\Pi} K(Z_2,n-2)$ be the fibration induced by h. If $v\in PH^{2n-3}(Y,Z_2)$ is any class satisfying $j^*v = \Sigma a_i\sigma^*\iota_{n-2+|b_i|}$, then $c(v)$ can be represented by $\Pi^*\iota_{n-2}\otimes\Pi^*\iota_{n-2}$.

Proof: $Y = \Omega Y'$ and $v = \sigma^*v'$ with $\sigma^*\otimes\sigma^*\overline{\mu}_{Y'}{}^*v' = \Pi^*\iota_{n-2}\otimes\Pi^*\iota_{n-2}$, and 3.2 can be applied.

Let φ_{4n} be a non-stable secondary operation associated with the relation $e(Sq^2Sq^{4n-1} + Sq^{4n}Sq^1) = e(Sq^{4n+1}) > 4n$ (e--the excess). If X,μ is an H-space with $H^*(X,Z_2)$ being an exterior algebra on odd dimensional generators, then $\varphi_{4n}: PH^{4n-1}(X,Z_2) \longrightarrow PH^{8n-1}(X,Z_2)/Sq^2$ (where A/Sq^2 denotes A/Sq^2A for an $\mathfrak{a}(2)$ module A). If X,μ is homotopy commutative, then $c: PH^*(X,Z_2) \longrightarrow H^*(X,Z_2)\otimes H^*(X,Z_2)/(1+T^*)$ induces a morphism

$c/Sq^2: PH^*(X,Z_2)/Sq^2 \longrightarrow [H^*(X,Z_2)/Sq^2 \otimes H^*(X,Z_2)/Sq^2]/(1+T^*)$.

3.4 Theorem: Let X,μ be a homotopy commutative and homotopy associative H-space, $H^*(X,Z_2)$--an exterior algebra on odd dimensional generators. Then φ_{4n} can be chosen so that one has a commutative diagram:

$$\begin{array}{ccc} PH^{4n-1}(X,Z_2) & \xrightarrow{\varphi_{4n}} & PH^{8n-1}(X,Z_2)/Sq^2 \\ \downarrow \rho & & \downarrow c/Sq^2 \\ H^{4n-1}(X,Z_2)/Sq^2 & \xrightarrow{\tilde{\varphi}} & (PH^*(X,Z_2)/Sq^2 \otimes PH^*(X,Z_2)/Sq^2)/\mathrm{im}(1+T^*) \end{array}$$

with $\tilde{\varphi}\rho(x) = [\rho(x)\otimes\rho(x)]$.

Proof: Let $x \in PH^{4n-1}(X,Z_2)$. Consider the following diagram:

$$\begin{array}{ccc} X' = X \times K(Z_2,4n-1) \times K(Z_2,8n-3) & \xrightarrow{\tilde{f}} & Y \\ \downarrow p_1 & & \downarrow \Pi \\ X & \xrightarrow{f} & K(Z_2,4n-1) \\ & & \downarrow h \\ & & K(Z_2,4n) \times K(Z_2,8n-2) \end{array}$$

where $\Pi: Y \longrightarrow K(Z_2,4n-1)$ is the $K(Z_2,4n-1) \times K(Z_2,8n-3)$ principal fibration induced by h, $h^*\iota_{4n} = Sq^1\iota_{4n-1}$, $h^*\iota_{8n-2} = Sq^{4n-1}\iota_{4n-1}$, $f^*\iota_{4n-1} = x$.

Now, by 1.1 and 1.2, $a(h) = 0 = c(h)$ and hence, $a(h\circ f) = \Omega h_{\#}a(f)$, $c(h\circ f) = \Omega h_{\#}c(f)$. Now $\Omega h_{\#} = (Sq^1, Sq^{4n-1})$,

$a(f) \in H^{4n-1}(X\wedge X\wedge X,Z_2) \oplus H^{8n-3}(X\wedge X\wedge X,Z_2)$

$c(f) \in H^{4n-1}(X\wedge X,Z_2) \oplus H^{8n-3}(X\wedge X,Z_2)$ hence, $\Omega h_{\#}a(f) = 0$, $\Omega h_{\#}c(f) = 0$ and thus, X' is a homotopy commutative and homotopy associative

H-space. It follows that the twisting function $w: X\wedge X \longrightarrow K(Z_2,4n-1) \times K(Z_2,8n-3)$ inducing the multiplication in X' satisfies $[w] \in \ker(1\otimes\overline{\mu}^* - \overline{\mu}^*\otimes 1) \cap \ker(1-T^*)$ and hence, $[w]$ represents an element in $P[\mathrm{Ext}^{2,*}_{H_*(X,Z_2)}(Z_2,Z_2)] = 0$ and, consequently, $[w] \in \mathrm{im}\,\overline{\mu}^*$. If $[w] = \overline{\mu}^*[g]$ where $g: X \longrightarrow K(Z_2,4n-1) \times K(Z_2,8n-3)$, then the mapping $\chi: X \longrightarrow X'$ given by $\chi(x) = x, g(x)$ is an H-mapping.

Let $f_1 = \tilde{f}\circ\chi: X \longrightarrow Y$. If $v \in PH^{8n-1}(Y,Z_2)$ restricts to $Sq^2(1\otimes\iota_{8n-3})$ in $H^*(K(Z_2,4n-1) \times K(Z_2,8n-3),Z_2)$ by 3.3 $c(v) = \Pi^*\iota_{4n-1}\otimes\Pi^*\iota_{4n-1}$. $(\Pi^*\iota_{4n-1}, y, v)$ is the universal example for φ_{4n} and as v can be altered by an element in $\Pi^*PH^*(K(Z_2,4n-1),Z_2)$, one may assume that $\sigma^*v = 1\otimes Sq^2\iota_{8n-4} \in H^*(\Omega Y,Z_2)$ $= H^*(K(Z_4,n-2),Z_2) \otimes H^*(K(Z_2,8n-4),Z_2)$.

Now: $(f_1^*\otimes f_1^*)c(v) = c(f_1^*(v)) + c(f_1)*\sigma^*v$, hence, $c(\varphi_{4n}(x)) = c(f_1^*(v)) = [x\otimes x]$ and 3.4 follows.

3.5 Corollary: If X,μ satisfies the hypothesis of 3.4, then $\ker \varphi_{4n} \subset \mathrm{im}\, Sq^2$.

4. Appendix.

2.1 and 3.5 yield sequences of primitive elements in $H^*(X,Z_2)$ of increasing order:

$$x_{4n+1} \longrightarrow Sq^{4n}x_{4n+1} \longrightarrow Sq^{8n}Sq^{4n}x_{4n+1}\cdots$$

$$\rho(x_{4n-1}) \longrightarrow \varphi_{4n}x_{4n-1} = \rho x_{8n-1} \longrightarrow \varphi_{8n}x_{8n-1} = \rho x_{16n-1} \longrightarrow \cdots$$

Further relations can be found between $\mathcal{G}(2)$ and φ_{4n} such as: $Sq^2\varphi_{4n}(x) = Sq^{4n}Sq^2x$. Much more involved relations can be obtained by comparing X to some universal homotopy commutative H-spaces such as SU.

The mod 2 chohomology of SU has the following properties:

(a) $Sq^2PH^{8n-1}(SU,Z_2) = PH^{8n+1}(SU,Z_2)$

(b) $Sq^4PH^{8n-3}(SU,Z_2) = PH^{8n+1}(SU,Z_2)$

(c) $Sq^{4,2}PH^{8n-5}(SU,Z_2) = PH^{8n+1}(SU,Z_2)$.

Consequently, one has:

4.1 Lemma: Let A be a (graded connected non-stable) $\mathcal{A}(2)$ module, $A^{2m} = 0$, $A^1 = 0$ and $Sq^{4n}: A^{4n+1} \longrightarrow A^{8n+1}$ is a monomorphism. Then any $\mathcal{A}(2)$ morphism $\theta: PH^*(SU, Z_2) \longrightarrow A$ which is a monomorphism in dimensions 3 and 5 is a monomorphism (in all dimensions).

Proof: Let $M = PH^*(SU, Z_2)$. One has the following diagram:

$$\begin{array}{ccc}
M^{4n+1} & \xrightarrow{\theta^{4n+1}} & A^{4n+1} \\
\approx \Big\downarrow Sq^{4n} & & 1\text{-}1 \Big\downarrow Sq^{4n} \\
M^{8n+1} & \xrightarrow{\theta^{8n+1}} & A^{8n+1} \\
\approx \Big\uparrow a_j & & \Big\uparrow a_j \\
M^{8n+1-2j} & \xrightarrow{\theta^{8n+1-2j}} & A^{8n+1-2j}
\end{array}$$

where $a_j = Sq^2, Sq^4$ or $Sq^4 Sq^2$ if $j = 1, 2,$ or 3 respectively.

If θ^{4n+1} is a monomorphism, so is θ^{8n+1} and, consequently, so is $\theta^{8n+1-2j}$, $j = 1, 2,$ or 3. As θ^5 is a monomorphism by induction, θ^j is a monomorphism for all j.

Let $SU(\{2\})$ be the mod 2 SU, i.e., there exist

$\psi_1: SU \longrightarrow SU(\{2\})$

$\psi_2: SU(\{2\}) \longrightarrow \prod_{n=1}^{\infty} K(Z, 2n+1) = SU_0$

so that ψ_i is an isomorphism of rational and mod 2 cohomology if $i = 1$ and of rational and mod p (p-odd) cohomology if $i = 2$.

4.2 Corollary: Let X, μ be an H-space satisfying the hypothesis of 2.1. If $Sq^2 PH^3(X, Z_2) \neq 0$, then there exists a mapping $h: X \longrightarrow SU(\{2\})$, h^* yielding a monomorphism of mod 2 cohomology.

Proof: The morphism $j: [X, SU(\{2\})] \longrightarrow H^3(X, Z)$ given by $j(f) = f^* w_3$ ($w_3 \in PH^3(SU(\{2\}), Z)$ is a generator) is an epimorphism. Thus, $h: X \longrightarrow SU(\{2\})$ satisfying $Sq^2 \rho_2 h^* w_3 \neq 0$ ($\rho_2: H^*(\ , Z) \longrightarrow H^*(\ , Z_2)$--the reduction). By 4.1, h^* is a monomorphism.

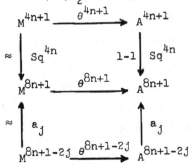

REFERENCES

1. Browder, W., Homotopy Commutative H-spaces, *Ann. of Math.* 75, 283-311 (1962).

2. Dyer, E. and Lashof, R.K., Homology of Iterated Loop Spaces, *Amer. J. of Math.* 84, 35-88 (1962).

3. Hubbuck, J., Automorphisms of Polynomial Algebras and Homotopy Commutativity in H-spaces, Mimeographed.

4. Kudo and Araki, S., Topology of H_n spaces and H-squaring operations, *Mem. Fac. Sci.*, Kyusyu Univ. Ser. A, 10, 85-120 (1956).

5. Stasheff, J.D., On Extensions of H-spaces, *Trans. Amer. Math. Soc.* 105, 126-135 (1962).

6. _________. Homotopy Associativity of H-spaces, I and II, *Trans. Amer. Math. Soc.* 108, 275-312 (1963).

7. Thomas, E., Steenrod Squares and H-spaces, *Ann. of Math.* 77, 306-317 (1963).

Lecture Notes in Mathematics

isher erschienen/Already published

ol. 1: J. Wermer, Seminar über Funktionen-Algebren. IV, 30 Seiten. 964. DM 3,80 / $ 1.10

ol. 2: A. Borel, Cohomologie des espaces localement compacts après. J. Leray. IV, 93 pages. 1964. DM 9,– / $ 2.60

ol. 3: J. F. Adams, Stable Homotopy Theory. Third edition. IV, 78 pages. 969. DM 8,– / $ 2.20

l. 4: M. Arkowitz and C. R. Curjel, Groups of Homotopy Classes. d. revised edition. IV, 36 pages. 1967. DM 4,80 / $ 1.40

l. 5: J.-P. Serre, Cohomologie Galoisienne. Troisième édition. I, 214 pages. 1965. DM 18,– / $ 5.00

l. 6: H. Hermes, Term Logic with Choise Operator. III, 55 pages. 70. DM 6,– / $ 1.70

. 7: Ph. Tondeur, Introduction to Lie Groups and Transformation oups. Second edition. VIII, 176 pages. 1969. DM 14,– / $ 3.80

. 8: G. Fichera, Linear Elliptic Differential Systems and Eigenvalue blems. IV, 176 pages. 1965. DM 13,50 / $ 3.80

9: P. L. Ivănescu, Pseudo-Boolean Programming and Applications. 50 pages. 1965. DM 4,80 / $ 1.40

10: H. Lüneburg, Die Suzukigruppen und ihre Geometrien. VI, Seiten. 1965. DM 8,– / $ 2.20

11: J.-P. Serre, Algèbre Locale. Multiplicités. Rédigé par P. Gabriel. onde édition. VIII, 192 pages. 1965. DM 12,– / $ 3.30

12: A. Dold, Halbexakte Homotopiefunktoren. II, 157 Seiten. 1966. 12,– / $ 3.30

13: E. Thomas, Seminar on Fiber Spaces. IV, 45 pages. 1966. 4,80 / $ 1.40

14: H. Werner, Vorlesung über Approximationstheorie. IV, 184 Sei- und 12 Seiten Anhang. 1966. DM 14,– / $ 3.90

15: F. Oort, Commutative Group Schemes. VI, 133 pages. 1966. 9,80 / $ 2.70

16: J. Pfanzagl and W. Pierlo, Compact Systems of Sets. IV, ages. 1966. DM 5,80 / $ 1.60

17: C. Müller, Spherical Harmonics. IV, 46 pages. 1966. ,– / $ 1.40

18: H.-B. Brinkmann und D. Puppe, Kategorien und Funktoren. 07 Seiten, 1966. DM 8,– / $ 2.20

9: G. Stolzenberg, Volumes, Limits and Extensions of Analytic ties. IV, 45 pages. 1966. DM 5,40 / $ 1.50

0: R. Hartshorne, Residues and Duality. VIII, 423 pages. 1966. 0,– / $ 5.50

1: Seminar on Complex Multiplication. By A. Borel, S. Chowla, Herz, K. Iwasawa, J.-P. Serre. IV, 102 pages. 1966. DM 8,– / $ 2.20

22: H. Bauer, Harmonische Räume und ihre Potentialtheorie. 5 Seiten. 1966. DM 14,– / $ 3.90

3: P. L. Ivănescu and S. Rudeanu, Pseudo-Boolean Methods for nt Programming. 120 pages. 1966. DM 10,– / $ 2.80

4: J. Lambek, Completions of Categories. IV, 69 pages. 1966. 60 / $ 1.90

5: R. Narasimhan, Introduction to the Theory of Analytic Spaces. pages. 1966. DM 10,– / $ 2.80

6: P.-A. Meyer, Processus de Markov. IV, 190 pages. 1967. – / $ 4.20

7: H. P. Künzi und S. T. Tan, Lineare Optimierung großer e. VI, 121 Seiten. 1966. DM 12,– / $ 3.30

: P. E. Conner and E. E. Floyd, The Relation of Cobordism to ries. VIII, 112 pages. 1966. DM 9,80 / $ 2.70

: K. Chandrasekharan, Einführung in die Analytische Zahlen- VI, 199 Seiten. 1966. DM 16,80 / $ 4.70

: A. Frölicher and W. Bucher, Calculus in Vector Spaces without , 146 pages. 1966. DM 12,– / $ 3.30

: Symposium on Probability Methods in Analysis. Chairman. appos. IV, 329 pages. 1967. DM 20,– / $ 5.50

: M. André, Méthode Simpliciale en Algèbre Homologique et Commutative. IV, 122 pages. 1967. DM 12,– / $ 3.30

: G. I. Targonski, Seminar on Functional Operators and s. IV, 110 pages. 1967. DM 10,– / $ 2.80

G. E. Bredon, Equivariant Cohomology Theories. VI, 64 pages. M 6,80 / $ 1.90

N. P. Bhatia and G. P. Szegö, Dynamical Systems. Stability nd Applications. VI, 416 pages. 1967. DM 24,– / $ 6.60

A. Borel, Topics in the Homology Theory of Fibre Bundles. ges. 1967. DM 9,– / $ 2.50

Vol. 37: R. B. Jensen, Modelle der Mengenlehre. X, 176 Seiten. 1967. DM 14,– / $ 3.90

Vol. 38: R. Berger, R. Kiehl, E. Kunz und H.-J. Nastold, Differentialrechnung in der analytischen Geometrie IV, 134 Seiten. 1967 DM 12,– / $ 3.30

Vol. 39: Séminaire de Probabilités I. II, 189 pages. 1967. DM 14,– / $ 3.90

Vol. 40: J. Tits, Tabellen zu den einfachen Lie Gruppen und ihren Darstellungen. VI, 53 Seiten. 1967. DM 6,80 / $ 1.90

Vol. 41: A. Grothendieck, Local Cohomology. VI, 106 pages. 1967. DM 10,– / $ 2.80

Vol. 42: J. F. Berglund and K. H. Hofmann, Compact Semitopological Semigroups and Weakly Almost Periodic Functions. VI, 160 pages. 1967. DM 12,– / $ 3.30

Vol. 43: D. G. Quillen, Homotopical Algebra. VI, 157 pages. 1967. DM 14,– / $ 3.90

Vol. 44: K. Urbanik, Lectures on Prediction Theory. IV, 50 pages. 1967. DM 5,80 / $ 1.60

Vol. 45: A. Wilansky, Topics in Functional Analysis. VI, 102 pages. 1967. DM 9,60 / $ 2.70

Vol. 46: P. E. Conner, Seminar on Periodic Maps. IV, 116 pages. 1967. DM 10,60 / $ 3.00

Vol. 47: Reports of the Midwest Category Seminar I. IV, 181 pages. 1967. DM 14,80 / $ 4.10

Vol. 48: G. de Rham, S. Maumary et M. A. Kervaire, Torsion et Type Simple d'Homotopie. IV, 101 pages. 1967. DM 9,60 / $ 2.70

Vol. 49: C. Faith, Lectures on Injective Modules and Quotient Rings. XVI, 140 pages. 1967. DM 12,80 / $ 3.60

Vol. 50: L. Zalcman, Analytic Capacity and Rational Approximation. VI, 155 pages. 1968. DM 13.20 / $ 3.70

Vol. 51: Séminaire de Probabilités II. IV, 199 pages. 1968. DM 14,– / $ 3.90

Vol. 52: D. J. Simms, Lie Groups and Quantum Mechanics. IV, 90 pages. 1968. DM 8,– / $ 2.20

Vol. 53: J. Cerf, Sur les difféomorphismes de la sphère de dimension trois (Γ_4= O). XII, 133 pages. 1968. DM 12,– / $ 3.30

Vol. 54: G. Shimura, Automorphic Functions and Number Theory. VI, 69 pages. 1968. DM 8,– / $ 2.20

Vol. 55: D. Gromoll, W. Klingenberg und W. Meyer, Riemannsche Geometrie im Großen. VI, 287 Seiten. 1968. DM 20,– / $ 5.50

Vol. 56: K. Floret und J. Wloka, Einführung in die Theorie der lokalkonvexen Räume. VIII, 194 Seiten. 1968. DM 16,– / $ 4.40

Vol. 57: F. Hirzebruch und K. H. Mayer, O (n)-Mannigfaltigkeiten, exotische Sphären und Singularitäten. IV, 132 Seiten. 1968. DM 10,80 / $ 3.00

Vol. 58: Kuramochi Boundaries of Riemann Surfaces. IV, 102 pages. 1968. DM 9,60 / $ 2.70

Vol. 59: K. Jänich, Differenzierbare G-Mannigfaltigkeiten. VI, 89 Seiten. 1968. DM 8,– / $ 2.20

Vol. 60: Seminar on Differential Equations and Dynamical Systems. Edited by G. S. Jones. VI, 106 pages. 1968. DM 9,60 / $ 2.70

Vol. 61: Reports of the Midwest Category Seminar II. IV, 91 pages. 1968. DM 9,60 / $ 2.70

Vol. 62: Harish-Chandra, Automorphic Forms on Semisimple Lie Groups X, 138 pages. 1968. DM 14,– / $ 3.90

Vol. 63: F. Albrecht, Topics in Control Theory. IV, 65 pages. 1968. DM 6,80 / $ 1.90

Vol. 64: H. Berens, Interpolationsmethoden zur Behandlung von Approximationsprozessen auf Banachräumen. VI, 90 Seiten. 1968. DM 8,– / $ 2.20

Vol. 65: D. Kölzow, Differentiation von Maßen. XII, 102 Seiten. 1968. DM 8,– / $ 2.20

Vol. 66: D. Ferus, Totale Absolutkrümmung in Differentialgeometrie und -topologie. VI, 85 Seiten. 1968. DM 8,– / $ 2.20

Vol. 67: F. Kamber and P. Tondeur, Flat Manifolds. IV, 53 pages. 1968. DM 5,80 / $ 1.60

Vol. 68: N. Boboc et P. Mustată, Espaces harmoniques associés aux opérateurs différentiels linéaires du second ordre de type elliptique. VI, 95 pages. 1968. DM 8,60 / $ 2.40

Vol. 69: Seminar über Potentialtheorie. Herausgegeben von H. Bauer. VI, 180 Seiten. 1968. DM 14,80 / $ 4.10

Vol. 70: Proceedings of the Summer School in Logic. Edited by M. H. Löb. IV, 331 pages. 1968. DM 20,– / $ 5.50

Vol. 71: Séminaire Pierre Lelong (Analyse), Année 1967 – 1968. VI, 190 pages. 1968. DM 14,– / $ 3.90

Bitte wenden / Continued

Vol. 72: The Syntax and Semantics of Infinitary Languages. Edited by J. Barwise. IV, 268 pages. 1968. DM 18,– / $ 5.00

Vol. 73: P. E. Conner, Lectures on the Action of a Finite Group. IV, 123 pages. 1968. DM 10,– / $ 2.80

Vol. 74: A. Fröhlich, Formal Groups. IV, 140 pages. 1968. DM 12,– / $ 3.30

Vol. 75: G. Lumer, Algèbres de fonctions et espaces de Hardy. VI, 80 pages. 1968. DM 8,– / $ 2.20

Vol. 76: R. G. Swan, Algebraic K-Theory. IV, 262 pages. 1968. DM 18,– / $ 5.00

Vol. 77: P.-A. Meyer, Processus de Markov: la frontière de Martin. IV, 123 pages. 1968. DM 10,– / $ 2.80

Vol. 78: H. Herrlich, Topologische Reflexionen und Coreflexionen. XVI, 166 Seiten. 1968. DM 12,– / $ 3.30

Vol. 79: A. Grothendieck, Catégories Cofibrées Additives et Complexe Cotangent Relatif. IV, 167 pages. 1968. DM 12,– / $ 3.30

Vol. 80: Seminar on Triples and Categorical Homology Theory. Edited by B. Eckmann. IV, 398 pages. 1969. DM 20,– / $ 5.50

Vol. 81: J.-P. Eckmann et M. Guenin, Méthodes Algébriques en Mécanique Statistique. VI, 131 pages. 1969. DM 12,– / $ 3.30

Vol. 82: J. Wloka, Grundräume und verallgemeinerte Funktionen. VIII, 131 Seiten. 1969. DM 12,– / $ 3.30

Vol. 83: O. Zariski, An Introduction to the Theory of Algebraic Surfaces. IV, 100 pages. 1969. DM 8,– / $ 2.20

Vol. 84: H. Lüneburg, Transitive Erweiterungen endlicher Permutationsgruppen. IV, 119 Seiten. 1969. DM 10,– / $ 2.80

Vol. 85: P. Cartier et D. Foata, Problèmes combinatoires de commutation et réarrangements. IV, 88 pages. 1969. DM 8,– / $ 2.20

Vol. 86: Category Theory, Homology Theory and their Applications I. Edited by P. Hilton. VI, 216 pages. 1969. DM 16,– / $ 4.40

Vol. 87: M. Tierney, Categorical Constructions in Stable Homotopy Theory. IV, 65 pages. 1969. DM 6,– / $ 1.70

Vol. 88: Séminaire de Probabilités III. IV, 229 pages. 1969. DM 18,– / $ 5.00

Vol. 89: Probability and Information Theory. Edited by M. Behara, K. Krickeberg and J. Wolfowitz. IV, 256 pages. 1969. DM 18,– / $ 5.00

Vol. 90: N. P. Bhatia and O. Hajek, Local Semi-Dynamical Systems. II, 157 pages. 1969. DM 14,– / $ 3.90

Vol. 91: N. N. Janenko, Die Zwischenschrittmethode zur Lösung mehrdimensionaler Probleme der mathematischen Physik. VIII, 194 Seiten. 1969. DM 16,80 / $ 4.70

Vol. 92: Category Theory, Homology Theory and their Applications II. Edited by P. Hilton. V, 308 pages. 1969. DM 20,– / $ 5.50

Vol. 93: K. R. Parthasarathy, Multipliers on Locally Compact Groups. III, 54 pages. 1969. DM 5,60 / $ 1.60

Vol. 94: M. Machover and J. Hirschfeld, Lectures on Non-Standard Analysis. VI, 79 pages. 1969. DM 6,– / $ 1.70

Vol. 95: A. S. Troelstra, Principles of Intuitionism. II, 111 pages. 1969. DM 10,– / $ 2.80

Vol. 96: H.-B. Brinkmann und D. Puppe, Abelsche und exakte Kategorien, Korrespondenzen. V, 141 Seiten. 1969. DM 10,– / $ 2.80

Vol. 97: S. O. Chase and M. E. Sweedler, Hopf Algebras and Galois theory. II, 133 pages. 1969. DM 10,– / $ 2.80

Vol. 98: M. Heins, Hardy Classes on Riemann Surfaces. III, 106 pages. 1969. DM 10,– / $ 2.80

Vol. 99: Category Theory, Homology Theory and their Applications III. Edited by P. Hilton. IV, 489 pages. 1969. DM 24,– / $ 6.60

Vol. 100: M. Artin and B. Mazur, Etale Homotopy. II, 196 Seiten. 1969. DM 12,– / $ 3.30

Vol. 101: G. P. Szegö et G. Treccani, Semigruppi di Trasformazioni Multivoche. VI, 177 pages. 1969. DM 14,– / $ 3.90

Vol. 102: F. Stummel, Rand- und Eigenwertaufgaben in Sobolewschen Räumen. VIII, 386 Seiten. 1969. DM 20,– / $ 5.50

Vol. 103: Lectures in Modern Analysis and Applications I. Edited by C. T. Taam. VII, 162 pages. 1969. DM 12,– / $ 3.30

Vol. 104: G. H. Pimbley, Jr., Eigenfunction Branches of Nonlinear Operators and their Bifurcations. II, 128 pages. 1969. DM 10,– / $ 2.80

Vol. 105: R. Larsen, The Multiplier Problem. VII, 284 pages. 1969. DM 18,– / $ 5.00

Vol. 106: Reports of the Midwest Category Seminar III. Edited by S. Mac Lane. III, 247 pages. 1969. DM 16,– / $ 4.40

Vol. 107: A. Peyerimhoff, Lectures on Summability. III, 111 pages. 1969. DM 8,– / $ 2.20

Vol. 108: Algebraic K-Theory and its Geometric Applications. Edited by R. M. F. Moss and C. B. Thomas. IV, 86 pages. 1969. DM 6,– / $ 1.70

Vol. 109: Conference on the Numerical Solution of Differential Equations. Edited by J. Ll. Morris. VI, 275 pages. 1969. DM 18,– / $ 5.00

Vol. 110: The Many Facets of Graph Theory. Edited by G. Chartrand and S. F. Kapoor. VIII, 290 pages. 1969. DM 18,– / $ 5.00

Vol. 111: K. H. Mayer, Relationen zwischen charakteristischen Zahlen III, 99 Seiten. 1969. DM 8,– / $ 2.20

Vol. 112: Colloquium on Methods of Optimization. Edited by N. N. Moiseev. IV, 293 pages. 1970. DM 18,– / $ 5.00

Vol. 113: R. Wille, Kongruenzklassengeometrien. III, 99 Seiten. 1970 DM 8,– / $ 2.20

Vol. 114: H. Jacquet and R. P. Langlands, Automorphic Forms on GL (2 VII, 548 pages. 1970. DM 24,– / $ 6.60

Vol. 115: K. H. Roggenkamp and V. Huber-Dyson, Lattices over Orders XIX, 290 pages. 1970. DM 18,– / $ 5.00

Vol. 116: Séminaire Pierre Lelong (Analyse) Année 1969. IV, 195 page 1970. DM 14,– / $ 3.90

Vol. 117: Y. Meyer, Nombres de Pisot, Nombres de Salem et Analy Harmonique. 63 pages. 1970. DM 6,– / $ 1.70

Vol. 118: Proceedings of the 15th Scandinavian Congress, Oslo 196 Edited by K. E. Aubert and W. Ljunggren. IV, 162 pages. 1970. DM 12,– / $ 3.30

Vol. 119: M. Raynaud, Faisceaux amples sur les schémas en group et les espaces homogènes. III, 219 pages. 1970. DM 14,– / $ 3.90

Vol. 120: D. Siefkes, Büchi's Monadic Second Order Success Arithmetic. XII, 130 Seiten. 1970. DM 12,– / $ 3.30

Vol. 121: H. S. Bear, Lectures on Gleason Parts. III, 47 pages. 19 DM 6,– / $ 1.70

Vol. 122: H. Zieschang, E. Vogt und H.-D. Coldewey, Flächen ebene diskontinuierliche Gruppen. VIII, 203 Seiten. 1970. DM 16, $ 4.40

Vol. 123: A. V. Jategaonkar, Left Principal Ideal Rings. VI, 145 pa 1970. DM 12,– / $ 3.30

Vol. 124: Séminare de Probabilités IV. Edited by P. A. Meyer. IV, pages. 1970. DM 20,– / $ 5.50

Vol. 125: Symposium on Automatic Demonstration. V, 310 pages. 1 DM 20,– / $ 5.50

Vol. 126: P. Schapira, Théorie des Hyperfonctions. XI, 157 pages. 1 DM 14,– / $ 3.90

Vol. 127: I. Stewart, Lie Algebras. IV, 97 pages. 1970. DM 10,– / $

Vol. 128: M. Takesaki, Tomita's Theory of Modular Hilbert Alge and its Applications. II, 123 pages. 1970. DM 10,– / $ 2.80

Vol. 129: K. H. Hofmann, The Duality of Compact Semigroups C*- Bigebras. XII, 142 pages. 1970. DM 14,– / $ 3.90

Vol. 130: F. Lorenz, Quadratische Formen über Körpern. II, 77 S 1970. DM 8,– / $ 2.20

Vol. 131: A Borel et al., Seminar on Algebraic Groups and Re Finite Groups. VII, 321 pages. 1970. DM 22,– / $ 6.10

Vol. 132: Symposium on Optimization. III, 348 pages. 1970. DM 2 $ 6.10

Vol. 133: F. Topsøe, Topology and Measure. XIV, 79 pages. DM 8,– / $ 2.20

Vol. 134: L. Smith, Lectures on the Eilenberg-Moore Spectral Seq VII, 142 pages. 1970. DM 14,– / $ 3.90

Vol. 135: W. Stoll, Value Distribution of Holomorphic Maps into pact Complex Manifolds. II, 267 pages. 1970. DM 18,– / $

Vol. 136: M. Karoubi et al., Séminaire Heidelberg-Saarbrücken buorg sur la K-Théorie. IV, 264 pages. 1970. DM 18,– / $ 5.00

Vol. 137: Reports of the Midwest Category Seminar IV. Ed S. MacLane. III, 139 pages. 1970. DM 12,– / $ 3.30

Vol. 138: D. Foata et M. Schützenberger, Théorie Géométriq Polynômes Eulériens. V, 94 pages. 1970. DM 10,– / $ 2.80

Vol. 139: A. Badrikian, Séminaire sur les Fonctions Aléatoires aires et les Mesures Cylindriques. VII, 221 pages. 1970. DM $ 5.00

Vol. 140: Lectures in Modern Analysis and Applications II. Ed C. T. Taam. VI, 119 pages. 1970. DM 10,– / $ 2.80

Vol. 141: G. Jameson, Ordered Linear Spaces. XV, 194 pages DM 16,– / $ 4.40

Vol. 142: K. W. Roggenkamp, Lattices over Orders II. V, 388 1970. DM 22,– / $ 6.10

Vol. 143: K. W. Gruenberg, Cohomological Topics in Group XIV, 275 pages. 1970. DM 20,– / $ 5.50